Glacial Geology

Glacial Geology

Ice Sheets and Landforms

MATTHEW R. BENNETT

School of Earth Sciences, University of Greenwich, UK

NEIL F. GLASSER

School of Biological and Earth Sciences,
Liverpool John Moores University, UK

JOHN WILEY & SONS

Chichester • New York • Brisbane • Toronto • Singapore

Other Wiley Editorial Offices

John Wiley & Sons, Inc., 605 Third Avenue,
New York, NY 10158-0012, USA

Jacaranda Wiley Ltd, 33 Park Road, Milton,
Queensland 4064, Australia

John Wiley & Sons (Canada) Ltd, 22 Worcester Road,
Rexdale, Ontario M9W 1L1, Canada

John Wiley & Sons (Asia) Pte Ltd, 2 Clementi Loop #02-01,
Jin Xing Distripark, Singapore 0512

Library of Congress Cataloging-in-Publication Data

Bennett, Matthew (Matthew R.)
    Glacial geology : ice sheets and landforms / Matthew R. Bennett,
Neil F. Glasser.
        p. cm.
    ISBN 0-471-96344-5 (cloth). — ISBN 0-471-96345-3 (paper)
    1. Glacial landforms. 2. Glaciers.   I. Glasser, Neil F.
II. Title.
GB581.B45   1996
551.3′1—dc20   95-46353
                                                         CIP
British Library Cataloguing in Publication Data

A catalogue record for this book is available from the British Library

ISBN 0-471-963445 (Cl)
      0-471-963453 (Pr)

Typeset in 10/12pt Palatino by Saxon Graphics Ltd, Derby
Printed and bound in Great Britain by Bookcraft (Bath) Ltd

This book is printed on acid-free paper responsibly manufactured from sustainable forestation, for
which at least two trees are planted for each one used for paper production.

Contents

Preface

This book is the product of two things, an enthusiasm for glacial geology and a perceived need for a student text with which to stimulate this enthusiasm in others. Its aim is therefore simple: to provide an account of glacial geology which is accessible to the undergraduate and uncluttered from unnecessary detail. We hope that by reading our book you will share some of our enthusiasm for glacial geology.

Matthew R. Bennett
and Neil F. Glasser
Chatham Maritime, 1995

Acknowledgements

The writing of any text inevitably draws upon the accumulated wisdom of innumerable colleagues. This book is written for the undergraduate and consequently is broad in approach, with little space for the explanation of detail. We apologise if this appears to neglect the valuable work of many glacial geologists. We alone are responsible for the style, content and any errors or omissions.

There are many people to whom we are indebted for help, advice, encouragement and forbearance in producing this book. Our views on glacial geology have been shaped by Geoffrey Boulton, Murray Gray and David Sugden, who not only introduced us to the subject but have shared their enthusiasm and knowledge with us. We would also like to acknowledge all those with whom we have spent both many happy and at times miserable hours in the field, usually in the pouring rain. In particular, we would like to thank Geoffrey Boulton, Chalmers Clapperton, Mike Hambrey, Richard Hodgkins, David Huddart, David Sugden and Charles Warren for valuable field discussion. We also thank our colleagues for their support in times of crisis. In particular, Alistair Baxter at the University of Greenwich has tolerated our often forceful demands for departmental resources. Neil Glasser would also like to thank his former colleagues at English Nature.

Drafts of this book have been reviewed at various stages by Alistair Baxter, Andy Bussell, Peter Doyle, John Gordon, Murray Gray, Richard Hindmarsh, Andy Kerr, Martin Kirkbride, Jeff Warburton and Alistair Wells. Their comments have done much to improve the clarity and accuracy of our message. Technical support at the University of Greenwich was provided by Pat Brown and Martin Gay. The illustrations were prepared by Hilary Foxwell, Angela Holder and Gecko Ltd. Jane Anderton assisted in the preparation of the index, for which we are grateful. Nick Dobson and his library team at Greenwich deserve special mention for providing an almost constant stream of books, journals and other reference material during a particularly difficult year. We would also like to thank all those who have supplied photographs for this book, in particular Ron Biggs and Andrew Bennett deserve special mention. Amanda Hewes and Nicky Christophers at John Wiley guided this project from the start and did much to make the job of preparing the manuscript relatively painless. Neil Glasser would like to acknowledge the support and patience of Mary Harrison who had the unfortunate privilege of living with the book as it was written. Finally, Matthew Bennett would like to thank Peter and Julie Doyle for their advice and friendship throughout this project.

Illustrations

Many of the illustrations within this book are not original. Where they are reproduced exactly as they were first published, permission has been obtained from the
copyright holder, and is indicated in the figure caption by the word *from* followed
by the source. Every effort has been made to obtain permission from the relevant
copyright holders, but where no reply was received we have assumed that there
was no objection to our using the material. Where we have significantly altered an
illustration, we have acknowledged this and signalled the change in the figure
caption by using the phrase *modified from*.

We are grateful to the following for permission to reproduce copyright figures
and photographs: Cambridge Air Photo Library (Figures 9.18 and 9.33); Balkema
(Figures 8.6 and 12.1); Blackwell Science Ltd (Figure 3.20); Chapman and Hall
(Figures 2.3, 2.5, 3.6, 3.9, 3.15 and 7.1); Geological Society Publishing House
(Figures 10.8, 10.9, 11.7 and 11.8); Institute of Arctic and Alpine Research,
University of Colorado (Figure 4.5); Institute of British Geographers (Box 2.3);
Longman (Figure 3.14); Martin Sharp (Figures 4.4 and 8.16); Metheun & Co (Figure
6.10); Open University Press (Figure 2.7); Royal Scottish Geographical Society (Box
2.3); Scandinavian University Press (Box 6.5); John Wiley & Sons (Figures 2.1, 2.2,
2.6 and Box 2.1).

1
Introduction

1.1 WHAT IS GLACIAL GEOLOGY AND WHY IS IT IMPORTANT?

Glacial geology is the study of the landforms and sediments created by glaciers, both past and present. Glaciers may range in size from small valley glaciers, such as those found in the European Alps, to large ice sheets such as that in Antarctica (Box 1.1).

Within Earth history glaciers have grown and decayed many times. The present landscape is a function of the glaciers which grew and decayed during the Cenozoic Ice Age. During the Cenozoic—the last 65 million years—the world's climate changed dramatically. The Antarctic ice sheet grew and was followed by ice sheets in Greenland and elsewhere in the Arctic north. Later, large mid-latitude ice sheets developed in North America, Scandinavia, Europe and Patagonia. These ice sheets changed and shaped the landscape beneath them and have left a record of their presence in the form of landforms and sediments. This record shows that these ice sheets are not only a consequence of oscillations in global climate, which have driven their growth and decay with amazing regularity during the last two million years, but that they have also helped to drive climate change by modifying and interacting with the atmosphere. We need, therefore, to study these glaciers if we are to understand the mechanisms of global climate change not only during the Cenozoic but within the whole of Earth history.

The glaciers of the Cenozoic have created a distinct landscape of landforms and sediments. This glacial landscape survives today and influences the way we build roads, railways, factories and houses (Boxes 1.2 and 1.3). The aesthetic appeal of this landscape is also the product of these glaciers. The spectacular mountain scenery of many areas is the result of glacial erosion, while glacial deposition often produces a gently rolling landscape. If we are to understand the form and texture

BOX 1.1: THE MORPHOLOGY OF GLACIERS

Glaciers range in size from small **valley glaciers**, such as those in the European Alps, to large **ice sheets**, such as that in Antarctica, which today contains 91% of all ice on Earth. As illustrated below, glaciers can be classified according to their relationship with topography and according to their size. There are two categories: (1) ice bodies that are unconstrained by topography; and (2) ice bodies that are constrained by topography. Within each of these categories a range of different glaciers can be defined in terms of their size.

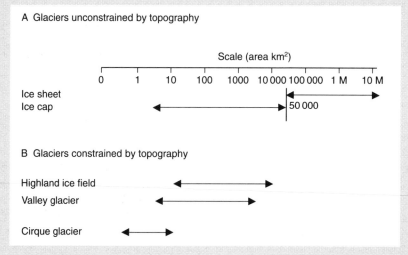

Ice sheets and ice caps are composed of a number of smaller elements. In the centre of an ice sheet or ice cap there is one or more **ice domes**. These domes form the central topography of the ice sheet. Ice flows from their centre, the **ice divide**, towards the ice sheet margin. The margin of an ice sheet may be characterised by **outlet glaciers**. These flow through peripheral mountains, such as the famous Beardmore and Shackleton outlet glaciers which flow from the central dome of the Antarctic Ice Sheet through the Transantarctic Mountains, or may simply appear as crevassed linear depressions flowing from the ice dome. Large fast-flowing outlet glaciers are often referred to as **ice streams**. The margin of an ice sheet may also be marked by an **ice shelf** if it terminates in the sea. An ice shelf is a floating slab of ice which deforms under its own weight. Where pieces of the ice shelf break off into the sea, **icebergs** are produced—a process known as **calving**.

BOX 1.2: CENOZOIC GLACIAL SEDIMENTS: AN ENGINEERING LEGACY

During the Cenozoic Ice Age approximately 30% of the Earth's land surface was glaciated, and as a consequence over 10% of it is now covered by glacial sediments—tills, silts, sands and gravels. In a country like Britain this proportion is even higher. Any form of construction on or in these sediments must consider their engineering properties. At what angle will the sediment stand if excavated? How will sediments respond when loaded? How variable are they? How permeable are they? These questions can only be answered with a detailed knowledge of the sediments and of the processes that deposited them: the contribution of the glacial geologist.

A good example is provided by a recent proposed development at Hardwick Air Field in Norfolk. In 1991, Norfolk County Council applied for planning permission to build a waste hill (landraise) 10 m high to dispose of 1.5 million m³ of domestic waste over 20 years. Crucial to their proposal was that the area was underlain by glacial till, rich in clay, which would act as a natural impermeable barrier to the poisonous fluids (leachate) generated within the decomposing waste. Normally an expensive containment liner is required to prevent contamination of the groundwater by the leachate. This proposal became the subject of local debate and as a consequence the planning application was called to a public planning inquiry in 1993. At this inquiry the objectors used a detailed knowledge of glacial till to argue that it was inadequate as an impermeable barrier in its natural state. Till contains fissures and pockets of sand through which the leachate may pass. Partly on the basis of this evidence the proposal was

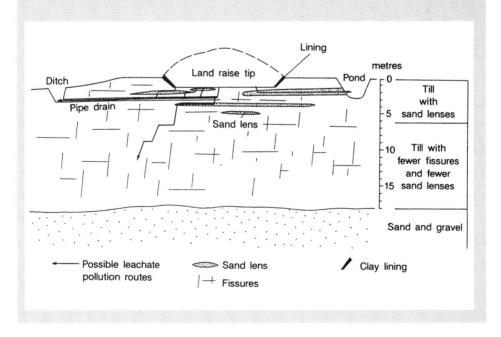

rejected. The example illustrates how a knowledge of glacial sediments is vital to engineering within glaciated terrains.

Source: Gray, J. M. 1993. Quaternary geology and waste disposal in south Norfolk, England. *Quaternary Science Reviews*, **12**, 899–912. [Diagram modified from: Gray, J. M. (1993) *Quaternary Science Reviews*, **12**, Fig. 9, p. 905]

BOX 1.3: CENOZOIC SAND AND GRAVEL: A RESOURCE

Within Britain approximately 280 million tonnes of aggregate, sand and gravel is used each year. This is used for concrete, as ballast and as road stone. About 39% of this is derived from natural deposits of sand and gravel, while the rest comes from crushed stone. Britain has a large reserve of sand and gravel, a legacy of earlier phases of the Cenozoic Ice Age.

Traditionally, reserves of sand and gravel are found by systematically mapping the sediment cover within a search area using regularly spaced excavations and drill holes. In areas of glacial terrain, however, these traditional methods can be replaced if one applies a knowledge of glacial depositional landforms. Different glacial depositional landforms are deposited in different ways and are composed of different sediments. By mapping and identifying these landforms one can identify areas in which potential sand and gravel reserves are likely to occur. These areas can then be targeted for further investigation. In this way exploration can proceed at much faster and therefore more economical rates. Expensive excavations and drill holes are restricted to those areas most likely to contain a valuable reserve of sand and gravel.

This approach was used by Crimes *et al.* (1992) in the search for sand and gravel on the Llyn Peninsula in North Wales. They calculated that by using these methods approximately 6 km^2 could be surveyed per day, in comparison with only 1 km^2 per day using the traditional methods. Understanding the glacial landscape can therefore have important economic benefits.

Sources: Merritt, J. W. 1992. A critical review of methods used in the appraisal of onshore sand and gravel resources in Britain. *Engineering Geology* **32**, 1–9. Crimes, T. P., Chester, D. K. & Thomas, G. S. P. 1992. Exploration of sand and gravel resources by geomorphological analysis in the glacial sediments of the Eastern Llyn Peninsula, Gwynedd, North Wales. *Engineering Geology* **32**, 137–156.

produces a gently rolling landscape. If we are to understand the form and texture of this glacial landscape we must understand the glaciers that produced it.

The landforms and sediments left by these glaciers are the clues from which the glaciers can be reconstructed and their behaviour studied. By studying glacial landscapes and reconstructing the glaciers that created them we can examine the way in which glaciers grow, decay and interact with climate. From such research we can begin to predict what will happen when the mid-latitude ice sheets next return, because although the present is optimistically termed the Postglacial period, there is no reason to suppose that large glaciers or ice sheets will not return to the mid-latitudes in years to come (Box 1.4).

BOX 1.4: ICE SHEETS, CLIMATE AND THE FUTURE

Today there is growing concern about what will happen to the world's ice sheets as a consequence of global warming. The key question is: how will the ice sheets respond? Equally, scientists are being asked to predict the effects of future glaciations, because the Cenozoic Ice Age is not over. In Britain, for example, any underground disposal site for the storage of high-grade nuclear waste may have to survive beneath a future ice sheet, since the radioactivity may take thousands of years to decay. Therefore, in designing such facilities we need to be able to predict the possibility of ice sheets returning to Britain, their impact on groundwater flow and the likely pattern of erosion and deposition within any such future ice sheet. The map below shows a computer prediction of a future European ice sheet 70 000 years from now.

The best way to understand present and future ice sheets is by examining the past: using the premise that the past is the key to the future. For example, at the

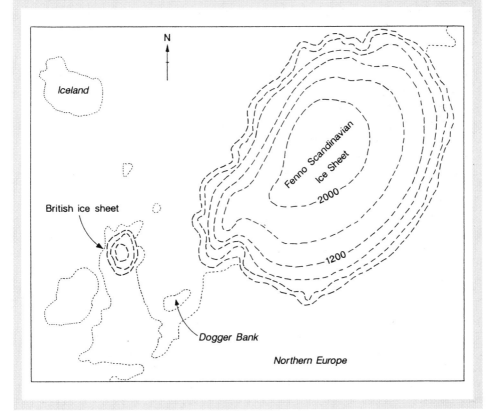

close of the last glacial cycle, air temperatures in the northern mid-latitudes rose by at least 1 °C every 10 years. By reconstructing how ice sheets responded to such dramatic changes, by examining the landform and sediment record they left, one can provide an insight into our present concern about how today's ice sheets may respond to a rise of only 1 °C every 100 years. Similarly, by understanding the behaviour and dynamics of former ice sheets we can constrain theoretical models with which to predict the size and dynamics of future ice sheets.

Source: Boulton, G. S. 1992. The spectrum of Quaternary global change—understanding the past and anticipating the future. *Geoscientist* 2, 10–12. [Diagram modified from: Boulton G. S. (1992) *Geoscientist* 2, Fig. 4, p. 11]

1.2 THE AIM AND STRUCTURE OF THIS BOOK

Glaciers are the scribes of the Cenozoic Ice Age and they have etched its story upon the landscape. It is a story that has been written repeatedly upon the same page with the successive growth and decay of each glacier. Each glacier has destroyed, remoulded or buried the evidence of earlier phases of glaciation. Deciphering the story of this complex geological record is therefore difficult and requires careful detective work. The landforms and sediments left by former glaciers provide the clues from which to reconstruct their form, mechanics and history. This book shows you how to interpret these clues.

First we must understand how glaciers form, how they work, and how the processes of glacial erosion and deposition operate within them to produce glacial landforms and sediments. The aim of this book is to show how glaciers work and how the processes which operate within them are recorded in the glacial landscape. With this knowledge you can begin to read the story of the Cenozoic Ice Age etched into the landscape by its glaciers.

As a starting point we first examine the history of ice on Earth and discuss the causes of ice ages. This is followed by two chapters which introduce the glacial system. If we are to interpret glacial landforms we must understand how the glaciers that produced them work. In short, we discover how glaciers grow, flow and decay. In Chapters 5 and 6 we explore the processes of glacial erosion and consider the landforms which they create; landforms which can be seen in the landscape today and which provide evidence about the dynamics of the glaciers that formed them. In Chapters 7–11 we tackle the processes of glacial sedimentation and landform development which provide some of the most important evidence of former glacier activity. The final chapter examines how the large-scale pattern of glacial erosion and deposition within a former ice sheet is reflected in the landscape.

2
The History of Ice on Earth

In this chapter we examine the history of ice on Earth. We review the occurrence of ice ages within the geological record, and the debate which surrounds their cause, before concentrating on the Cenozoic Ice Age. We describe the rhythm of glaciation in the Cenozoic and discuss the nature of the landform and sediment record which it produced.

2.1 ICE AGES AND EARTH HISTORY

The rocks and fossils of the geological record provide us with evidence of global climatic change throughout much of Earth history. From this record it is apparent that during the last 600 million years, the Phanerozoic, the Earth's climate has varied between two states: one of global warming, the **'greenhouse' state**, and one of global cooling, the **'ice-house' state** (Figure 2.1). During periods of global refrigeration, or **ice ages**, glaciers dominated the high-latitude areas of the globe. During 'greenhouse' periods the Earth was usually ice-free and there was generally a period of global warmth. The number of oscillations between these two states recognisable in the geological record depends largely on the scale or the detail with which this record is viewed. Over the largest timescale it is possible to recognise seven 'ice-house'–'greenhouse' oscillations within the Phanerozoic (Figure 2.1). A similar pattern of oscillations may have occurred during the Proterozoic and Archaean eons of the Precambrian, but the palaeoclimatic record from this very ancient time is poor (Figure 2.1).

In examining each 'ice-house' state in the Phanerozoic, we can identify several similarities. Each 'ice-house' state appears to start with a long period of global cooling which culminates in the growth of high-latitude ice sheets at the Earth's poles. These periods of global refrigeration are normally terminated by a sudden episode of climatic warming. The development of ice sheets in the mid-latitudes is uncom-

8

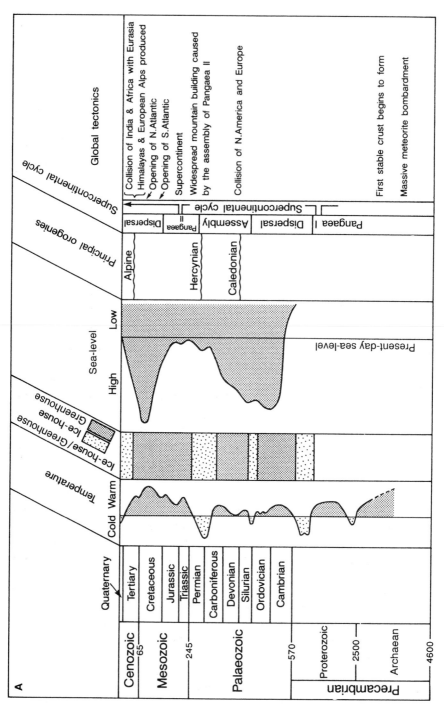

Figure 2.1 The chronostratigraphical scale and the variation in climate, sea-level and plate tectonics during Earth history. [Diagram reproduced with permission from: Doyle et al. (1994) Key to Earth History, John Wiley, Fig. 10.8, p. 168]

mon. Mid-latitude ice sheets are, however, a feature of the most recent or Cenozoic Ice Age, during which they expanded and contracted with a regular periodicity. Mid-latitude ice sheets may also have grown during both the Carboniferous and Late Proterozoic Ice Ages (Figure 2.1). In contrast, 'greenhouse' conditions appear to start rapidly and are normally terminated in a more gradual fashion via a period of cooling which take places at different rates in different geographical locations. 'Greenhouse' conditions are loosely correlated with periods of geological time when: (1) the atmosphere possessed a higher concentration of greenhouse gases, such as carbon dioxide; (2) sea-level was high and therefore large areas of the continents were flooded; and (3) the Earth's ocean basins were very large and/or well connected.

This record of climate change indicates a natural oscillation between periods of global cooling and global warming. The cause of these oscillations is not certain but we can identify four broad types of climatic forcing which may be responsible: (1) variation in the amount and distribution of solar radiation received at the top of the atmosphere; (2) variation in the composition of the atmosphere; (3) the nature of the Earth's surface including the distribution of land and sea; and (4) extra-terrestrial causes.

1. **Radiation variations.** The first explanation for climate change involves variations in the amount of solar radiation received in the upper part of the Earth's atmosphere. Variation in the input of solar radiation to the Earth occurs at two broad scales. First, it varies at a high frequency due to **Milankovitch radiation variations**. As we shall see later, these determine the climatic pattern within an 'ice age'—the cyclic growth and decay of ice sheets—but cannot explain the onset of an 'ice-house', since Milankovitch radiation variations occur continuously in both 'ice-house' and 'greenhouse' periods. The second scale of radiation variation is of much lower frequency, but higher magnitude. The solar system rotates around the centre of the galaxy once every 300 million years, a period known as the **galactic year**, and in so doing it passes through two stationary 'clouds' of hydrogen-rich particles which may cause fluctuations in the amount of solar radiation received by the Earth. Consequently, every 150 million years the Earth may receive less solar radiation. It has been suggested that the record of 'ice age', or 'ice-house' states has an approximate periodicity of 150 million years. This is an interesting hypothesis, but at present the geological record of glaciation and palaeoclimate on these long timescales is not precise enough to allow this to be rigorously tested.

2. **Variations in atmospheric carbon dioxide.** The budget of atmospheric carbon dioxide helps to determine the Earth's climatic state. Carbon dioxide is a 'greenhouse gas', that is one that helps insulate the Earth. The proportion of carbon dioxide within the atmosphere will influence global temperatures: the more carbon dioxide, the greater the greenhouse effect and the warmer the climate. At the simplest level, atmospheric carbon dioxide concentrations are dependent on the rate of volcanic outgassing, which is primarily a function of plate tectonics, and on changes in the rate of continental weathering. Sea-floor spreading, and the associated volcanic activity, provides the main input of car-

bon dioxide (volcanic gas) to the atmosphere, while chemical weathering of silicate rocks removes carbon dioxide. Carbon dioxide is removed by chemical weathering because rain dissolves carbon dioxide within the atmosphere to produce an acidic solution that breaks down silicate rocks and locks the carbon molecules into the weathered rock. Periods of continental rifting and rapid sea-floor spreading may therefore be associated with warmer climates, while periods of uplift and erosion, which reveal fresh silicate rocks for weathering, may have lower carbon dioxide contents and be cooler. The Cretaceous 'greenhouse' period has been correlated with high carbon dioxide values produced by the rapid rate of continental rifting and sea-floor spreading associated with the opening of the Atlantic Ocean during the late Mesozoic (Figure 2.1). The high sea-levels of this time would also have reduced the area of silicate rocks available for weathering. The atmospheric budget of carbon dioxide may therefore help to explain the variation between 'ice-house' and 'greenhouse' states recorded in the geological record. However, it is important to note that the carbon dioxide budget is much more complex than the above discussion might suggest.

3. **The distribution of land and sea.** The nature of the Earth's surface is an important control on climate, particularly through the location of the continental plates which determine the distribution of land and sea. There is a clear correlation between 'ice-house' conditions and the times when plate tectonics produced a concentration of continental masses in high latitudes. There are several reasons why this may happen. The presence of high-latitude land provides a surface for the accumulation of snow. Snow increases the reflective qualities of land, its **albedo**, leading to an increase in the reflection of solar radiation back to the atmosphere. As a consequence, less radiation is available to warm the land surface. Further, high-latitude landmasses may block the poleward penetration of warm ocean currents from the equator and therefore limit the poleward transport of heat. This has played an important role in developing the Cenozoic 'ice-house'. Periods dominated by 'ice-house' conditions may therefore be explained by the concentration of continents in high-latitude areas through plate tectonics. In addition to helping to explain 'ice-house' conditions, the distribution of continents may also explain 'greenhouse' states. Most 'greenhouse' periods are associated with large well-connected oceans, allowing the poleward transport of heat. For example, the mid-Cretaceous 'greenhouse' phase was one of the warmest periods in the Earth's history, and it occurred at a time when ocean basins were well connected and the resulting ocean circulation pattern may have allowed efficient heat transfer towards the poles. In addition, high sea-levels and extensive continental flooding also appear to be loosely correlated with periods of 'greenhouse' conditions (Figure 2.1). Oceans and continental seas can absorb much more heat than land, while evaporation from their surfaces increases cloud cover and therefore helps retain heat. The high sea-levels of the Cretaceous 'greenhouse' phase may also, therefore, be of considerable importance to the pronounced warmth of this period (Figure 2.1).

The importance of continental distribution in determining global climate has been examined in detail through the use of **general circulation models**. These are three-dimensional computer models of the Earth's climate system. For example, one particular set of experiments examined the global climate associated with different continental distributions, in particular with: (1) a large polar continent without an ice cap; (2) a large polar continent with an ice cap; and (3) a large tropical continent. The results are illustrated in Figure 2.2. The experiment with a polar continent (without an ice sheet) produced a globally averaged surface temperature that was about 12 °C cooler than when the model was run using a tropical continent (Figure 2.2). The presence or absence of an ice cap on the polar continent was also found to be very important in determining global temperature. The increased reflectivity of the ice cap enhances the rate at which the polar continent is cooled (Figure 2.2). These experiments help demonstrate the importance of the distribution of land and sea in determining the Earth's climatic state.

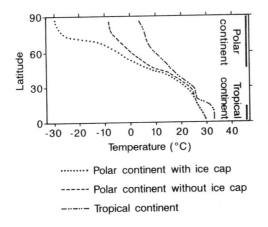

········ Polar continent with ice cap

------ Polar continent without ice cap

—··—··— Tropical continent

Figure 2.2   *A plot of average annual air temperature against latitude for three different continental distributions as predicted by a general circulation model. [Diagram reproduced with permission from: Doyle et al. (1994) Key to Earth History, John Wiley, Box 10.4, p. 157]*

4. **Extra-terrestrial causes.** The impact of meteorites or 'bolides' from space has been put forward as a possible explanation for the change between 'ice-house' and 'greenhouse' states on Earth. The impact of a large meteorite could potentially cause global cooling through the production and injection of masses of dust into the Earth's upper atmosphere, blocking off the sun's radiation. Equally, the impact of a large extra-terrestrial body into an ocean might also promote higher temperatures through the vaporisation of large amounts of water. The increased cloud cover which would result from this would also insulate the Earth, preventing heat loss and thereby causing global warming. Although such ideas are currently fashionable, for the most part they remain untested.

The quest for an explanation for these large-scale climatic oscillations is a complex one and is made difficult by the increasing lack of precise palaeoclimatic data with increasing geological age. It is also unclear whether a single causal mechanism explains all events or whether each episode was triggered by a different combination of factors. Despite this, however, changes in palaeogeography and in atmospheric carbon dioxide levels through time provide some of the most plausible explanations for the oscillation between 'ice-house' and 'greenhouse' climates in Earth history.

Most of our information about global climate during 'ice-house' conditions comes from the Cenozoic Ice Age. In comparison with the record for earlier periods of glaciation, a large amount is known about this period of 'ice-house' conditions due to the quality of the environmental record. More importantly, it is the ice sheets of this period which have provided us with the legacy of sediments and landforms we see today. In the next section we introduce the story of the Cenozoic Ice Age.

2.2 THE CENOZOIC ICE AGE

The events of the Cenozoic Ice Age are recorded by numerous different lines of evidence. These can, however, be broadly classified as: (1) terrestrial landforms and sediments; (2) the terrestrial floral and faunal (biological) record; and (3) ocean sediments, flora and fauna. Terrestrial landforms and sediments not only record the growth and decay of ice sheets, but also the changing environment in areas far beyond the limit of the ice. For example, many of the sand dunes within today's deserts are relict landforms which reflect ancient wind directions dominant during the Cenozoic Ice Age. Changes in climate are also recorded by changes in vegetation and in the faunal assemblage present within an area. Fossil remains of fauna and pollen from vegetation accumulate in lakes and peat bogs to provide a record of these changes. However, both the terrestrial landform and biological record of the Cenozoic Ice Age are incomplete due to frequent breaks or hiatuses within the depositional record. In the case of glacial landforms these breaks occur because each successive glacier erodes the evidence of earlier glaciers. Lakes and bogs in which faunal and floral remains accumulate are subject to periodic desiccation or become infilled, and rarely contain very long sedimentary records. In contrast, however, sedimentation within ocean basins has been almost continuous throughout the Cenozoic and consequently these sediments contain the best record of the Cenozoic Ice Age. Here the changing isotopic composition of the oxygen within sea-water, a function of global ice volume, is recorded in the calcareous skeletons and shells of micro-organisms which accumulate on the ocean floor. Variations in the isotopic composition of the oceans provide a record of global ice volume (Box 2.1), which allows us chart the history of the Cenozoic Ice Age (Figure 2.3).

At the start of the Cenozoic, 65 million years ago, the Earth was ice-free. Forests grew, even at high latitudes. The world's continents were slowly moving apart, having been part of a single supercontinent, Pangaea, during the Jurassic some 100

BOX 2.1: GLACIAL HISTORY AND THE OXYGEN ISOTOPE RECORD

The isotopic composition of ocean water changes with the growth and decay of ice sheets. A record of this isotopic variation and therefore a record of global ice volume is stored in shell carbonate within deep-sea sediments.

Stable isotopes are elements which have more than one stable atomic state. The nucleus of an atom contains positively charged mass particles called **protons** and mass particles with no electrical charge known as **neutrons**. Orbiting the nucleus at various distances are negatively charged particles known as **electrons** which control the chemical properties. Certain elements may possess the same number of protons, but a different number of neutrons; these are called **isotopes**. For example, oxygen has two isotopes. Both have the same number of protons (eight), but one has eight neutrons and the other ten. This is indicated by the **atomic mass number** (protons + neutrons), all oxygen has eight protons, but one isotope has eight neutrons and the other ten giving ^{16}O and ^{18}O. Some isotopes are stable, like oxygen, while others are unstable and will alter through time to obtain a more stable form—a process known as **radioactive decay**.

On average, the ratio of ^{18}O to ^{16}O in today's ocean water (H_2O) is about 1:500, or alternatively 0.2% of all oxygen is ^{18}O. The oxygen locked up in the calcium carbonate ($CaCO_3$) of marine shells reflects the ratio existing at the time of the shells' growth. This ratio varies with the growth and decay of the Earth's ice sheets. When sea-water evaporates, a process of natural fractionation occurs and more water molecules with ^{16}O are evaporated than those with ^{18}O because the atomic mass of ^{16}O is less (i.e. the molecule is lighter). Atmospheric water (clouds and rain) are therefore enriched in ^{16}O. In a non-glacial environment, the balance of ^{18}O to ^{16}O is maintained because rain-water falling on land quickly returns to the ocean via rivers. In contrast, during a glacial period the oceanic balance of ^{18}O to ^{16}O is upset because atmospheric moisture is not returned

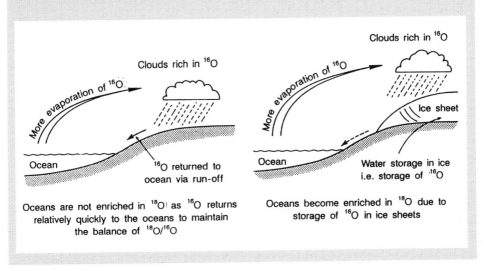

Oceans are not enriched in ^{18}O as ^{16}O returns relatively quickly to the oceans to maintain the balance of $^{18}O/^{16}O$

Oceans become enriched in ^{18}O due to storage of ^{16}O in ice sheets quickly to the oceans but falls as snow and is stored in ice sheets, as shown in the diagram. Consequently the oceans are enriched in ^{18}O during periods of glaciation. A record of the Earth's glacial history during the Cenozoic Ice Age can therefore be obtained from an analysis of oxygen isotopic composition of deep-ocean sediments: horizons in which the shells and skeletons of marine organisms are enriched in ^{18}O correspond to periods when the Earth's total ice volume was large, while those horizons with less ^{18}O correspond to periods with a smaller global ice volume.

Source: Imbrie, J. & Imbrie, K. P. 1979. *Ice Ages: Solving the Mystery.* Harvard University Press, Cambridge, Massachusetts. [Diagrams modified from: Doyle *et al.* (1994) *Key to Earth History,* John Wiley, Figure 5.14, p. 84]

million years before (Figure 2.1). As the continents dispersed, the pattern of ocean circulation also underwent gradual change (Figure 2.4). It is these changes in ocean circulation which appear to have played an important role in global cooling at the onset of the Cenozoic Ice Age. At the start of the Cenozoic, major equatorial currents encircled the globe and the transfer of heat from the equator to the poles was relatively efficient. This kept the high latitudes warm and global climate free from temperature extremes. From this point on, major changes in the southern and northern hemispheres had a dramatic effect upon the Earth's climatic system. About 50 million years ago Australia began to move northwards, breaking free of Antarctica and opening a Southern Ocean in which westerly winds were able to establish a major ocean current known as the **Antarctic Circumpolar Current** (Figure 2.4). Today this current completely encircles the Antarctic continent and is partly responsible for the extreme cold of these latitudes. The current acts as a barrier to warm water from low latitudes reaching the Antarctic coast, and prevents the efficient transfer of heat from the relatively warm equatorial and mid-latitude oceans to the high southern latitudes.

With the development of the Antarctic Circumpolar Current, progressive cooling of the southern high latitudes proceeded in a stepped fashion (Figure 2.3). One major cooling episode took place about 38 million years ago at the end of the Eocene, which marked the onset of freezing conditions at sea-level around Antarctica. This sudden cooling probably reflects the growth of ice bodies on the Antarctic continent. As the ice sheets grew, the albedo of the land surface would have increased, since ice reflects heat more effectively; consequently less solar radiation would be absorbed, leading to further cooling. This feedback system accelerated the rate of cooling of southern latitudes and the growth of the Antarctic ice sheet. The impact of the Antarctic Circumpolar Current increased when Antarctica finally broke away from South America about 25 million years ago, creating an oceanic gap between them (Figure 2.4). The current now became truly circumpolar and the further cooling led to further development and expansion of the Antarctic Ice Sheet (Figure 2.3).

In the northern hemisphere the lack of land over the pole and the absence of a strong circumpolar current meant that cooling occurred much later and it was not

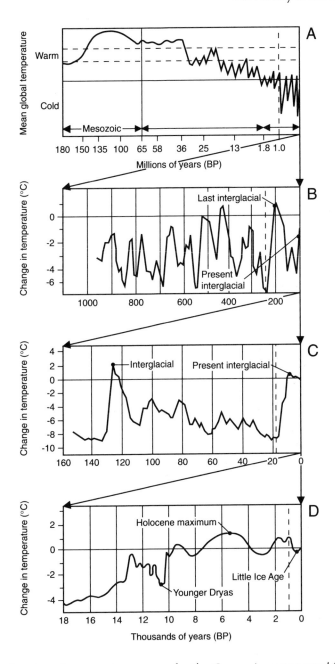

Figure 2.3 *A series of temperature curves for the Cenozoic constructed from information obtained from oxygen isotopes and from historical data. [Diagram reproduced with permission from: Boulton (1993) In: Duff (Ed.)* Holmes' Principles of Physical Geology, *Chapman & Hall, Fig. 21.18, p. 446]*

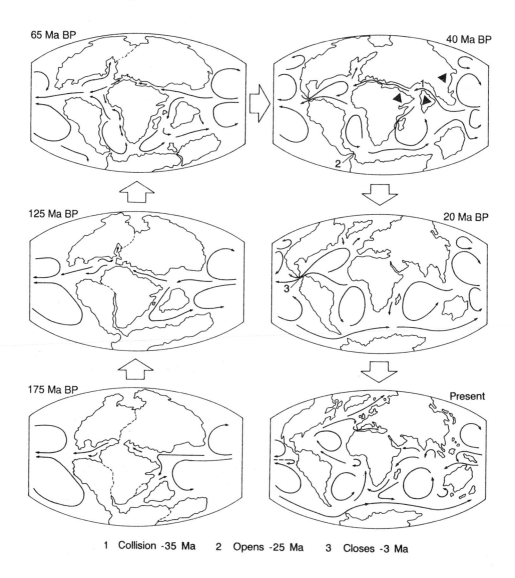

65 Ma BP

40 Ma BP

125 Ma BP

20 Ma BP

175 Ma BP

Present

1   Collision -35 Ma   2   Opens -25 Ma   3   Closes -3 Ma

Figure 2.4   *Schematic reconstruction of the pattern of major ocean currents during the Mesozoic and Cenozoic.1, Collision (35 Ma BP); 2, opens (25 Ma BP); 3, closes (3 Ma BP). [Modified from: Williams* et al. *(1993)* Quaternary Environments, *Edward Arnold, Fig. 2.2, p. 18]*

until the late Pliocene, around three million years ago, that conditions had cooled sufficiently for significant ice bodies to develop in Greenland, Arctic Canada and Siberia. The closure of the seaway between North and South America, approximately three million years ago, may have been significant in precipitating the onset of glaciation in the northern hemisphere, strengthening the North Atlantic currents and in turn delivering snow to the northern high latitudes. By 2.4 million years ago, moderately sized ice sheets had grown in the northern hemisphere in response to these changes. From this time onwards the mid-latitudes saw the regular growth and decay of ice sheets and glaciers. During cold phases, known as **glacials**, ice sheets and glaciers reached their maximum and covered much of the northern mid-latitudes. These cold periods were interspersed with warmer periods, called **inter-glacials**, in which the distribution of ice was similar to that of today (Table 2.1). A **glacial cycle** consists of both a glacial and an interglacial period.

Table 2.1 *The present-day distribution of glacier ice on Earth. Note that most of the volume of glacier ice is located in the Antarctic Ice Sheet.*

	Area (10^6 km^2)	Volume (10^6 km^3)
Antarctica	12.6 (84.3%)	31.0 (91.4%)
Greenland	1.8 (12.1%)	2.8 (8.3%)
Other	0.54 (3.6%)	0.12 (0.3%)

The oscillation between periods of warm and cold appears initially to have had a periodicity of 41 000 years. After about 900 000 years ago this changed to a periodicity of 100 000 years (Figure 2.3). This change was reflected by the growth of much larger mid-latitude ice sheets during the cold phases. It is these large ice sheets which have left the dramatic imprint on the landscape of the northern mid-latitudes we see today.

The regular oscillation between cold and warmth, and therefore between glacial and interglacial episodes, is driven by regular changes in the amount of solar radiation received by the Earth from the Sun. These regular variations are caused by cyclic changes in the Earth's orbit around the Sun. The magnitude and frequency of the changes in solar radiation received by the Earth as a consequence of these orbital changes was first calculated by the astronomer **Milutin Milankovitch** in 1941. He identified three orbital processes which would control these changes: axial tilt, eccentricity and precession of the equinoxes (Figure 2.5).

1. **Axial tilt.** The Earth's axis of rotation is inclined at present at an angle of about 23.5°. However, this angle is not constant and varies between 24.5° and 21.5° every 40 000 years. When the angle of tilt is greatest there is the greatest difference in seasonal heating at any latitude.
2. **Eccentricity of the orbit.** The Earth's orbit gradually changes from an elliptical orbit to a circular one every 100 000 years.
3. **Precession of the equinoxes.** The Earth wobbles slightly on its axis due to the gravitational pull of the Sun and Moon. Consequently the time of the year at which the Earth is nearest to the Sun changes through time in a cycle which takes about 23 000 years.

These three orbital cycles cause subtle variations in the amount of solar radiation received at different latitudes through time. It is these variations which have pro-

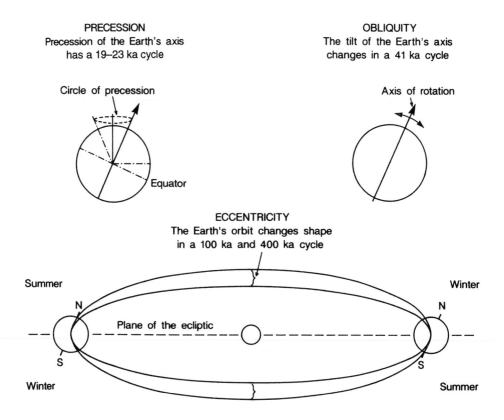

Figure 2.5   *Orbital oscillations causing fluctuation in the solar radiation received by Earth deduced by Milankovitch in 1941. [Diagram reproduced with permission from: Boulton (1993) In: Duff (Ed.) Holmes' Principles of Physical Geology, Chapman & Hall, Fig. 21.9, p. 447]*

vided the rhythm of glacial expansion and contraction during the Cenozoic Ice Age (Figure 2.3) and probably that during earlier ice ages in Earth history.

One of the most significant aspects of the Cenozoic Ice Age is the growth of mid-latitude ice sheets, and the greater apparent intensity of the Cenozoic 'ice-house' than some earlier 'ice-house' episodes is hard to explain. One hypothesis which explains this is the impact on atmospheric circulation of Cenozoic mountain building. Within the northern hemisphere there are two regions of extensive highland: (1) the Tibetan Plateau and the Himalayan mountains of Asia, and (2) the Western Cordilleras of North America. Both these areas are broad, plateau-like bulges on which narrower mountain ranges are superimposed and they have only reached their present elevation within the last million years. The large-scale circulation of the atmosphere is strongly influenced by these areas of uplift. They produce large-scale standing waves (**Rosby waves**) within the upper atmosphere of the northern hemisphere (Figure 2.6). These waves occur in the lee of the western edge of the

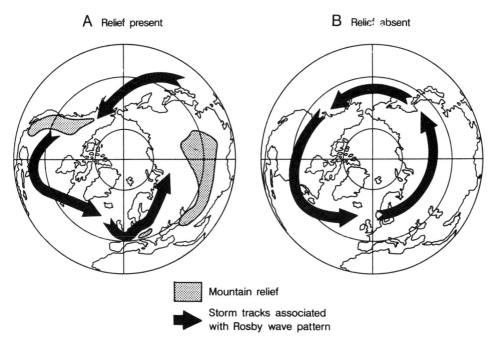

A Relief present   **B** Relief absent

Mountain relief

Storm tracks associated
with Rosby wave pattern

Figure 2.6 *The pattern of atmospheric circulation in the northern hemisphere with and without mountain relief present, illustrating the role of mountain relief in modifying atmospheric circulation. The relief causes a large wave-like structure (Rosby waves) to develop within the jet stream which control storm tracks and helps transfer cold polar air into the mid-latitudes. [Diagram reproduced with permission from: Doyle et al. (1994) Key to Earth History, John Wiley, Box: 10.3, p. 155]*

North American Cordilleras, to the east of the mountains of the European continent, and in the lee of the Tibetan plateau. Without these mountains the pattern of atmospheric circulation would be much smoother (Figure 2.6B). Consequently the Late Cenozoic uplift of these mountainous areas produced a series of waves and troughs within the pattern of atmospheric circulation of the northern hemisphere. There is therefore a strong north–south component to the circulation, which during a period of Milankovitch cooling would tend to draw down polar air into mid-latitudes, thereby intensifying the cooling episode in these areas. The occurrence and distribution of mountains within an ice age may therefore have important consequences in determining its intensity and may help to explain why some are associated with large mid-latitude ice sheets and others are not.

2.3 THE NATURE OF THE GLACIAL RECORD

Our understanding of the history of the Cenozoic Ice Age is based primarily on the oxygen isotope record from the ocean sediments. This, however, can only provide an indication of the variation in the total global ice volume and it is clear that not all glaciers grew and decayed synchronously across the globe. Instead, there is a complex pattern of regional variation which is masked by the growth and decay of the large North American Ice Sheet which, because of its size, dominates the oxygen isotope record. In order to unravel this regional variation we must turn to the direct evidence of glaciation. Glaciers may be reconstructed in several different ways using a variety of different types of data, including:

1. **Glacial landforms and sediments.** Accurate mapping of landforms or sediments formed by a particular ice sheet can be used to reconstruct its morphology and dynamics.
2. **Numerical models.** Theoretical ice sheets may be reconstructed using computer-generated models based on equations designed to replicate the physical behaviour of ice and upon inferences about the climate that drives the growth and decay of the ice sheet. These models may be of two types: (1) ice-sheet models generated independently of field evidence, which are then compared or tested against field evidence; and (2) ice-sheet models which are constrained by field evidence and the input variables are modified until the modelled ice sheet replicates the field evidence present.
3. **Raised shorelines.** As ice sheets grow, their weight depresses the crust; as they decay, the crust rises. This is known as **isostatic depression**. The crust responds relatively slowly to the change in the load placed upon it and consequently it continues to rise long after the ice sheet has melted. Parts of North America and Scandinavia are still experiencing **isostatic rebound** following the last glacial cycle. This history of isostatic depression can be reconstructed from evidence such as raised shorelines in coastal areas. The highest raised shorelines occur in areas close to the centres of the former ice sheet, where the ice was thickest and the crustal depression greatest; the lowest occur close to the ice-sheet margins. Raised shorelines should therefore slope away from the centre of an ice sheet. Lines of equal isostatic uplift, **isobases**, can be drawn between raised shorelines of similar elevations and used to locate the centre of former ice sheets. The degree of uplift following deglaciation may also provide estimates of ice-sheet thickness.

Of these different lines of evidence the most direct, and most frequently used, is to examine sediments and landforms left by the glaciers themselves. It is this evidence which forms the glacial landscape we see today. These landscapes range from areas within the interior of former ice sheets, which have been deeply eroded and scoured, to the more marginal areas, where the landscape is dominated by depositional landforms and thick layers of glacial sediment.

The record of landforms and sediments left by these ice sheets is, unfortunately, often fragmentary and incomplete. Glaciers are erosive and recent glaciers tend to destroy the evidence left by earlier ones. To confuse things further, a glacier may not always destroy all the evidence of earlier episodes of glaciation, particularly if it covers a less extensive area. As a consequence, glacial landscapes may contain landforms and sediments formed by different glaciers at different times. This is particularly true of the erosional landscape, which is more robust than depositional landforms, the latter being relatively easily modified or removed by subsequent ice action. For example, in Britain the magnificent mountain scenery of Scotland, Wales and the Lake District is the product of several phases of glacial erosion which cannot be distinguished from each other. In contrast, few, if any, glacial sediments or depositional landforms survive in these areas which do not date from the final phase of glaciation. However, in the more lowland areas of Britain, towards the periphery of the former ice sheets, where the intensity of glacial erosion was less, glacial landforms and sediments formed by earlier phases of glaciation have survived. Areas such as Britain, which have been glaciated many times during the Cenozoic Ice Age, may possess glacial landscapes that contain a complex and fragmentary record of successive ice sheets and periods of glaciation.

2.4 GLACIAL STRATIGRAPHY

Unravelling the history and interpreting the significance of glacial landscapes is an exercise in stratigraphy. Stratigraphy involves determining the relative order of sediments or landforms and interpreting them as events or episodes within Earth history. It allows the relative order, size and character of ice sheets to be established from the evidence left within the glacial landscape. Glacial stratigraphy involves two distinct stages: (1) establishing the relative order of landforms and sediments; and (2) interpreting those landforms and sediments in terms of the glacial environments in which they formed. The relative order of a sequence of landforms or sediments is established through the application of a series of simple tools which form the 'stratigraphical tool kit'. This tool kit includes:

1. **The principle of superposition.** In any undisturbed sequence of sediments or landforms the layer or landform at the bottom of the sequence was formed first.
2. **Cross-cutting relationships.** If a landform cuts across or truncates another it must post-date the truncated landform (Box 2.2).
3. **Lithostratigraphy.** It is possible to correlate sediments on the basis of their lithological character. For example, glacial sediments deposited by a single ice sheet may contain a similar range of rock types or lithologies.
4. **Biostratigraphy.** When glacial sediments are found in association with organic or fossil remains this may allow them to be correlated. For example, during the early part of the Quaternary many small mammals underwent rapid evolution, and some species existed only for very short periods of time. Sediments that contain fossil mammals at the same evolutionary stage must have formed at a similar time and can therefore be correlated. Similarly, most of the interglacial periods in Britain had different assemblages of flora and fauna and can be distinguished on this basis. Consequently any glacial sediments found in association with them can be placed into the appropriate period of glaciation.
5. **Geochronology.** A wide range of techniques exist by which sediments and landforms may be dated either relative to one another or with reference to an absolute timescale. Absolute dating might involve the application of radiometric dating techniques which rely on the time-dependent decay of unstable isotopes through the release of radioactive particles.

BOX 2.2: THE APPLICATION OF CROSS-CUTTING RELATION-SHIPS IN THE STUDY OF ICE SHEET BEHAVIOUR

Boulton and Clark (1990a,b) studied satellite images from most of the area of the Canadian mainland once covered by the North American (Laurentide) Ice Sheet. On these they were able to recognise and map a complex pattern of superimposed and cross-cut subglacial bedforms (see Section 9.2.1). The largest of these bedforms are mega-scale glacial lineations—low, broad ridges of sediment typically between 200 and 1300 m wide and between 8 and 70 km long—which form subglacially and parallel to the direction of ice flow. Not only do these giant lineations cross-cut or truncate one another but they have drumlins superimposed upon them. Drumlins also form subglacially and are elongated in the direction of ice flow. These superimposed drumlins are often orientated parallel to a different ice flow direction from the main lineation. By mapping out these lineations and interpreting them in terms of ice flow directions, a complex map of cross-cutting ice-flow direction is obtained, as shown below. In the diagram below, each line is a summary of the ice flow deduced from the sub-

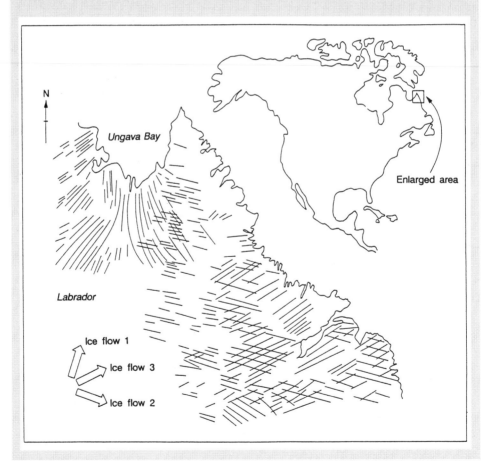

glacial bedforms present. The sequence of ice flows which make up this pattern can be deduced by the study of aerial photographs and with the application of the principles of superposition and cross-cutting relationships: the youngest lineations, and therefore most recent ice flow, are either superimposed on, or truncate, the older lineations. Boulton and Clark (1990a,b) interpreted the sequential pattern of ice flows in terms of a shifting pattern of ice divides within the North American Ice Sheet. Ice flows from the centre of an ice sheet, its ice divide, towards its margin; if the ice divide moves then the pattern of ice flow within the ice sheet will change and subglacial bedforms may become superimposed. This example illustrates the application of very simple stratigraphical tools in determining the glacial history of one of the largest ice sheets that existed during the Cenozoic Ice Age.

Sources: Boulton, G. S. & Clark, C. D. 1990a. A highly mobile Laurentide ice sheet revealed by satellite images of glacial lineations. *Nature* **346**, 813–817. Boulton, G. S. & Clark, C. D. 1990b. The Laurentide ice sheet through the last glacial cycle: the topology of drift lineations as a key to the dynamic behaviour of former ice sheets. *Transactions of the Royal Society of Edinburgh* **81**, 327–347. [Diagram modified from: Boulton & Clark (1990b) *Transactions of the Royal Society of Edinburgh* **81**, Figure 8, p. 335]

Having established the relative order of the landforms and sediments it is necessary to interpret them in terms of the glaciers which formed them, a task that is not always easy (Box 2.3). They provide clues not only about the extent and surface morphology of a glacier but more importantly about its dynamics and behaviour. This behaviour is primarily a function of climate, which allows inferences about palaeoclimate to be made from former glaciers. It is this interpretation of glacial landforms and sediments which is crucial to understanding the significance of a glacial landscape and it is their interpretation which forms the focus of this book.

Interpretation involves the application of the principle of **uniformitarianism** or **actualism**. This states that the present is the key to the past. By examining the behaviour of contemporary glaciers and their products, we can interpret the landforms and sediments produced by ancient glaciers. In the contemporary environment the aim is, therefore, to understand the processes which operate to create the landforms and sediment (the product). In interpreting ancient environments the reverse is true: process can be deduced through the study of the product by reference to contemporary models or analogues. The glacial products—landforms and sediments—are used to infer the processes which operated within the former glacier that produced them. The logic involved in this process is illustrated in Figure 2.7. It rests upon the assumption that modern glaciers are suitable analogues for those of the past. There is no reason to question such an assumption since the physical properties of ice have always been the same and as we will see in the next chapter, the glacier system behaves in a relatively predictable way in response to variation in climate.

BOX 2.3: THE INTERPRETATION OF GLACIAL LANDFORMS

Different glacial geologists may interpret the geomorphological record differently depending upon the model with which the landscape is viewed. This point is illustrated by the moraines of Glen Geusachan in the Cairngorms, Scotland. Within the Cairngorms there are numerous glacial deposits of a variety of different ages. Some relate to the deglaciation of the Scottish Ice Sheet at the close of the last glacial cycle, while others were deposited during a brief return to glacial conditions at the close of this cycle, an event known as the Younger Dryas. At this time an ice-field existed in the Western Highlands and valley glaciers in the Cairngorms.

The moraines of Glen Geusachan were first mapped by Sugden (1970) who interpreted them in terms of the prevailing model of the time, which stressed the importance of glaciofluvial landforms and of ice sheet stagnation. He argued that most of the landforms pre-dated the Younger Dryas. This view was vigorously challenged in a series of papers by Sissons, who finally published a new geomorphological interpretation of Glen Geusachan in 1979. Sissons dismissed the earlier glaciofluvial landform interpretations and suggested that the moraines were best interpreted as hummocky moraine produced by the stagnation of a small valley glacier at the close of the Younger Dryas. In turn, this interpretation was challenged by Bennett and Glasser (1991). They agreed that the moraines were the product of a Younger Dryas valley glacier, but challenged Sissons' interpretation as hummocky moraine formed during ice stagnation. Instead they mapped individual ice-marginal landforms formed at an actively retreating ice margin.

Three sets of glacial geologists have therefore examined the same evidence and come up with three different interpretations, primarily because each entered the field with a different objective and was guided by different models. This illustrates some of the difficulty involved in interpreting the glacial landscape and making correct inferences about the glaciers which formed it. In the map extracts below, A is from Bennett and Glasser (1990), B is from Sissons (1979) and C is from Sugden (1970).

Sources: Bennett, M. R. & Glasser, N. F. 1991. The glacial landforms of Glen Geusachan, Cairngorms: a reinterpretation. *Scottish Geographical Magazine* **107**, 116–123. Sissons, J. B. 1979. The Loch Lomond Stadial in the Cairngorm mountains. *Scottish Geographical Magazine* **95**, 66–82. Sugden, D. E. 1970. Landforms of deglaciation in the Cairngorms, Scotland. *Transactions of the Institute of British Geographers* **51**, 201–219. [Diagrams reproduced with permission from: Bennett & Glasser (1991) *Scottish Geographical Magazine* **107**, Figure 3, p. 118. Sissons (1979) *Scottish Geographical Magazine* **95**, Figure 3, p. 70. Sugden (1970) *Transactions of the Institute of British Geographers* **51**, Figure 1, p. 203]

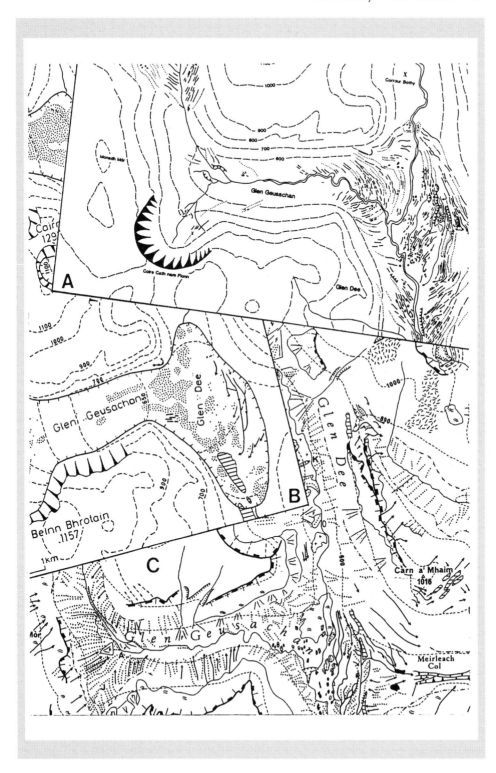

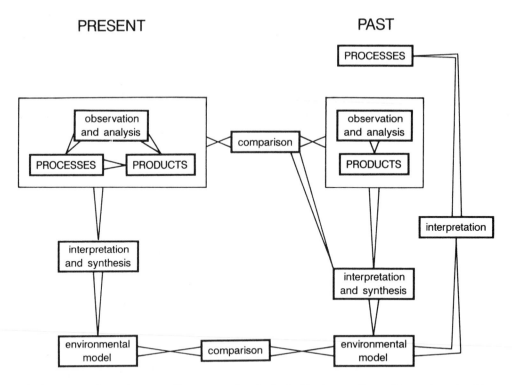

Figure 2.7   *Flow diagram illustrating the logic behind actualism. [Reproduced with permission from:* Earth History 1, Science Foundation Course Unit 26, *Open University, p. 6]*

This book revolves around this concept. We shall first explore how glaciers work today before examining their products. By understanding modern glaciers we can interpret the record of the ancient glaciers that have shaped the glacial landscape.

2.5 SUMMARY

Global climate has oscillated between 'greenhouse' and 'ice-house' periods throughout much of Earth history. The most plausible explanation for this oscillation in global climate involves changes in palaeogeography—the distribution of land and sea—and variation in the carbon dioxide content of the atmosphere. Within an 'ice-house' period climatic oscillations, marked by the growth and decay of glaciers, are controlled by variations in the amount of solar radiation received by the atmosphere. This variation is caused by cyclic changes in the Earth's orbit known as Milankovitch cycles. During the Cenozoic (the last 65 million years) the Earth has experienced a period of 'ice-house' conditions. Ice sheets have grown and decayed in both polar and mid-latitude locations at regular intervals throughout this time. These ice sheets have etched their story into the landscape and have left a legacy of glacial landforms and sediments. These landforms and sediments provide the clues by which glacial geologists can reconstruct these ice sheets and read the story of the Cenozoic Ice Age. In order to read this story glacial geologists must first establish the chronology of glacial events within an area through the application of glacial stratigraphy. The landforms and sediments can then be used to reconstruct the morphology and dynamics of the ice sheet that formed them. Interpreting the glacial landscape relies upon the application of uniformitarianism or actualism: the principle that modern glaciers can be used as analogues with which to interpret the landforms and sediments left by former glaciers.

2.6 SUGGESTED READING

A simple overview of palaeoclimate through Earth history can be obtained from Doyle *et al.* (1994). A more detailed account of the Earth's palaeoclimatic history is contained within Van Andel (1994) and within Frakes *et al.* (1992). The occurrence of ice ages within the geological record and their significance is discussed in several contributions, including Hambrey & Harland (1981), Young (1991), and Hambrey (1992). The history of the Cenozoic Ice Age is reviewed by Dawson (1992). The rhythm of the Cenozoic Ice Age and the role of Milankovitch radiation variations is discussed by Imbrie & Imbrie (1979), while the onset of glaciation in the northern hemisphere and the role of mountain building is discussed by Ruddiman & Kutzbach (1990), Raymo & Ruddiman (1992) and Raymo (1994). The methods by which glaciers can be reconstructed are reviewed by Andrews (1982) and examples can be obtained from Ballantyne (1989) and Boulton *et al.* (1985). Doyle *et al.* (1994) cover the basic principles of stratigraphy necessary to interpret the relative order of events within a glacial landscape. The book by Lowe & Walker (1984) also contains important information about the reconstruction of Cenozoic environments.

Andrews, J. T. 1982. On the reconstruction of Pleistocene ice sheets: a review. *Quaternary Science Reviews* **1**, 1–30.

Ballantyne, C. K. 1989. The Loch Lomond Readvance on the Isle of Skye, Scotland: glacier reconstruction and palaeoclimatic implications. *Journal of Quaternary Science* **4**, 95–108.

Boulton, G. S., Smith, G. D., Jones, A. S. & Newsome, J. 1985. Glacial geology and glaciology of the last mid-latitude ice sheets. *Journal of the Geological Society of London* **142**, 447–474.

Dawson, A. G. 1992. *Ice Age Earth.* Routledge, London.

Doyle, P., Bennett, M. R. & Baxter, A. N. 1994. *The Key to Earth History.* John Wiley & Sons, Chichester.

Frakes, L. A., Francis, J. E. & Syktus, J. I. 1992. *Climate Modes of the Phanerozoic.* Cambridge University Press, Cambridge.

Hambrey, M. J. 1992. Secrets of a tropical ice age. *New Scientist* (1 February), 42–49.

Hambrey, M. J. & Harland, W. B. 1981. *The Earth's Pre-Pleistocene Glacial Record.* Cambridge University Press, Cambridge.

Imbrie, J. & Imbrie, K. P. 1979. *Ice Ages: Solving the Mystery.* Harvard University Press, Cambridge, Massachusetts.

Lowe, J. J. & Walker, M. J. C. 1984. *Reconstructing Quaternary Environments*. Longman, London.

Raymo, M. E. 1994. The initiation of Northern Hemisphere glaciation. *Annual Review of Earth and Planetary Sciences* **22**, 353–383.

Raymo, M. E. & Ruddiman, W. F. 1992. Tectonic forcing of Late Cenozoic climate. *Nature* **359**, 117–122.

Ruddiman, W. F. & Kutzbach, J. E. 1990. Late Cenozoic plateau uplift and climatic change. *Transactions of the Royal Society of Edinburgh: Earth Sciences* **81**, 301–314.

Van Andel, T. H. 1994. *New Views on an Old Planet* (Second Edition). Cambridge University Press, Cambridge.

Young, G. M. 1991. The geologic record of glaciation: relevance to the climatic history of Earth. *Geoscience Canada* **18**, 100–108.

3
Mass Balance and the
Mechanism of Ice Flow

In this chapter we shall examine the fundamental mechanics of glaciers. We consider the formation of glaciers and of glacier ice, how glaciers flow and why some glaciers flow faster and are more active than others.

3.1 ANNUAL MASS BALANCE

A glacier will form whenever a body of snow accumulates, compacts, and turns to ice. This can occur in any climatic zone where the input of snow exceeds the rate at which it melts. The length of time required to form a glacier will depend on the rate at which the snow accumulates and turns to ice. If this rate of **accumulation** is high and the loss due to melting is low then a glacier will form quickly. Once established, its survival will depend on the balance between accumulation and melting (**ablation**). This balance between accumulation and ablation is known as the **mass balance** of the glacier and is largely dependent on climate.

Figure 3.1 shows how the total amount of accumulation and the total amount of ablation each year defines the mass balance of a glacier. Ablation will tend to dominate in the warm summer months and accumulation in the winter months. If the amount of ablation equals the amount of accumulation over a year the **net balance** of the glacier will be zero and its size will remain constant (Figure 3.1). On the other hand, if there is more accumulation than ablation then the net balance will be positive and the glacier will grow and expand. If it is has a negative mass balance then the glacier will gradually disappear.

The study of glacier mass balance is therefore the study of inputs and outputs to the glacial ice system. Inputs to a glacier's mass balance include snow, hail, frost, avalanched snow and rainfall. If these inputs survive summer ablation they will

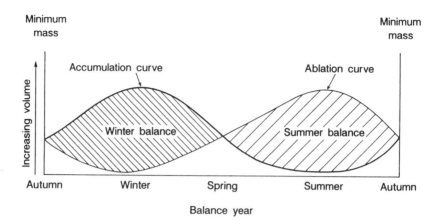

Figure 3.1   *Diagram to illustrate how curves for accumulation and ablation define a balance year for a glacier. The winter balance is positive and the summer balance negative. If the winter and summer balances equal one another the mass balance for the glacier will be zero and it will neither grow nor decay. [Modified from: Sugden & John (1976)* Glaciers and Landscapes, *Edward Arnold, Fig. 3.2, p. 37]*

begin a process of transformation into glacier ice. The term **firn** or **néve** is used for snow which has survived a summer melt season and has begun this transformation. The transformation involves: (1) compaction; (2) the expulsion of air; and (3) the growth of an interlocking system of ice crystals. Dry fresh snow is about 97% air by volume and has a density of 0.1 mg m^{-3}, while glacier ice has almost no air within it and a density of 0.9 mg m^{-3}. The rate at which this transformation takes place is dependent on climate. If snow fall is high and significant melting occurs, then the process can be rapid, since older snow is quickly buried by fresh accumulation which compacts the firn, while alternate melting and refreezing encourages the growth of new ice crystals. In contrast, where accumulation rates are low and little melting occurs, the transformation can be extremely slow. For example, in the interior of the Eastern Antarctic ice sheet, where there is very little accumulation and even less melt, the transformation may take up to 3500 years. By contrast, on the Seward glacier in Alaska the transformation is achieved in as little as 3 to 5 years, due to the high accumulation and melt rates.

Outputs from the mass balance system are collectively termed **ablation**. This ablation can occur in three ways: by ice melt, iceberg calving and sublimation. **Glacial meltwater** is derived from direct melting of ice on the surface of, or within, a glacier. On the surface this is a function of solar radiation received, while within and at the base of the glacier heat is supplied by: (1) friction due to ice flow, and (2) heat derived from the Earth's crust beneath the glacier (**geothermal heat**). Melting can therefore occur both on the surface of the glacier and within the glacier itself. Surface melting is primarily a result of warm air temperatures and is therefore highly seasonal, whereas melting within the glacier is not. It is important to emphasise that melting is not simply confined to the ice margin but may occur across the whole of the glacier surface.

Where glaciers terminate in water, either in the sea or in a lake, blocks of ice will break from the front (**snout** or **terminus**) of the glacier as icebergs (Figure 3.2). This process is known as **iceberg calving**. It can be a particularly rapid way of losing mass from a glacier. In very cold and dry environments, mass may also be lost through **sublimation**, which is the direct evaporation of ice.

Figure 3.2 *A calving glacier margin on the Upsalla glacier in Argentina. Note the steep ice front, dense crevasses and icebergs. [Photograph: N. F. Glasser]*

In summary, it is the relative balance between inputs and outputs to a glacier that determines its mass balance and therefore whether it will expand, contract or remain unchanged. So far we have considered the whole of a glacier and the total balance between accumulation and ablation over its surface. However, accumulation and ablation do not occur equally over the whole surface of a glacier. Ablation dominates at the terminus of a glacier, where it may calve icebergs and where the climate is relatively warm, while accumulation dominates in its upper regions, where temperature and precipitation are suitable for snowfall. It is this spatial imbalance between accumulation and ablation that creates the surface slope on a glacier that drives ice flow.

3.2 THE MASS BALANCE GRADIENT: THE GLACIAL DRIVING MECHANISM

Most ablation will occur at the snout or terminus of a glacier, which is usually its point of lowest elevation, where air temperatures will be highest and where ice-

berg calving may occur. The rate of mass loss (ablation) will decrease with elevation, as atmospheric air temperature falls with altitude. Accumulation may be
more uniform over the surface of a glacier, but will tend to increase with elevation.
A glacier can, therefore, be divided into two areas: (1) an **accumulation zone**
where accumulation exceeds ablation; and (2) an **ablation zone** where ablation
exceeds accumulation (Figure 3.3). The line between the accumulation zone and
the ablation zone is known as the **equilibrium line**: along this line accumulation is
balanced by ablation. This is sometimes referred to as the **snowline**.

On a valley glacier or ice sheet, mass is added at the top in the accumulation
zone and taken away through melting and calving at the terminus in the ablation
zone. Consequently the surface profile of the glacier will steepen with increasing
accumulation (Figure 3.4). In this way the surface slope will steepen until sufficient
stress builds up within the ice to cause it to flow. Ice flow transfers mass from the
accumulation zone to the ablation zone, reducing the surface slope of the glacier
and therefore the stress imposed on the ice. This transfer maintains the glacier
slope at a constant or equilibrium angle. It is therefore the gradient between accumulation and ablation across a glacier which causes it to flow. The larger this gradient, the greater the glacier flow required to maintain an equilibrium slope. This

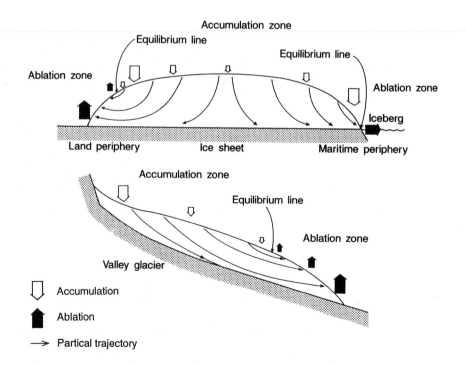

Figure 3.3 *Schematic diagram of an ice sheet and valley glacier showing the location
of the equilibrium line, the accumulation zone and the ablation zone. Principal flow
paths are also shown. [Modified from: Sugden & John (1976) Glaciers and Landscapes,
Edward Arnold, Fig. 4.8, p. 63]*

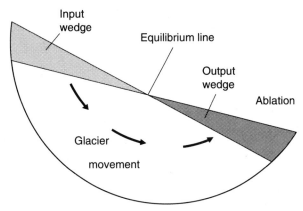

Figure 3.4   *Idealised glacier with net accumulation or input 'wedge' and net ablation or output 'wedge'. Glacier flow from the accumulation zone to the ablation zone is necessary to maintain a constant slope. [Reproduced with permission from: Sugden & John (1976) Glaciers and Landscapes, Edward Arnold, Fig. 3.5, p. 41]*

gradient is known as the **net balance gradient** and is defined as the increase in net balance (accumulation minus ablation) with altitude. It is composed of the sum of the rate of increase in accumulation and the rate of decrease in ablation with altitude. The higher the net balance gradient, the thicker the wedges of accumulation and ablation shown in Figure 3.4 will be, and as a result, the more rapid the glacier flow must be to maintain a constant or equilibrium slope. The net balance gradient will be high on glaciers that have high rates of accumulation and large amounts of ablation. It will therefore be steepest on glaciers that experience warm damp maritime climates and lowest for those in cold dry continental areas (Figure 3.5). Glaciers located in continental areas will therefore flow more slowly than those in warm maritime areas. The rate of glacier flow varies from one glacier to the next and much of this variation is due to differences in the net balance gradient between different glaciers.

3.3 MECHANISMS OF ICE FLOW

A glacier flows because the ice within it deforms in response to gravity. This gravitational force is derived from the fact that a glacier slopes towards its terminus, which is a result of the imbalance between accumulation and ablation across a glacier. If there is no surface slope, that is no imbalance between accumulation and ablation, the glacier will not flow. The force per unit area set up within a mass of ice by gravity which causes it to deform is known as the **shear stress**. The level of shear stress experienced within an ice mass at any point is dependent upon both

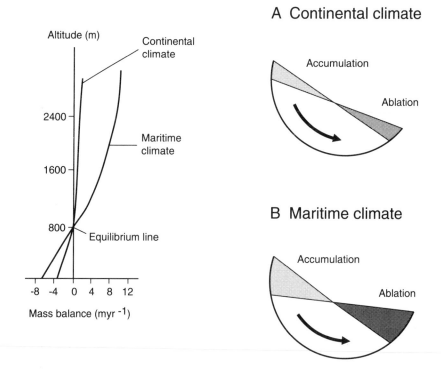

Figure 3.5   *Net balance gradients and climatic regime. Two idealised glaciers with net accumulation and ablation 'wedges' are shown. One of these glaciers has a maritime climate while the other has a continental climate. The mass balance curves illustrate the difference in the net balance gradient between the two glaciers. The larger amounts of accumulation and ablation, and the steeper net balance gradient, on the maritime glacier results in greater ice flow. [Modified from: Kerr (1993) Terra Nova **5**, Fig. 2, p. 333]*

the overlying weight of ice and the surface slope of the glacier. This can be summarised in the equation

$$\tau = \rho g(s - z) \sin \alpha \qquad (3.1)$$

where τ = shear stress at a point within the glacier, ρ = the density of ice, g = the acceleration due to gravity, α = the surface slope of the glacier, s = surface elevation, and z = elevation of a point within the glacier. At the base of a glacier the product of $(s - z)$ will be equal to the ice thickness so that the shear stress at the base of a glacier, its **basal shear stress**, is given by

$$\tau = \rho g h \sin \alpha \qquad (3.2)$$

where τ = shear stress, ρ = the density of ice, g = the acceleration due to gravity, h = the thickness of the glacier, and α = the surface slope of the glacier. This equation predicts that the basal shear stress will vary with glacier thickness and with surface slope, and that high basal shear stress values occur where the ice is thick and steep.

Different materials can withstand different values of shear stress before they will deform and/or fracture. In the case of ice, deformation occurs under relatively low values of shear stress and the values obtained from beneath glaciers are remarkably constant. In most situations shear stress at the base of a glacier flowing over bedrock varies between about 50 and 100 kilopascals (kilopascals = kpa; 100 kpa = 1.0 bars or kg cm^{-2}). At levels of less than 50 kpa, ice does not normally deform, while ice cannot usually withstand a shear stress of more than 150 kpa. Therefore glaciers flow because the imbalance between accumulation and ablation leads to an increase in the surface slope of a glacier, which increases its shear stress until it deforms or flows.

Values of basal shear stress may be much lower where the glacier flows over a bed that is not rigid but is composed of a **deformable sediment**. In this situation a large proportion of the forward movement of the glacier may be produced by flow within this deformable sediment as the stress within the glacier is transferred to the bed. Consequently, movement may not be primarily controlled by the proper-ties of the ice but by the mechanical properties of the sediment. As these tend to be weaker, sediment and glacier coupled flow may be initiated at much lower shear stress values (see Section 3.3.3).

The relatively constant values of shear stress found beneath glaciers flowing over rigid bedrock are an important characteristic and explain why most ice bodies have a parabolic profile; that is, a glacier slope which is steep at the margin and flattens off towards the centre. If shear stress is constant it follows from the above equation (3.2) that a large ice thickness must be associated with a small surface slope, and a small ice thickness with a large surface slope. The longitudinal profile of a glacier will therefore have a parabolic form, with high slopes at the margin and low slopes in the accumulation zone. The slope of this profile may be reduced or modified if the glacier flows over a layer of deformable sediment (Figure 3.6). This consistency of ice surface profile makes it possible to reconstruct the form of Cenozoic ice sheets that have long since disappeared if their former margin is

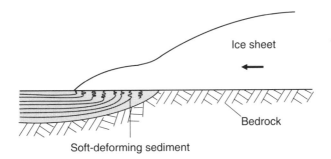

Figure 3.6 *Schematic cross-section through an ice sheet to show how its profile may be modified by the presence of an area of soft deformable sediment. [Diagram repro-duced with permission from: Boulton (1993) In: Duff (Ed.) Holmes' Principles of Physical Geology, Chapman & Hall, Fig. 20.23, p. 418]*

known (Box 3.1). The consistency of basal shear stress values is also of importance in providing a way of critically testing glacier reconstructions based on geomorphological evidence. Glacial reconstructions should obey the same physical laws as glaciers today and should therefore have basal shear stress values of between about 50 and 100 kpa over hard substrates (Box 3.2).

Ice can flow in response to the shear stress applied to it through three different mechanisms: (1) by **internal deformation**, (2) by **basal sliding**, and (3) by **subglacial bed deformation**.

BOX 3.1: THE PREDICTION OF ICE SHEET PROFILES

It is a reasonable approximation to regard ice as perfectly plastic with a yield stress of 100 kpa, that is ice will deform plastically if a stress of 100 kpa is applied. Nye (1952) used this assumption to derive a simple equation with which to calculate ice surface profiles. On a horizontal bed the altitude of the ice surface at any point inland from a known margin can be found from the formula

$$h = \sqrt{2h_o s}$$

$$h_o = \frac{\tau}{\rho g} = 11$$

where h = ice altitude in metres, τ = basal shear stress, s = horizontal distance from the margin in metres, ρ = density of ice, and g = acceleration due to gravity.

From comparison with real profiles, it seems that Nye's profile slightly overestimates the slope near the centre of an ice sheet, as shown below. Despite this, Nye's simple equation has been used widely in reconstructing former ice sheets.

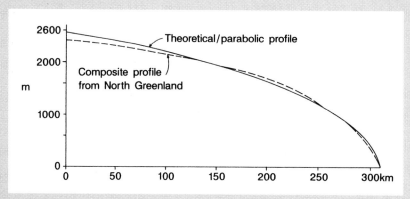

Source: Nye, J. F. 1952. A method of calculating the thickness of ice sheets. *Nature* **169**, 529–530. [Diagram modified from: Sugden & John (1976) *Glaciers and Landscapes*, Arnold, Figure 4.4, p.60]

BOX 3.2: ESTIMATES OF BASAL SHEAR STRESS AND THE ACCURACY OF GLACIAL RECONSTRUCTIONS

During the last glacial period the Yellowstone National Park in North America was covered by a number of small mountain ice-fields, one of which is referred to as the Pinedale Ice Mass. From a careful field study of the landforms and sediments within this area, Pierce (1979) was able to reconstruct the morphology of this ice mass. Any reconstruction of a former ice-field on the basis of geomorphological evidence, is likely to be speculative in some respect. How accurate, therefore, is the reconstruction? Pierce (1979) argued that glaciology offers a method of independently evaluating glacier reconstructions: former glaciers should obey the same physical laws as modern glaciers. In particular, the calculation of basal shear stress provides a powerful tool. The two parameters that together specify the shape of a reconstructed glacier—ice thickness and surface slope—are also the primary variables that determine basal shear stress. Empirical observations suggest that the vast majority of modern glaciers, flowing over undeformable substrata, have basal shear stresses between 50 and 150 kpa. Pierce (1979) used this to test the validity of his reconstruction of the ancient Pinedale ice-field: if the reconstruction is valid it should give basal shear stress values of this order.

Pierce (1979) calculated the basal shear stress of 50 reaches, each 5–15 km in length, along flow lines within his reconstructed ice-field. The results are displayed below on a graph with the ice thickness on one axis and the angle of the former glacier slope on the other. All the values of basal shear stress fall between 60 and 150 kpa, which is fairly consistent with modern glaciers. The high values of basal shear stress tend to occur in areas likely to have experienced extending flow, while the lower values are typical of areas of compressive or decelerating flow. Pierce (1979) concluded that his glacier reconstruction had values of basal

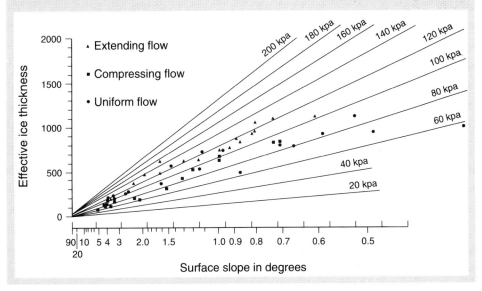

shear stress that are both internally consistent and consistent with the values of basal shear stress obtained for modern glaciers. This strongly supports the validity of his reconstruction and illustrates how basic glaciological principles can be used to test the accuracy of glacier reconstructions based on geomorphological evidence.

Source: Pierce, K. L. 1979. History and dynamics of glaciation in the Northern Yellowstone National Park area. *US Geological Survey Professional Paper 729-F*. [Diagram modified from: Pierce (1979) *US Geological Survey Professional Paper 729-F*, Figure 48, p. 71]

3.3.1 Internal Deformation

The internal deformation of ice is achieved in two ways: (1) by the process of **creep**, and (2) by large-scale folding and faulting.

Creep involves both the deformation of ice crystals and, at higher temperatures, the mutual displacement of ice crystals relative to one another in response to the shear stress placed upon the ice. The rate of ice creep is a function of the shear stress applied: the greater the shear stress, the greater the rate of ice creep. This relationship is known as **Glen's flow law** and emphasises the sensitivity of ice creep to shear stress: simply doubling the shear stress will increase the rate of creep by a factor of eight. This explains why most creep takes place in the basal layers of a glacier where shear stress is greatest, because of the greater ice thickness. The rate of deformation is also controlled by temperature, because ice is more plastic at higher temperatures.

The rate of glacier creep may vary down-glacier depending upon whether the glacier is experiencing accelerating (**extensional**) or decelerating (**compressional**) flow. The distribution of zones of extending or compressional flow varies with scale. On a glacier scale, compressive flow tends to occur where the ice thickness decreases down-glacier (in the ablation zone), and extending flow tends to occur where thickness increases down-glacier (in the accumulation zone). At a small-scale, extending flow tends to occur on slopes beneath ice that steepens down-glacier, while compressive flow tends to occur where basal slopes shallow down-glacier. The pattern of surface fractures (**crevasses**; Figure 3.7) on the glacier reflects the type of flow—extending or compressional—that takes place (Figure 3.8).

Under certain conditions, creep cannot adjust sufficiently fast to the stresses set up within the ice, and faults and/or folds are formed. The type of faults that form depend upon whether the ice is experiencing a zone of longitudinal extension or one of compression (Figure 3.8). Areas of compressive flow are also controlled by the temperature of the basal ice, as discussed in Section 7.3.

3.3.2 Basal Sliding

There are two main processes by which ice sheets can slide over their beds: (1) enhanced basal creep, and (2) regelation slip.

Figure 3.7 *Crevasses on the Upsalla glacier in Argentina. Crevasses open in areas of extensional stress within a glacier. They may subsequently close if the ice flows into an area of compressional stress. [Photograph: N. F. Glasser]*

Enhanced basal creep is an extension of the normal ice-creep process. It explains how basal ice deforms around irregularities on the ice–bed interface. A glacier bed is not smooth but will contain irregularities, such as bedrock bumps or lodged boulders, which protrude into the bottom of the moving glacier. Basal ice pressure within the ice increases on the upstream side of such obstacles (see

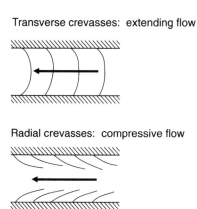

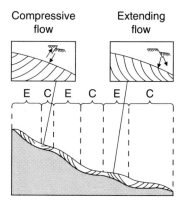

Figure 3.8 *Compressive and extending flow within glaciers. The pattern of surface crevasses associated with each is different. Compressive stress is associated with a decrease in the subglacial slope, while extentional stresses occur where the glacier bed steepens*

Section 4.3), and this increases the rate of ice deformation at this point, allowing the ice to flow more efficiently around the obstacle. The larger the obstacle, the higher the increase in basal pressure and the greater the rate of deformation. Consequently, the process is most efficient for larger obstacles.

Regelation slip occurs when ice at its pressure melting point moves across a series of irregularities or bumps. On the upstream side of each obstacle the basal ice pressure is higher as the ice moves against the obstacle (see Section 4.3). The melting point of ice falls as pressure rises and, as a consequence, basal ice melts on the upstream side of obstacles. The meltwater produced will flow around the bump to the downstream side where the pressure is lower and consequently it will refreeze to form **regelation ice**. This process is most effective for small obstacles because the higher temperature gradients across them can drive the heat flux generated by the freezing of the meltwater (release of latent heat) from the downstream side of the obstacle through the rock bump to assist ice melting on the upstream side.

There are therefore two processes of basal sliding. One of these, regelation slip, operates best in passing small obstacles, while enhanced basal creep works best for larger obstacles. In between the two there is a critical obstacle size range where neither process is particularly effective. A bed with obstacles in this size range will therefore pose the greatest resistance to basal sliding.

The efficiency of basal sliding depends not only upon the size of obstacles, but also more generally upon the overall level of bed friction. Bed friction is a function of the number of points of contact between the ice and the bed: if these are few then the friction resisting basal sliding will be low. The number of points of contact depends primarily on the amount of water present at the ice–bed interface and its pressure (see Section 4.3). A film of water only a few millimetres thick will reduce friction sufficiently to increase the rate of basal sliding. The presence of water-filled

basal cavities at the ice–bed interface can also dramatically increase the rate of basal sliding (see Section 4.3). The rate of basal sliding is also affected by the amount of debris within the basal ice; if the debris content is very large the rate of basal sliding may be reduced (see Section 5.1.3).

3.3.3 Subglacial Bed Deformation

When an ice sheet flows over unfrozen sediment it may cause this sediment to deform beneath the weight of the ice. This deformation occurs when the water pressure in the pores or spaces between the sediment grains increases sufficiently to reduce the resistance between individual grains. This allows them to move or flow relative to one another as a slurry-like mass. In response to the shearing force imposed by the overriding glacier, this slurry forms a continuously deforming layer on which the glacier moves (Box 3.3). This process can be dramatic. For example, 90% of the forward motion of the Breiðamerkurjökull in south-east Iceland may be due to subglacial bed deformation.

3.4 THE PRINCIPLES OF BASAL THERMAL REGIME

The principal control on which combination of flow processes—creep or basal sliding—operates beneath a given glacier is the temperature of the basal ice. Some glaciers are frozen to their beds. No meltwater is present at the ice–bed interface and basal sliding does not occur (Figure 3.9). Such glaciers are composed of **cold ice**. In contrast, other glaciers are composed of **warm ice**, where basal ice is constantly melting and the ice–bed interface is therefore lubricated with meltwater. In such situations, basal sliding is an important component of flow (Figure 3.9). An ice sheet with a warm base has a much greater potential for fast flow and therefore greater potential to modify its bed by erosion than one which is frozen to it. Basal ice temperature (**basal thermal regime**) is therefore one of the most important controls on the geomorphological impact of a glacier since it controls the pattern of erosion and deposition within it. Not only does basal thermal regime vary between glaciers, but it also varies within a particular ice body.

The temperature at the base of a glacier is determined by the balance between: (1) the heat generated at the base of the glacier; and (2) the temperature gradient within the overlying ice, which determines the rate at which the basal heat is drawn away by conduction from the ice–bed interface. Heat is generated at the base of a glacier in three ways: (1) by geothermal heat entering the basal ice from the Earth's crust; (2) by frictional heat produced by sliding at the base of the glacier; and (3) by frictional heat produced by the internal deformation of the glacier. These three heat sources combine to warm the base of the glacier. The rate at which this heat is conducted away from the base of a glacier depends upon the temperature gradient within the overlying ice. This temperature gradient depends on: (1) the temperature at the base of the ice; (2) the temperature of the glacier sur-

BOX 3.3: THE OBSERVATION OF SUBGLACIAL DEFORMATION

There are very few direct observations of deforming sedimentary layers beneath glaciers, due to the problems of access. Some of the most comprehensive data to date have been obtained by Boulton and Hindmarsh (1987) from beneath Breiðamerkurjökull in Iceland. A series of tunnels were excavated within the basal ice of Breiðamerkurjökull from which a series of strain markers (small cylinders) were inserted through drill holes into the subglacial till beneath the tunnels, along with a number of other instruments. The experiment lasted for 136 hours, during which time glacier velocity and the subglacial water pressure were monitored. At the end of the experiment, water was pumped from the subglacial bed beneath the ice tunnels, before a section through the bed was dug to study the strain markers. The results are shown in the diagram below. Displacement of the strain markers clearly illustrates the deformation of an upper horizon of till over a relatively undeformed lower layer. It was estimated that approximately 90% of the forward movement of the glacier was due to this subglacial deformation. These observations along with the other measurements were used to derive a mathematical flow law with which to describe the deformation of the till. This flow law has been used to predict and model the behaviour of deforming sediments beneath glaciers and is critical to theories about the formation of subglacial landforms.

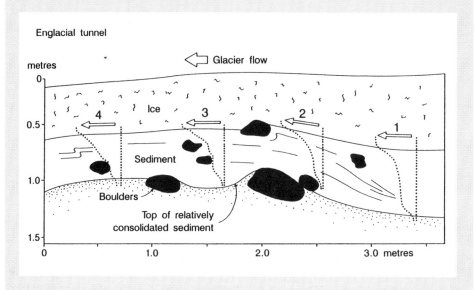

Source: Boulton, G. S. & Hindmarsh, R. C. A. 1987. Sediment deformation beneath glaciers: rheology and geological consequences. *Journal of Geophysical Research* **92**, 9059–9082. [Diagram modified from: Boulton & Hindmarsh (1987) *Journal of Geophysical Research* **92**, Figure 2, p. 9062]

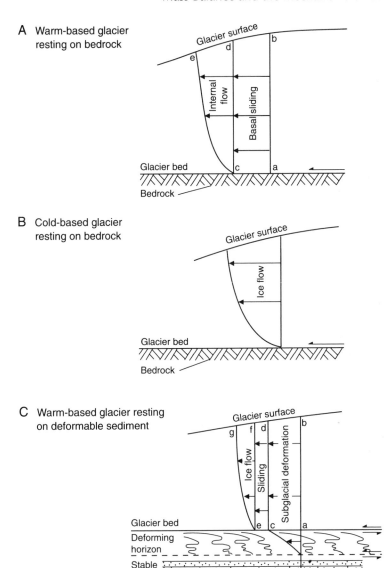

Figure 3.9 *The velocity distribution within three glaciers with different thermal regimes. If we insert a vertical line within a glacier (a–b) at a given time, this line will be displaced as shown.* **A:** *Warm-based glacier resting on hard bedrock (the line a–b is displaced to c–d by basal sliding and then deformed to follow the line e–c by internal deformation).* **B:** *Cold-based glacier resting on hard bedrock.* **C:** *Warm-based glacier resting on deformable sediment (the line a–b is displaced to c–d by subglacial deformation, to e–f by basal sliding and to e–g by internal flow).* [*Diagram reproduced with permission from: Boulton (1993) In: Duff (Ed.)* Holmes' Principles of Physical Geology, *Chapman & Hall, Fig. 20.20, p. 416*]

face; (3) the thickness of the ice; and (4) the thermal conductivity of the ice. The temperature of the basal ice is therefore a function of the amount of heat generated at the base of the glacier and the rate at which it is conducted away along the thermal gradient within the overlying ice. Three basal ice conditions can be defined.

1. **Boundary Condition A.** Net basal melting (warm ice). In this case more heat is generated at the base of the glacier than can be removed by conduction in the direction of the temperature gradient.

    Basal heat generated   >   Heat conducted away

    Heat input   >   Heat output

2. **Boundary Condition B.** Equilibrium between melting and freezing. In this case the heat generated at the base of the glacier is equal to that conducted away along the direction of the temperature gradient.

    Basal heat generated   =   Heat conducted away

    Heat input   =   Heat output

3. **Boundary Condition C.** Net basal freezing (cold ice). In this case all the heat generated at the base of the glacier is quickly removed away from the bed along the direction of the temperature gradient and the ice remains frozen to its bed.

    Basal heat generated   <   Heat conducted away

    Heat input   <   Heat output

These three boundary conditions or basal temperature states can be thought of as separate thermal zones, each of which has its own attributes in terms of the types of basal flow (Figure 3.9). Each is also dominated by different processes and geomorphological activity, as we will see in later chapters.

In the discussion so far, heat transfer within the glacier has simply been assumed to be a function of conduction along the temperature gradient within the ice. In practice, heat is also transferred through **advection**. Advection is the transfer of heat energy in a horizontal or vertical direction by the movement of ice or snow. For example, the downward movement of cold snow or firn within the accumulation area of a glacier will lead to glacier cooling. The rate of advection is therefore partly a function of accumulation rates: high accumulation rates result in strong heat fluxes due to the passage of cold ice through the glacier system. Listed below are the principal variables that help determine the basal ice temperature of a glacier:

1. **Ice thickness.** Increasing ice thickness will have the effect of increasing the basal ice temperature. This is due to the insulating effect of ice: more ice equals more insulation.

2. **Accumulation rate.** Basal ice temperature is also affected by accumulation rates through the process of advection. In a glacier, advection occurs as the result of the accumulation of fresh snow on the surface gradually moving down through the glacier to the base. If this snow is cold it will cause the glacier to cool, and conversely, if it is warm it may cause an increase in temperature. For example, in the interior of an ice sheet, where accumulation rates may be very low and where snow accumulates at low temperatures, advection will favour basal freezing. Towards the margin of this ice sheet, where accumulation rates may be higher and where the snow accumulates at higher temperatures, advection may favour basal melting. The incorporation of large amounts of cold dry snow also has the effect of improving the temperature gradient within the ice and therefore the rate of heat conduction, which also helps reduce basal ice temperatures. Advection may also warm a glacier through the upward movement of warm ice in the ablation zone.

3. **Ice surface temperature.** An increase in the surface temperature of a glacier will reduce the temperature gradient within the ice and is therefore likely to increase basal ice temperature. There is a direct relationship between surface temperature and basal temperature: a fall of 1 °C in surface temperature causes a fall of 1 °C in basal temperature. This may be achieved by incorporating large amounts of wet snow and/or through summer melting. The percolation of meltwater through an ice body will have the effect of raising its temperature, because as it refreezes, latent heat is liberated. For every gram of meltwater that freezes, the temperature of 160 grams of ice is raised by 1 °C.

4. **Geothermal heat.** An increase in the flux of geothermal heat will increase the basal ice temperature.

5. **Frictional heat.** An increase in ice velocity will increase the amount of frictional heat generated and in turn increase the basal ice temperature. A glacier flow of 20 m per year produces the same amount of heat as that produced by the average geothermal heat flux. As we will see in Section 3.5, ice flow within a glacier varies spatially, increasing from zero beneath the ice divide of an ice sheet towards the equilibrium line before it decreases towards the ice margin. Consequently, the heat generated by friction will also increase to a maximum close to the equilibrium line. This variation in the amount of heat generated by friction is particularly important in determining the spatial variation of ice temperature within a glacier.

The above allows us to predict the situations likely to produce cold- and warm-based ice. **Cold-based glaciers** are likely to occur where the glacier is thin, slow moving, and where there is little or no surface melting in summer and the surface layers of ice are cooled severely each winter. A typical temperature profile through a cold glacier is shown in Figure 3.10A. The increase in temperature with depth is due to the insulating effect of the overlying layer of ice and the increase in pressure with depth. The temperature gradient is positive (i.e. warmer at the base than at the surface) and therefore the heat will flow from the glacier bed to the surface. Heat will only flow along a negative gradient, that is from warm to cold. Any heat

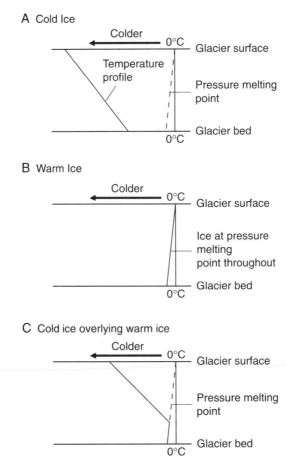

Figure 3.10   *Idealised temperature profiles through three glaciers. **A:** Temperature profile found within cold-based ice. **B:** Temperature profile found in warm-based ice. **C:** Temperature profile found where cold ice overlies warm ice at depth. [Modified from: Chorley et al. (1984) Geomorphology, Methuen, Fig. 17.4, p. 435]*

generated at the base of a glacier will in this case be conducted quickly away from the glacier base through the ice. Boundary Condition C will therefore prevail.

In contrast, **warm-based glaciers** are likely to occur where the ice is thick, fast moving and where summer melting is high. The percolation of meltwater through the glacier body will warm the ice as it refreezes through the release of latent heat, and the ice temperature may be close to its melting point. The melting point of ice varies with depth due to the change in pressure. Beneath 2000 m of ice, within an ice sheet, the melting point will be −1.6 °C instead of 0 °C. This is termed the **pressure melting point**. Within an ice mass close to its melting point, the temperature profile will look like that in Figure 3.10B. In this case the profile is positive, that is colder at the base than at the surface; consequently, any basal heat generated will not be able to escape and boundary Condition A will therefore apply. The cases outlined above represent two extremes of a continuum. In rare instances glaciers may be entirely warm- or cold-based but in practice basal boundary conditions vary both in space and in time within a single glacier. For example, Figure 3.10C shows a situation where cold ice overlies warm ice at the glacier bed.

3.4.1 Spatial Variation of Basal Thermal Regime within a Glacier

Figure 3.11 shows just one way in which the three boundary conditions or thermal zones can be combined within a single glacier. At its simplest, this can be thought of as a continuum between cold and warm ice (Figure 3.11A–E). Boundary Condition B has been split into two zones, one in which there is slight net freezing (B1) and one in which there is slightly more melting (B2). This reflects the position of this thermal zone within the transition between the two extreme types of basal boundary condition. In Figure 3.11A, boundary Condition C occurs throughout the glacier and the underlying ground is frozen (permafrost). Since the glacier is frozen to its bed, no basal sliding occurs. In the next diagram the central part of the glacier lies in thermal balance and is neither predominantly melting nor freezing (Figure 3.11B). In this central zone basal sliding will occur. This will cause compres-

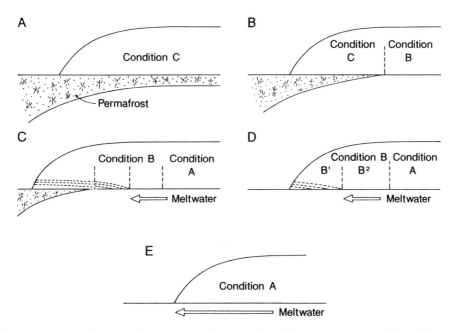

Figure 3.11 *Schematic diagrams to illustrate the different patterns of basal thermal regime that might exist within a glacier. See text for the precise definition of each thermal condition. Condition A = warm; Condition B = thermal equilibrium; Condition C = cold. [Modified from: Boulton (1972) In: Price & Sugden (Eds) Polar Geomorphology, Institute of British Geographers, Special Publication 4, Fig. 1, p. 4]*

sion toward the transition into the outer zone of the glacier which is still frozen to its bed and will not therefore experience basal sliding. Situations like this result in large compressive zones near glacier margins in which basal ice is often thrust toward the glacier surface (see Section 7.3). Figure 3.11C shows a more complex pattern of thermal zones, in which the centre of the ice sheet or glacier experiences basal melting. Meltwater passes out from this zone under hydrostatic (water) pressure, then freezes to the bed beneath the cold ice in the intermediate zone (B1) and new basal ice is formed. Beyond this the glacier is frozen to its bed. Figures 3.11D and E show the other end of the continuum, moving toward a glacier that is melting at its base and one which will therefore experience basal sliding throughout.

In practice, the pattern or order of these thermal boundaries within a glacier may vary dramatically. For example, irregular basal topography or climatic variation across a glacier produce more complex localised patterns. Figure 3.12A shows a cross-section through a large mid-latitude ice sheet. The pattern of basal thermal regime within it is one idea of what the pattern may have looked like in one of the large mid-latitude ice sheets of the Cenozoic Ice Age. This pattern of basal thermal regime is based primarily upon the effects of ice thickness and climate, and ignores the effects of advection. Climatic variation across the ice sheet results in the contrast between the northern and southern margins: a contrast between a cold continental and a warm maritime climate. Figure 3.12B shows how the presence of a deformable bed can be exploited only where the glacier is warm based. The pattern of basal thermal regime can be further complicated by introducing basal topography (Figure 3.12C). A deep trench beneath a zone of the ice sheet that would normally be cold at the base may induce melting at the glacier sole. Equally, a raised mountain area may experience basal freezing in what would otherwise be a zone of melting (Figure 3.12C).

An alternative view of the basal temperature distribution within the former mid-latitude ice sheets of the Cenozoic Ice Age is obtained when the role of advection is emphasised. In this model the cooling effect of incorporating cold snow (advection cooling) in the ice sheet centre causes it to be cold based in the middle, while the increase in frictional heating towards the ice margin and the upward transfer of warm ice by compressive flow (advection warming) in the ablation zone cause the ice sheet to be warm based towards its margin (Figure 3.13).

Whatever the pattern of basal thermal regime within an ice sheet, it has a profound effect upon the work done by the ice sheet and on the geomorphology produced. This point will be returned to in later chapters but is well illustrated by the control of the basal regime on the mechanisms of ice flow (Figure 3.9). Where the ice is frozen to its bed, irrespective of the nature of the substrate, flow may only occur by internal deformation and movement is concentrated above the bed. As a consequence, the potential to modify the bed is small. In contrast, in zones of thermal melting, flow may occur by basal sliding and by subglacial deformation where the substrate is appropriate (Figure 3.9), and the ice sheet has a more profound effect on its bed.

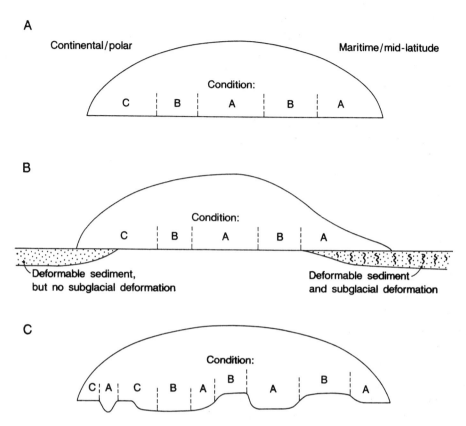

Figure 3.12   *Schematic cross-sections through three ice sheets showing the pattern of basal thermal regime within them and the role of lithology and topography in modifying the pattern present. See text for the precise definition of each thermal condition. Condition A = warm; Condition B = thermal equilibrium; Condition C = cold*

3.4.2 Temporal Variation in Basal Thermal Regime within a Glacier

Just as the basal ice temperature may vary within a glacier, it may also vary through time. In particular, the pattern of basal thermal regime within an ice sheet is not static. As an ice sheet grows and decays the pattern of basal ice temperature within it will also evolve. Consequently, the pattern of processes controlled by it will also vary through time as the pattern of thermal regime changes. Figure 3.13 shows a hypothetical cross-section through a mid-latitude ice sheet, showing the evolution of the pattern of basal thermal regime within it. This diagram is highly schematic and in reality the patterns are likely to have been much more complex. It serves to illustrate, however, that at any location the temperature of the ice above may change as the ice sheet grows and decays, and consequently the processes operating upon it will also change. This pattern results primarily due to

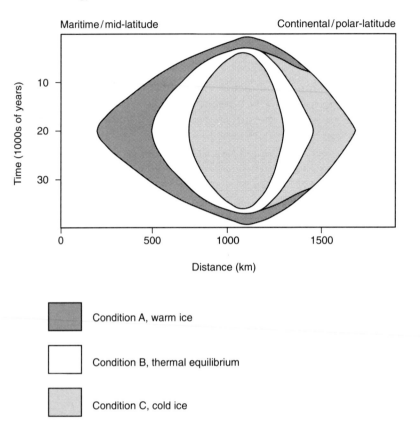

Figure 3.13 *A schematic time–distance diagram for a single span or cross-section through an ice sheet. It shows the growth and decay of this ice sheet through time and the evolution of its basal thermal regime. Note that as the ice sheet grows and decays, a zone of warm ice moves across the landsurface. The pattern of the basal thermal regime at the ice sheet maximum can be seen by taking a time slice through the diagram at approximately 20 000 years*

two main factors: (1) the increase in frictional heating towards the equilibrium line maintains a warm outer ring to the ice sheet; and (2) cooling by advection causes the central part of the ice sheet to remain cold.

   At a much smaller scale the pattern of basal thermal regime, and in particular the boundaries between one thermal zone and another, may change as a result of local or regional fluctuations in: (1) ice velocity, (2) accumulation, (3) geothermal heat flux, and (4) ice thickness. On a longer timescale, glacial erosion may modify basal topography, causing variations in the thermal regime. Any of these variables may change the temperature of the basal ice and therefore the dynamic nature of the processes operating at the ice–bed interface.

3.5 PATTERNS AND RATES OF ICE FLOW

Within a glacier, flow usually follows the direction of the surface slope. Figure 3.3 shows a cross-section through both an ice sheet and a valley glacier. In the ice sheet, ice flows in two opposite directions from the summit or **ice divide** (Figure 3.3). In the accumulation area, flow takes place downwards into the ice, counteracting the upward growth of the surface through accumulation. In the ablation zone, the surface is lowered by ablation which causes ice to effectively rise towards the surface. Ice may also rise towards the surface in the ablation zone due to compressional flow at the glacier margin. Flow within a valley glacier or channel shows a similar pattern (Figure 3.3). Ice flow is maximum in the centre of the channel or valley and near the surface, that is the point furthest from the frictional resistance of the valley sides (Figure 3.14 and Box 3.4).

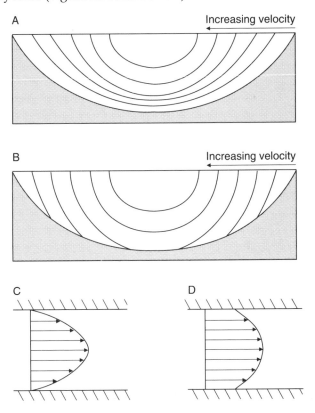

Figure 3.14 *Pattern of ice flow within glaciated valleys.* **A:** *The flow pattern in cross-section where no basal sliding is present.* **B:** *The flow pattern in cross-section where basal sliding is present.* **C:** *The flow pattern in plan where no basal sliding is present.* **D:** *The flow pattern in plan where basal sliding is present. [Diagram reproduced with permission from: Summerfield (1990)* Global Geomorphology, *Longman, Fig. 11.5, p. 265]*

BOX 3.4: THE FIRST MEASUREMENTS OF GLACIER FLOW BY JAMES D. FORBES IN 1842

In 1842 James Forbes set up a series of glacial experiments on the Mer de Glace above Chamonix in the French Alps. These experiments were to prove to be one of the most important and original contributions to glaciology. Through detailed mapping and sequential survey of fixed points on the glacier during the summer of 1842, Forbes was able to: (1) demonstrate the rate of glacier flow; (2) show that glaciers moved in a slow and regular fashion; and (3) that glaciers flowed fastest along their centre line. On the basis of these simple measurements and his detailed observations he was able to make a series of inferences. First he was able to dismiss one of the early theories of glacier flow which envisaged that ice flowed as a consequence of the expansion of water within crevasses as it froze each night, a process which would not give the regular pattern of glacier flow measured. Forbes distinguished correctly between internal deformation of ice and basal sliding; he concluded, correctly in the case of warm-based ice, that movement varied with the weather and with the season; and his deduction that movement diminished steadily from the ice surface to the glacier bed has also been shown to be substantially correct. Forbes also suggested that some glaciers might slide over their beds, while others in colder conditions might not if they were frozen to them. Although his measurements may appear obvious in terms of current knowledge, at the time they represented a major step forward for glaciology. More importantly, Forbes' inferences from his measurements reveal a profound understanding of the mechanics of glaciers. The tradition of detailed observation and experimentation in glaciology established by Forbes in 1842 has been followed by generations of glaciologists since.

Sources: Forbes, J. D. 1843. *Travels through the Alps of Savoy and Other Parts of the Pennine Chain with Observations on the Phenomena of Glaciers.* Black, Edinburgh. Cunningham, F. H. 1989. James David Forbes on the Mer de Glace in 1842: early quantification in glaciology. In: Tinkler, K. J. (Ed.) *History of Geomorphology,* The Binghampton Symposia in Geomorphology: International Series **19**, Unwin Hyman, Boston, 109–126.

In an ideal ice sheet the rate of flow will tend to increase from the ice divide towards the equilibrium line, where it will reach a maximum, before decreasing towards the terminus. This observation can be explained schematically with reference to Figure 3.15. If we assume a constant average velocity through the whole thickness of the ice sheet, then at a point at a given distance (x) from the ice divide the horizontal flow velocity (u) must be sufficient to remove by flow all the accumulation (a) that occurs up-ice of that point. The amount of ice or discharge which can pass through this point is given by

$$Discharge = uh \tag{3.3}$$

where u = average ice flow velocity (m s^{-1}), and h = ice thickness (m). This discharge must equal the accumulation rate up-ice of that point if the ice sheet is to maintain a steady state (i.e. size). Therefore the amount of ice to be discharged by the glacier is given by

$$\text{Discharge} = xa \qquad (3.4)$$

where x = distance from point x to the ice divide (m), and a = average accumulation (m yr^{-1}). It follows therefore that average ice velocity (u) can be calculated from

$$uh = \text{Discharge} = xa \qquad (3.5)$$

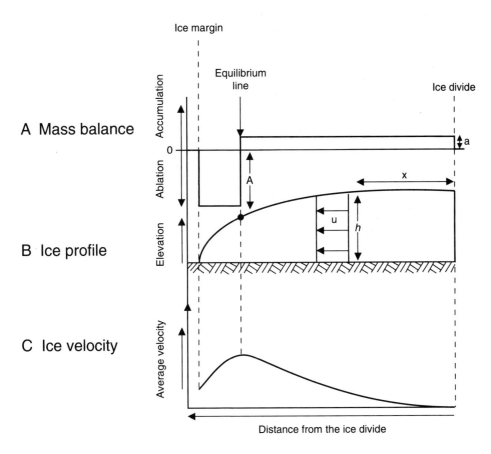

Figure 3.15 *Pattern of ice flow within an ice sheet. **A:** A simplified mass balance pattern on the glacier surface. **B:** The ice sheet in profile. **C:** The ice flow velocity within the ice sheet. Ice flow velocity rises from zero at the ice divide to a maximum near to the equilibrium line, from where it falls. [Diagram reproduced with permission from: Boulton (1993) In: Duff (Ed.) Holmes' Principles of Physical Geology, Chapman & Hall, Fig. 20.14, p. 412]*

This can be rearranged to give

$$u = \frac{xa}{h}$$

(3.6)

where u = average ice flow volocity (m s^{-1}), h = ice thickness (m), x = distance from point x to the ice divide (m), and a = average accumulation (m yr^{-1}).

Using this simple equation, ice-flow velocity increases from zero at the ice divide ($x=0$) to a maximum at the equilibrium line, thereafter it decreases. Clearly the assumptions made in calculating the velocity pattern in Figure 3.15 are considerable, but the pattern produced holds as a general model or first approximation of the velocity pattern within an ideal ice sheet. This pattern is significant because it implies that: (1) little or no geomorphological work will take place beneath ice divides due to low ice velocities; and (2) that most geomorphological work will be done beneath the equilibrium line, which is usually located relatively close to the ice margin.

Most glaciers have velocities in the range of 3–300 m per year, but their velocity can reach 1–2 km per year in steep terrain or where there is a high mass balance gradient. A few glaciers flow at speeds that are much higher. These are commonly associated with large outlet glaciers from ice sheets such as that in Greenland or Antarctica. In these **ice streams** flow is channelled down valleys and velocities may reach as much as 7–12 km per year. These ice streams may drain significant areas of an ice sheet and because they are fed by a large accumulation area their velocities are not normally limited by the supply of ice (i.e. the rate of accumulation).

Some glaciers may also experience periodic **surges** in ice flow often 10–100 times greater than previous ice velocities. Surges are usually limited by the amount of ice available in the accumulation zone so that increased flow rates cannot be sustained. Not all glaciers are prone to surges, and those that do, appear to surge at regular intervals. It has been suggested that only about 4% of all glaciers surge, although they tend to be concentrated in certain geographical areas, such as in Svalbard (Figure 3.16).

It has been suggested that there are two modes of glacier flow, one 'normal' and one 'fast' (Figure 3.17). In an ice stream the 'fast' flow mode is maintained because of the availability of ice within the large accumulation zone of the ice sheet which it drains. In the case of surging valley glaciers the pulse of 'fast' flow is limited by the amount of ice available in the much smaller accumulation zone. Once this ice has been discharged the flow must return to 'normal'. It is possible to conceptualise a surge as the product of an excess of accumulation of ice above that which 'normal' flow can discharge. This excess accumulation may be stored in the accumulation area until it reaches a critical level when it may trigger a pulse of 'fast' flow. The periodic nature of a surge is explained in this model by the time necessary between pulses of 'fast' flow to build up the excess of ice or the stress necessary to trigger the event. This will vary from one glacier to next, which explains why different glaciers surge with different periodicities. The exact nature of the instability that generates a surge and the mechanisms of 'fast' flow are not however well understood at present (see Section 4.3).

Figure 3.16   *The snout of a small surging glacier (Arebreen) in Reindalen, Svalbard. Note the steep ice front. [Photograph: N. F. Glasser]*

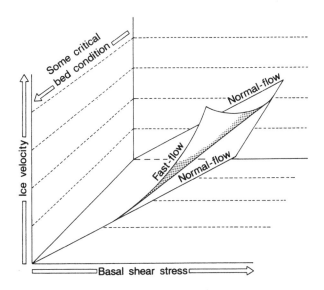

Figure 3.17   *Modes of glacier flow. Two flow states exist: 'fast flow' and 'normal flow'. The existence of 'fast flow' is dependent on some critical bed condition without which it cannot occur. A surging glacier is able to switch between fast and normal flow rates*

3.6 GLACIER RESPONSE TO CLIMATE CHANGE

The size of a glacier is determined primarily by climate. Changes in climate will cause the glacier's margins to expand or contract since climate controls a glacier's mass balance. If the balance is positive (i.e. more accumulation than ablation) it will expand and grow. If it is negative then the glacier will contract. Glaciers are constantly responding to changes in their mass balance budget, adjusting to annual variations as well as to long-term trends (Figure 3.18).

To understand how the glacier system responds to climate change we need to consider what happens when a glacier thins or thickens due to changes in its mass balance. If climate deteriorates (i.e. temperatures fall and/or snowfall increases) every part of the glacier is likely to thicken. This thickening will result from either an increase in accumulation or a reduction in ablation and may cause the ice margin to advance. Conversely, a climatic warming will lead to overall thinning of the glacier and retreat of its margin. Glacier response to such changes in mass balance may be either stable or unstable. A stable response is one in which the glacier thickens or thins in proportion to the size of the change in mass balance. An unstable response is one in which the glacier thickens or thins out of proportion with the size of the change in mass balance which triggers it. The type of response depends on whether the flow is extending or compressive (Figure 3.8). Unstable behaviour occurs with compressive flow. This can be illustrated in relation to Figure 3.19. In this figure, AA and BB are two cross-sections in a glacier subject to compressive flow. The discharge (Q, i.e. volume per unit time) of ice through cross-section BB is less than that through cross-section AA by the amount of ablation (ab) which occurs on the glacier surface between points A and B ($Q_{AA} = Q_{BB} + ab$). If we now place a uniform layer of snow over the glacier surface it can be

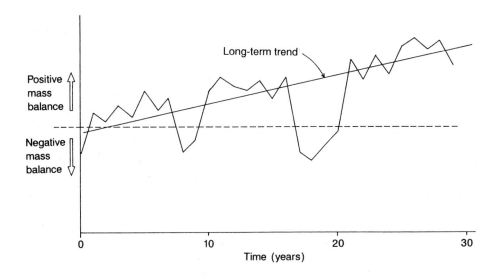

Figure 3.18   *Schematic diagram to illustrate both short-term and long-term mass balance trends*

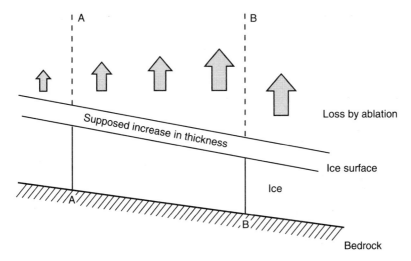

Figure 3.19   *Diagram to explain why the response to an increase in accumulation is unstable in an area of compressive flow. See text for details. [Modified from: Sugden & John (1976)* Glaciers and Landscapes, *Edward Arnold, Fig. 3.12, p. 48]*

shown mathematically that the increase in discharge caused by the layer of snow is proportional to the original flow through the section. Consequently, since the flow through cross-section AA is larger than that through BB, the increase in discharge due to the new layer of snow will be greater at AA than at BB ($Q_{AA} > Q_{BB} + ab$). Therefore the ice must thicken between points A and B to accommodate this extreme volume. With continued flow this effect will be accentuated down-ice and the glacier will progressively thicken towards its snout, eventually causing it to advance.

Changes in mass balance are propagated down-glacier by means of **kinematic waves**. A kinematic wave can be viewed as a bulge in the glacier surface. The greater thickness of ice in the bulge will locally increase the basal shear stress and therefore the rate of ice deformation and consequently the glacier velocity. The bulge will therefore move down-ice faster than the thinner ice on the up-ice and down-ice side of the bulge. It is important to stress that it is the bulge which moves and not the ice: a boulder on the surface would temporally move at a higher velocity while it was on top of the bulge and then return back to its normal velocity as the bulge passed by. Kinematic waves originate in the vicinity of the equilibrium line. In general, the equilibrium line represents the junction between predominantly extending flow in the accumulation zone and the compressive flow typical of the ablation zone. An increase in ice thickness, caused by a deterioration in climate, will therefore cause a stable thickening of the glacier above the equilibrium line and an unstable response below it. It is this difference in thickening response which initiates a kinematic wave. The wave, once formed, travels down the glacier at a rate faster than the ice velocity. When the wave reaches the snout, often years later, it may initiate an advance of the ice margin. The passage of a kinematic wave is similar to the passage of a flood wave within a river. It is important to emphasise that kinematic waves are rarely visible at the glacier surface. They are simply the mechanism by which mass balance changes are propagated throughout the glacier system; the means by which it adjusts to changes in climate by extension or contraction of the ice margin.

The rate at which a kinematic wave passes through a glacier is highly variable. Some glaciers respond quickly to changes in mass balance while others do not. The length of time for glacier adjustment to a change in mass balance, the **response time**, depends on the sensitivity of the particular glacier to change and upon the nature of the change. For example, an excess of ablation at the glacier snout one year may cause a glacier snout to retreat almost immediately; by contrast an advance in the snout due to an excess of accumulation must first feed through the whole glacier before it has an effect. The response time is also controlled by the sensitivity of the glacial system, which depends on a variety of parameters, including glacier morphology and activity (i.e. the mass balance gradient). At a simple level, the larger the ice body, the slower the response time. Large ice sheets will only respond to large and sustained changes in mass balance, while small cirque or valley glaciers will often respond quickly to minor fluctuations. The rate of response of these smaller ice bodies is influenced by their morphology, as illustrated in Figure 3.20. Here two valley glaciers with different slopes are affected differently when the equilibrium line is raised by an identical amount. The ratio of accumulation to ablation area of valley glacier B is much more sensitive than that of valley glacier A and will therefore be more responsive to climatic change. In a similar manner, glaciers that are very active, that is they have large mass balance gradients, will respond much more quickly to changes in mass balance than those which are less active.

The link between climate, mass balance and glacier response is complex and may not always be apparent. This is particularly the case where glaciers terminate or calve into water. On land a glacier can react to changes in input by extending or withdrawing its snout. An extension of the snout means that an increased surface area is exposed to ablation since the glacier advances into warmer areas at lower altitude. A glacier in a deep **fjord** (a glacial valley which has been drowned by the sea) cannot do this so easily. This can be illustrated by considering a tidewater glacier in an ideal fjord of constant width and depth (Figure 3.21). The ablation area of the glacier is limited to the lower reaches of the glacier and, more importantly, to the amount of ice that can be discharged as icebergs from the cross-sectional area of the snout. If there is a shift to a positive net balance the glacier will begin to advance. Since it cannot extend to lower altitudes to enhance ablation, it will continue to advance until it can spread out and increase the cross-sectional area exposed to melting and calving. This will only occur at the fjord mouth or at a point at which the fjord widens or deepens. Consequently, fjord glaciers are particularly sensitive to changes in mass balance and relatively minor climatic changes can cause spectacular variations in the position of snouts. The most stable positions for calving glaciers within fjords include: (1) fjord mouths; (2) fjord bifurcations; (3) points where fjords widen, narrow or are bordered by low ground; and (4) where they deepen or shallow. These locations are known as **pinning points**,

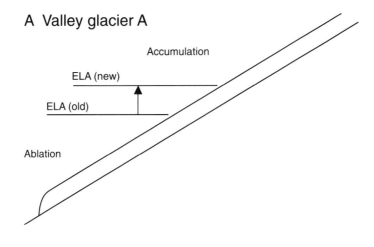

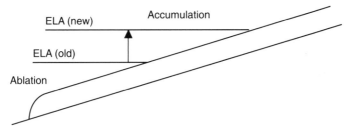

Figure 3.20   *Diagram to show the impact of a rise in equilibrium line altitude (ELA) on two valley glaciers with different gradients. The impact of the rise is greater on valley glacier B due to its shallower gradient. [Modified from: Kerr (1993) Terra Nova **5**, Fig. 4, p. 335]*

points where the ablation geometry of the ice margin may be changed, and which control the location of calving ice margins (Figure 3.21).

Understanding the interaction of climate, mass balance and glacier response is important not only in examining the response of glaciers to relatively minor climatic fluctuations but also to the more dramatic and long-term fluctuations associated with the onset of a glacial cycle. The growth and decay of large ice sheets is a complex problem. Traditionally, ice sheets are considered to grow through a sequence of larger and larger ice bodies (snow patches → cirque glaciers → valley glaciers → ice-fields → ice caps → ice sheets), developing first in upland areas and then expanding into lowland regions as the ice bodies merge and grow. More recently, computer models have suggested that the sequence of growth may follow a slightly different pattern: snow patches → cirque glaciers → valley glaciers →

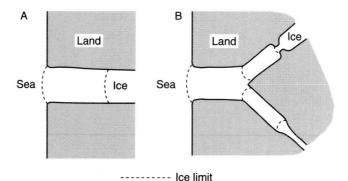

Figure 3.21   *Diagrams to illustrate the control of fjord geometry on the location of ice margins. **A**: In a fjord of constant width and depth a glacier experiencing a climatic deterioration will advance to the fjord mouth, because only there can the ice margin increase in size, thereby increasing the rate of calving and therefore ablation. **B**: Examples of pinning points*

piedmont lobes → small ice sheet. In this scenario valley glaciers first develop in mountainous areas and flow out into low-lying areas where the ice spreads out as large lobes, known as **piedmont lobes**. These lobes merge and thicken rapidly in an unstable fashion to give small ice sheets, which then merge to give larger ones. The piedmont lobes thicken and grow dramatically since they have low gradients and consequently low ice velocities and cannot, therefore, discharge the ice pouring into them from the fast-flowing valley glaciers which drain the steep mountainous areas behind. Topography may also play a very important part in the rate at which an ice sheet develops. This has recently been illustrated by a computer model of a former ice cap which existed in the Scottish Highlands at the close of the last glacial cycle, during a cold period known as the Younger Dryas (10 000 years ago). This computer model illustrates the sensitivity of this former ice cap to the mountain topography of the highlands, and showed that certain types of topography accelerate ice cap growth. As illustrated in Figure 3.22, for a given deterioration in climate, parts of the ice cap which advanced into large basins grew more dramatically than those centred on topographic ridges. Consequently, the location of large topographic basins has a dramatic effect on the rate at which ice sheets grow. In these locations small deteriorations in climate may have a dramatic effect on the size of the ice body.

Once established, an ice sheet will continue to grow provided that there is an excess of precipitation over that which the ice sheet can discharge and ablate. Growth will be driven by a variety of positive feedback systems. For example, the Earth's atmosphere is characterised by strong altitudinal gradients in temperature and precipitation. As an ice sheet grows, an increasing proportion of its area will lie at more favourable altitudes for accumulation facilitating further growth. Ice sheets will grow until: (1) their size is limited by the available space, such as the edge of the continental shelf and the presence of deep water; (2) precipitation starvation sets in when the interior of an ice sheet becomes so removed from sources

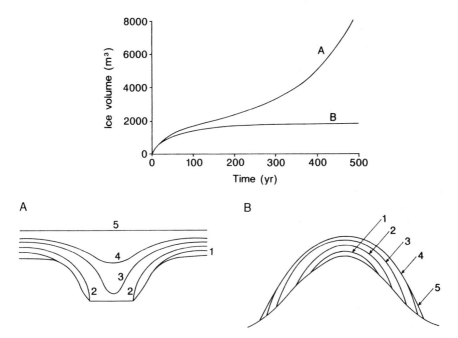

Figure 3.22   *The effects of topography on the rate of ice sheet growth. Ice advancing into topographic basins (**A**) may experience rapid non-linear growth in which a small deterioration in climate may have a dramatic effect on the size of the ice sheet. As the glaciers merge within the basin their ablation areas are reduced and rapid growth occurs as accumulation dominates over ablation (i.e. as the glacier grows in size its ablation area decreases in size). In contrast, glaciers advancing down the side of a mountain range (**B**) do not experience unstable growth as the ablation area increases in size as the glacier grows. [Diagram modified from: Payne & Sugden (1990)* Earth Surface Processes and Landforms, *15, Fig. 6, p.632]*

of precipitation around its margins that its rate of accumulation falls; and (3) climate changes, reducing accumulation or increasing ablation. Theoretically, ice sheets should reach an optimum or equilibrium size for the prevailing climate and topographic location. The ice sheet will exist until some change in this climate occurs to cause it to decay (**deglaciation**).

Deglaciation may be driven by either a decrease in precipitation and therefore accumulation, or an increase in ablation, or a combination of both. An increase in ablation may not only be achieved through a rise in air temperature but also through a rise in sea-level. Rising sea-level may increase the area subject to calving and therefore induce rapid ablation. Rapid calving will increase the discharge of ice towards these margins, a process which may initiate the development of large fast-flowing ice streams draining the ice sheet centre. It has been suggested that this process may cause the rapid 'down-draw' and deglaciation of an ice sheet. Traditionally there are two models used to explain how deglaciation occurs.

1. **Catastrophic down-wasting by areal stagnation.** In this model the equilibrium line rises quickly above the ice sheet, depriving it of accumulation. As a consequence, down-wasting is rapid and traditionally believed to be associated with little forward flow. This model has been used to explain the rapid decay of mid-latitude ice sheets during the Cenozoic Ice Age and the presence of extensive areas of stagnation-type landforms within the limits of these former ice sheets.
2. **Active deglaciation by regular ice-marginal retreat.** In this model ice sheets decay in a regular fashion through the contraction and retreat of the ice margin. Forward flow is maintained even as the margin of the glacier retreats. This mechanism can occur both slowly and quickly.

Although widely postulated as a mechanism of glacier decay, areal stagnation is now considered by many geologists to be of only regional or local importance. It is unlikely that ice sheets will be completely deprived of accumulation during decay. More significantly, ablation will always be maximum at the margin and minimum at the ice divide, maintaining the surface ice sheet gradient, basal shear stress and therefore forward flow. Stagnation on a local or regional scale may, however, occur where very low angled lobes of ice exist, a situation which may result after a glacier has surged. Glaciers decay actively through the regular retreat of their ice margin. This process may involve a fast, continuous and therefore rapid retreat or it may be punctuated by a series of still stands, seasonal readvances, surges or larger prolonged readvances. The pattern of deglaciation is therefore rarely a simple and uniform one.

3.7 SUMMARY

The mass balance—accumulation minus ablation over a year—of a glacier determines whether it will grow or decay over time. A positive balance (more accumulation than ablation) will result in glacier expansion. A negative mass balance (more ablation than accumulation) will result in glacier decay. Accumulation and ablation occur on different parts of the ice sheets. It is this spatial imbalance which drives glacier flow. The stronger this imbalance, the faster the glacier will flow. Glaciers flow through three different mechanisms: (1) by internal deformation; (2) by basal sliding; and (3) by subglacial deformation. The temperature distribution at the base of a glacier, the basal thermal regime, is important in determining the operation of these processes. Basal thermal regime is controlled by the influx of geothermal heat from the Earth and by frictional heat released during glacier movement and the rate at which heat is removed by the glacier. Glaciers may be either warm-based or cold-based. In warm-based glaciers, meltwater is released at the base of the glacier and ice flow occurs by all three processes: internal deformation, basal sliding and subglacial deformation. Cold-based glaciers are frozen to their beds and can therefore only flow by internal deformation, which restricts their geomorphological impact. The basal thermal regime of a glacier may vary both in space and time and as a consequence the geomorphological potential of a glacier will also vary. The rate or velocity of glacier flow is controlled primarily by the mass balance gradient of the glacier: the steeper the gradient, the faster the ice flow. Some glaciers may be subject to periods of rapid flow known as surges, which result from instability at the glacier bed. The growth and decay of glaciers is controlled by their mass balance budget, which is a function of climate.

3.8 SUGGESTED READING

Probably the most comprehensive and accessible treatment of the mechanics of glaciers is to be found in the book by Paterson (1994). The importance of mass balance in determining glacier activity is considered by Andrews (1972), and volume 66A of the journal *Geografiska Annaler* consists of papers about mass balance. The application of basal shear stress in testing glacier reconstructions is covered in Pierce (1979), Thorp (1991), and Bennett & Boulton (1993). There are numerous empirical and theoretical papers on glacier flow; of particular interest are Kambe & La Chapelle (1964), Weertman (1957, 1964, 1972), Kamb (1970), Iken (1981), Iken *et al.* (1983), and Iken & Bindschadler (1986). A whole issue of *Journal of Geophysical Research*, volume 92, is devoted to fast-flowing glaciers. Subglacial deformation is dealt with by Boulton & Hindmarsh (1987), Alley *et al.* (1986, 1987a,b), and Kamb (1991)

One of the most important papers on the role of basal thermal regime in determining a glacier's potential to erode and deposit is Boulton (1972). Several papers calculate the basal thermal regime of ancient glaciers, including Sugden (1977) and Glasser (1995). The glacial structure within surging glaciers is considered by Lawson *et al.* (1994), while the mechanisms of surging glaciers are reviewed in Clarke *et al.* (1984), Kamb *et al.* (1985), Kamb (1987), Raymond (1987), and Sharp (1988).

The behaviour of calving glaciers is considered by Mercer (1961) and Warren (1992). Finally the evolution of valley glaciers into ice sheets is considered by Payne & Sugden (1990) and by Hindmarsh (1990).

Alley, R. B., Blankenship, D. D., Bentley, C. R. & Rooney, S. T. 1986. Deformation of till beneath Ice Stream B, West Antarctica. *Nature* **322**, 57–59.

Alley, R. B., Blankenship, D. D., Bentley, C. R. & Rooney, S. T. 1987a. Till beneath Ice Stream B: 3. Till deformation: evidence and implications. *Journal of Geophysical Research* **92**, 8921–8929.

Alley, R. B., Blankenship, D. D., Rooney, S. T. & Bentley, C. R. 1987b. Till beneath Ice Stream B: 4. A coupled ice–till flow model. *Journal of Geophysical Research* **92**, 8931–8940.

Andrews, J. T. 1972. Glacier power, mass balance, velocities and erosion potential. *Zeitschrift für Geomorphologie* **13**, 1–17.

Bennett, M. R. & Boulton, G. S. 1993. A reinterpretation of Scottish 'hummocky moraine' and its significance for the deglaciation of the Scottish Highlands during the Younger Dryas or Loch Lomond Stadial. *Geological Magazine* **130**, 301–318.

Boulton, G. S. 1972. The role of thermal regime in glacial sedimentation. In: Price, R. J. & Sugden, D. E. (Eds) *Polar Geomorphology.* Institute of British Geographers, Special Publication **4**, 1–19.

Boulton, G. S. & Hindmarsh, R. C. A. 1987. Sediment deformation beneath glaciers: rheology and geological consequences. *Journal of Geophysical Research* **92**, 9059–9082.

Clarke, G. K. C., Collins, S. G. & Thompson, D. E. 1984. Flow, thermal structure, and subglacial conditions of a surge-type glacier. *Canadian Journal of Earth Sciences* **21**, 232–240.

Glasser, N. F. 1995. Modelling the effect of topography on ice sheet erosion, Scotland. *Geografiska Annaler* **77A**, 67–82.

Hindmarsh, R. C. A. 1990. Time-scales and degrees of freedom operating in the evolution of continental ice-sheets. *Transactions of the Royal Society of Edinburgh* **81**, 371–384.

Iken, A. 1981. The effects of the subglacial water pressure on the sliding velocity of a glacier in an idealised numerical model. *Journal of Glaciology* **27**, 407–421.

Iken, A. & Bindschadler, R. A. 1986. Combined measurements of subglacial water pressure and surface velocity of Findelengletscher, Switzerland: conclusions about the drainage system and sliding mechanism. *Journal of Glaciology* **32**, 101–119.

Iken, A., Röthlisberger, H., Flotron, A. & Haeberli, W. 1983. The uplift of Unteraargletscher at the beginning of the melt season—a consequence of water storage at the bed? *Journal of Glaciology* **29**, 28–47.

Kamb, B. 1970. Sliding motion of glaciers: theory and observation. *Review of Geophysics and Space Physics* **8**, 673–728.

Kamb, B. 1987. Glacier surge mechanism based on linked cavity configuration of the basal water conduit system. *Journal of Geophysical Research* **92**, 9083–9100.

Kamb, B. 1991. Rheological nonlinearity and flow instability in the deforming bed mechanism of ice stream motion. *Journal of Geophysical Research* **96**, 16 585–16 595.

Kamb, B. & La Chapelle, E. 1964. Direct observation of the mechanism of glacier sliding over bedrock. *Journal of Glaciology* **5**, 159–172.

Kamb, B., Raymond, C. F., Harrison, W. D., Englehardt, H., Echelmeyer, K. A., Humphrey, N., Brugman, M. M. & Pfeffer, T. 1985. Glacier surge mechanism: 1982–1983 surge.of Variegated Glacier, Alaska. *Science* **227**, 469–479.

Lawson, W. J., Sharp, M. J. & Hambrey, M. J. 1994. The structural geology of a surge type glacier. *Journal of Structural Geology* **16**, 1447–1462.

Mercer, J. H. 1961. The response of fjord glaciers to changes in the firn limit. *Journal of Glaciology* **3**, 850–858.

Paterson, W. S. B. 1994. *The Physics of Glaciers* (Third Edition). Pergamon, Oxford.

Payne, A. & Sugden, D. E. 1990. Topography and ice sheet growth. *Earth Surface Processes and Landforms* **15**, 625–639.

Pierce, K. L. 1979. History and dynamics of glaciation in the Northern Yellowstone National Park area. *US Geological Survey Professional Paper* **729-F**.

Raymond, C. F. 1987. How do glaciers surge? A review. *Journal of Geophysical* Research **92**, 9121–9134.

Sharp, M. 1988. Surging glaciers: behaviour and mechanisms. *Progress in Physical Geography* **12**, 349–370.

Sugden, D. E. 1977. Reconstruction of the morphology, dynamics and thermal characteristics of the Laurentide ice-sheet at its maximum. *Arctic and Alpine Research* **9**, 21–47.

Thorp, P. W. 1991. Surface profiles and basal shear stresses of outlet glaciers from a Lateglacial mountain ice field in western Scotland. *Journal of Glaciology* **37**, 77–88.

Warren, C. R. 1992. Iceberg calving and the glacioclimatic record. *Progress in Physical Geography* **16**, 253–282.

Weertman, J. 1957. On the sliding of glaciers. *Journal of Glaciology* **3**, 33–38.

Weertman, J. 1964. The theory of glacier sliding. *Journal of Glaciology* **29**, 287–303.

Weertman, J. 1972. General theory of water flow at the base of a glacier or ice sheet. *Reviews of Geophysics and Space Physics* **10**, 287–333.

4
Glacial Meltwater

In this chapter we shall examine the role of glacial meltwater within the glacial system, its characteristics and its significance. Glacial meltwater is an important part of most glaciers, because: (1) it is the main ablation product; (2) it is intimately involved in glacier sliding; and (3) it is responsible for removing debris from the ice–rock interface and carrying it beyond the confines of the glacier.

4.1 GLACIAL MELTWATER

Glacial meltwater is derived from the melting of ice on the surface, at the bed, or within a glacier. This melting will occur whenever there is sufficient heat to turn the ice back into water. Heat can be supplied by: (1) solar radiation, (2) friction due to ice flow, and (3) heat derived from the Earth's crust beneath the glacier (geothermal heat). As a consequence, melting can occur both on the surface of the glacier and within the glacier itself. Surface melting on the ice surface is probably the most important contribution to glacial meltwater. Since surface melting by solar radiation is a result of warm air temperatures, it is highly seasonal. The volume of water within the glacial meltwater system also depends on the amount contributed by rainfall, snow-melt and valley side streams, all of which can add significant quantities of externally derived water to the glacial system.

4.2 THE GLACIAL CHANNEL SYSTEM

Glacial meltwater finds its way through the glacier from its point of origin through a variety of different flow paths. Meltwater derived from surface melting and from basal melting will tend to follow different paths through a glacier, although both will invariably involve channel flow either within or on the glacier. The type of channel system within a particular glacier depends primarily on its thermal regime. On glaciers consisting of cold ice the meltwater is unable to penetrate without freezing and tends to be confined to the surface and ice margin. On glaciers consisting of warm ice, water may penetrate throughout the ice mass and water flow will occur in **supraglacial** (surface), **englacial** (within) and **subglacial** (beneath) channels. In glaciers with mixed thermal regimes, more complex flow patterns are to be expected and water may become trapped in small subglacial lakes at the junction between ice of different temperatures (see Section 3.4).

Supraglacial channels tend to be less than a few metres wide and may exploit structural weaknesses within the ice (Figure 4.1A). In plan they may adopt either meandering or straight courses. Velocities within these channels are usually high due to their smooth sides. On warm ice, supraglacial channels are usually short and are interrupted by vertical shafts which divert the water into the glacier. These shafts may simply be crevasses or alternatively they may be cylindrical vertical tunnels, known as **moulins** (Figure 4.1B).

The internal geometry of moulins has been studied in detail on the Storglaciären in Sweden. Detailed mapping on the surface of this glacier has shown that moulins exploit crevasses. As ice moves down-glacier it may encounter changes in the stress fields within the glacier, causing crevasses to open, deepen and then close. When a crevasse opens on a glacier surface it may intersect a meltwater stream. The crevasse will then fill with water, until it opens and deepens sufficiently to intersect englacial drainage passages, at which point the water within it will drain away. When glacier flow moves the crevasse into an area of compression the crevasse may close, but the heat carried into it by the meltwater may keep the drainage channel open and thereby form a moulin. The moulins of Storglaciären consist of near-vertical shafts, 30–40 m deep, which feed englacial tunnels that descend from the base of the moulin at an angle of between 0° and 45°. The orientation of this englacial tunnel is often controlled by the orientation of the original crevasse from which the moulin formed.

Within the body of a warm glacier, water flow occurs through a series of tunnels. The presence and size of these tunnels depends upon the balance between: (1) ice pressure, which will tend to close the tunnel through ice deformation; and (2) frictional melting of the tunnel walls by the flow of water and by the transfer of heat carried into the glacier by the water, which will tend to keep the tunnels open. The natural shape for a tunnel or conduit, well away from the bed, is circular due to these two opposing factors. More importantly, the size of conduit will vary with changes in discharge over a matter of weeks. As discharge, or water flow, increases, the conduit will grow in size.

The formation of these tunnels is an area of some uncertainty. It has been suggested that water first flows through the ice along the boundaries between individual ice crystals. This inter-crystal flow quickly forms a network of small tubes. The bigger tubes grow rapidly, at the expense of the smaller ones, due to the greater water flow and therefore frictional heating within them. In this way a three-dimensional dendritic network of conduits from small tubes through to larger tunnels will form. Moulins cut across this integrated channel network.

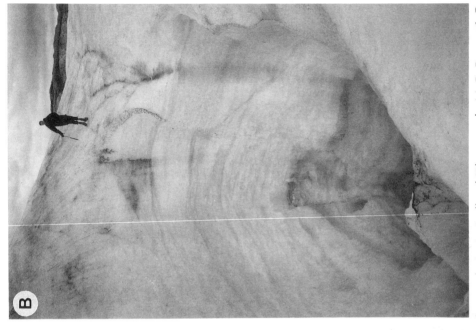

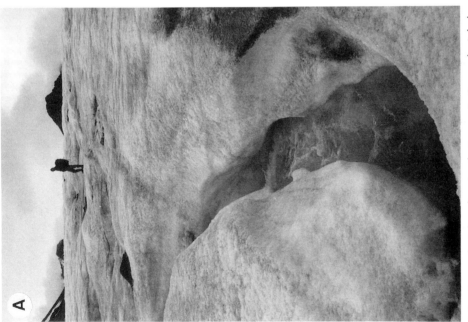

Figure 4.1 Supraglacial meltwater features. **A:** Supraglacial channel on a glacier in Svalbard [Photograph: M. R. Bennett]. **B:** Moulin on the Upsalla glacier in Argentina. [Photograph: N. F. Glasser]

The orientation of this network of conduits and tunnels is controlled by the water pressure gradient within the glacier. Water will flow down the pressure gradient from areas of high to low pressure. It is possible to determine the nature of this pressure gradient within a glacier and therefore the direction of water flow within it. Consider Figure 4.2, which shows a hypothetical water-filled tube beneath a glacier. The weight of ice above point A is equal to the weight of the water column BC which it forces up. A line between A and C defines a surface of equal potential pressure. Along this line the pressure due to the weight of the overlying ice is equal to the water pressure it generates. If we now move the tube towards the right, towards the ice margin, the weight of the ice above point A will fall, and consequently the water column BC will be lower. A new lower equipotential surface is defined. Water will flow at right angles to these equipotential surfaces from a surface of higher potential pressure to one of lower potential pressure. As a consequence, englacial conduits and tunnels will be orientated perpendicular to surfaces of equipotential pressure, as shown in Figure 4.3. Some moulins may be an exception to this, reflecting their origin as crevasses. The geometry of the equipotential surfaces within a glacier is determined by the variation in ice thickness, which is primarily controlled by the surface slope of the glacier and secondarily by the slope of the underlying topography. The surface of a glacier does not always slope in sympathy with the slope of the glacier bed. As a consequence, subglacial meltwater may not always flow directly down the maximum slope beneath the glacier and may in some cases even flow uphill. Under an ice sheet water flow will be approximately radial, in sympathy with the surface slope and the direction of ice flow, but will deviate around hills and bumps and be concentrated in topographic depressions such as valleys.

It is possible to calculate the water pressure potential at a series of points at the base of a glacier from a knowledge of the variation in ice thickness. These points can be contoured to define a surface known as the subglacial hydraulic potential surface (Figure 4.3). Provided that any subglacial tunnel is completely water-filled, then the tunnel should be orientated at right angles to this hydraulic surface. The

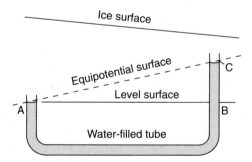

Figure 4.2   *Diagram to illustrate the hydraulic head which drives water flow within a glacier. The weight of ice above point A is equal to the elevation of the water column BC. The thinner the ice above point A the less the hydraulic head (BC); consequently, the hydraulic head or potential will fall towards the ice margin or in the direction of glacier slope. Water flows from areas of high hydraulic potential to areas of low potential*

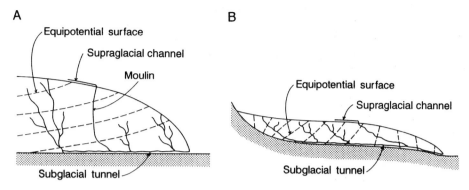

Figure 4.3   The pattern of equipotential surfaces within a glacier (i.e. surfaces of equal hydraulic potential). Water always flows from areas of high potential to areas of low potential and will therefore flow at right angles to the equipotential surfaces, as shown

ability to calculate this surface is a useful tool in the interpretation of the glacial landform record (Box 4.1). However, if the subglacial tunnel is not completely full of water, something which may occur at the ice margin, then the water flow and the orientation of the tunnel may be controlled by the underlying topography beneath the glacier and not by the subglacial water pressure surface. The presence of gravity-driven subglacial flow at the margin of a glacier has recently been examined using **dye-tracing** experiments (Box 4.2). Dye is added to meltwater on the surface of the glacier and the time taken for it to travel through the glacier to the snout is recorded. From this information, reconstructions of the internal drainage system within a glacier can be made and compared to that predicted by theory.

The nature of subglacial channel flow is still an area of some uncertainty. Subglacial channels may either be cut upwards into the ice (**Röthlisberger** or **R-channels**) or cut into bedrock (**Nye** or **N-channels**) (Figure 4.4). In the case of R-channels their direction is still determined by that of the water pressure gradient. In contrast, N-channels may exploit weaknesses in the underlying rock.

In general, subglacial drainage beneath glaciers flowing over hard substrates may be achieved through two types of channel system which may coexist: channels and linked cavities. First, drainage may be achieved though a series of large channels, discharging the bulk of the meltwater flow. The number of such channels will decrease towards the ice margin, and most of the meltwater discharge will occur through a single channel. Secondly, in addition to these major meltwater arteries, subglacial water may exist and flow within a series of **linked subglacial cavities** (Figure 4.5). As ice flows over bumps in its bed, water-filled cavities may form in their lee where the pressure drops. In certain situations these water-filled cavities appear to be linked by a series of small, irregular R- and N-channels (Figures 4.4 and 4.5). Flow in this system of linked cavities will tend to be orientated perpendicular to the direction of glacier flow, since the high pressures are generated at the down-glacier toe of the cavity as it closes. Within many glaciers there may therefore be two independent discharge routes for subglacial meltwater.

BOX 4.1: EQUIPOTENTIAL SURFACES: A TOOL IN GLACIAL RECONSTRUCTION

There is a long running controversy about the nature of the glacial history of part of Antarctica known as the Dry Valleys in Victoria Land. At present these valleys contain no ice. Some researchers argue that this area was once covered by part of the Antarctic Ice Sheet, while others have suggested that it was only ever covered by local valley glaciers. Sugden *et al.* (1991) set out to resolve this debate. Within the dry valleys there are a series of spectacular subglacial melt-water channels. In order to resolve whether these channels were formed under an extension of the Antarctic ice sheet or beneath local ice, Sugden *et al.* (1991) calculated the subglacial hydraulic potential surface that would have existed within the area had an ice sheet been present. The subglacial meltwater channels should, if formed beneath the ice sheet, run at right angles to the contours used to define this surface (i.e. they should run down the pressure gradient).

This involved first estimating the location of the ice margin on the assumption that the area had been covered by an ice sheet and then estimating its surface morphology. The surface morphology was estimated by using the profile equation of Nye (1952) (Box 3.1), which assumes a parabolic ice surface profile. This ice surface was then compared with the topography of the dry valley area to determine the variation in ice thickness, from which the subglacial hydraulic potential surface was calculated using a simple equation developed by R. L. Shreve. This equation states that

$$\phi = \phi_0 + p + p_w g h$$

Here ϕ is the subglacial water pressure potential; ϕ_0 is a constant; p is water pressure in a conduit at the base of the glacier, which is assumed to be equal to the pressure of the overlying ice; p_w is water density; g is acceleration due to gravity; and h is glacier thickness.

If the subglacial meltwater channels were formed beneath the ice sheet then they should run at right angles to the subglacial equipotential surface calculated from Shreve's equation. If they had formed beneath local ice the pattern would be different, reflecting the equipotential surface within local valley glaciers. The meltwater channels were found to run at right angles to the equipotential contours calculated for the ice sheet, which provides strong support for the contention that the area was once covered by an extension of the Antarctic Ice Sheet. This example illustrates how the application of simple principles in glaciology can resolve problems of interpretation in areas of former glaciation.

Source: Sugden, D. E., Denton, D. H. & Marchant, D. R. 1991. Subglacial melt-water channel system and ice sheet overriding of the Asgard range, Antarctica. *Geografiska Annaler* **73A**, 109–121.

BOX 4.2: DETERMINING THE GEOMETRY OF GLACIAL DRAINAGE SYSTEMS

The geometry of the englacial and subglacial drainage system of modern glaciers can be examined using dye-tracing. Dye is added to supraglacial meltwater as it disappears into the glacier. The time taken for this meltwater to reach the glacier snout and emerge is recorded. If this procedure is repeated at a large number of locations over a small valley glacier it is possible to obtain some idea of its internal drainage structure.

Sharp *et al.* (1993) did this for the Haut Glacier D'Arolla in Switzerland. A total of 342 dye-injection experiments were conducted using 47 moulins distributed widely across the glacier surface. These experiments were combined with detailed measurements of both the surface and subsurface topography of the glacier using conventional field survey techniques and radio echo sounding. The dye-injection experiments were used to reconstruct the pattern of internal drainage within the glacier. This was then compared with the basal water pressure gradients within the glacier calculated from R. L. Shreve's equation (Box 4.1). The drainage system should flow normal to the contours that define the subglacial hydraulic potential surface. A close approximation was found between the reconstruction and the pattern of contours or water pressure gradient. However, some discrepancies were noted, which might imply that the drainage close to the ice margin was at least seasonally driven by the slope of the subglacial topography and not by the surface slope of the glacier.

Source: Sharp, M., Richards, K., Willis, I, Arnold, N., Nienow, P., Lawson, W. & Tison, J. 1993. Geometry, bed topography and drainage system structure of the Haut Glacier D'Arolla, Switzerland. *Earth Surface Processes and Landforms* **18**, 557–571.

This picture of subglacial drainage does not hold for glaciers flowing over soft deformable beds. In some cases discharge may be achieved through porous flow within the deformable bed, or by water flow within the deforming sediment. However, in some cases this will not be sufficient to discharge all the basal meltwater. In these situations subglacial drainage probably takes the form of broad shallow channels cut into the surface of the deformable bed. There is, however, no field evidence, at present, to support this theoretical inference.

The pattern of subglacial drainage is also affected by the permeability of the bedrock over which the glacier is flowing. It is generally assumed that meltwater from the base of a glacier is discharged in a relatively thin subglacial zone. However, when glaciers flow over extensive aquifers or areas of permeable lithology this may not be the case. At these locations subglacial drainage will occur through the aquifer or permeable bedrock. Recently numerical models have suggested that high rates of meltwater discharge into an aquifer from continental-scale ice sheets could strongly influence the rate and pattern of groundwater flow within the aquifer. The aquifer would become integrated into the hydrological sys-

Figure 4.4   *N-channels in front of the Glacier de Tsanfleuron, Switzerland.*
[Photograph: M. Sharp]

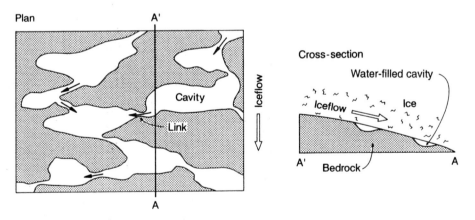

Figure 4.5   *A network of linked basal cavities in plan and cross-section. Each cavity is linked by N- or R-channels. [Reproduced with permission from: Hooke (1989) Arctic and Alpine Research **21**, Fig. 4, p. 226]*

tem of the ice sheet and flow within it would reflect the same hydraulic gradients that control the flow of water within the ice sheet. Consequently, flow within the aquifer would be reorganised and the local topographic variables which normally operate to control the rate and direction of groundwater flow would be overridden.

This is of particular importance because parts of the former mid-latitude ice sheets in North America and northern Europe advanced over extensive aquifers.

4.3 SUBGLACIAL WATER PRESSURE

Subglacial water pressure has an important role in many subglacial processes, through its control on the effective normal pressure beneath a glacier. Effective normal pressure is the force per unit area imposed vertically by a glacier on its bed. For a cold-based glacier it is effectively equal to the weight of the overlying ice: thick ice imposes a greater pressure than thin ice. This is summarised by

$$N = pgh \qquad (4.1)$$

where N = normal effective pressure, p = density of ice, g = acceleration due to gravity, and h = ice thickness.

If water is present at the glacier bed, however, the effective normal pressure is reduced by an amount equal to the subglacial water pressure. Put crudely, the greater the water pressure, the more it can support the weight of the glacier and thereby reduce the effective normal pressure. The equation is modified to

$$N = pgh - wp \qquad (4.2)$$

where N = effective normal pressure, p = density of ice, g = acceleration due to gravity, h = ice thickness, and wp = subglacial water pressure.

This is only true where the glacier has a flat bed. Effective normal pressure is modified by the flow of ice over obstacles (Figure 4.6). As ice flows against the upstream side of an obstacle, the effective normal pressure increases by an amount proportional to the rate of glacier flow against the obstacle. Effective normal pressure is also reduced in the lee or on the downstream side of the obstacle (Figure 4.6). The pressure fluctuation caused by the flow of ice against the obstacle is, therefore, positive on the upstream side and negative on the downstream side. The negative pressure fluctuation on the downstream side of an obstacle may cause a cavity to form in the lee of the obstacle if it exceeds the effective normal pressure at this point (Figure 4.7). Cavity formation is favoured by: (1) high basal water pressures which reduce effective normal pressure; and (2) high rates of basal sliding which give large pressure fluctuations over obstacles. Theoretical calculations show that cavities can open at sliding velocities of about 9 m per year beneath a thickness of 100 m of ice, while velocities of 35 m per year are required with ice thickness of the order of 400 m.

Basal water pressure is controlled by four variables: (1) glacier thickness (the greater the weight of the overlying ice, the greater the water pressure); (2) the rate of water supply (the input of large amounts of meltwater may increase the pressure); (3) the rate of meltwater discharge (an efficient subglacial drainage system will reduce water pressure); and (4) the nature of the underlying geology (permeable bedrock will reduce water pressure). Variations in the rate of water supply

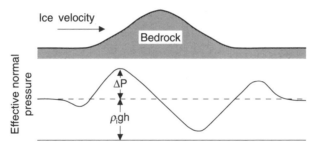

Figure 4.6   *Schematic diagram of the normal effective pressure distribution at the glacier bed as ice flows over a bedrock obstacle. [Modified from: Boulton (1974) In: Coates (Ed.)* Glacial Geomorphology, *George Allen & Unwin, Fig. 8, p. 55]*

Figure 4.7   *Subglacial cavity in the lee of a rock step beneath a glacier in Patagonia. The ice is flowing from right to left over a bedrock step. [Photograph: N. F. Glasser]*

and in the rate of meltwater discharge are responsible for much of the seasonal variation in water pressure observed at some glaciers. Early in the melt season, water pressure may be very high due to the abundance of meltwater and the relative inefficiency of the channel network (see Section 4.4). As the subglacial channel network develops during the ablation season, discharge becomes more efficient and the water pressure generally falls.

As we will see in Chapter 6, variations in water pressure and its influence on effective normal pressure and cavity formation are very important to the processes of glacial erosion. Basal water pressure is also important in determining the rate of basal sliding (see Section 3.3.2). Effective normal pressure helps determine the friction experienced between a glacier and its bed. If the water pressure rises, effective normal pressure will fall, thereby reducing basal friction and increasing basal sliding. This explains why the basal sliding velocity of many glaciers often increases during the summer melt season or after a large rainfall event. For example, field observations on the Unteraargletscher have shown that it moves vertically by 0.4 m at the start of the melt season due to increased water pressure. This is followed by a similar downward movement at a constant rate over the next three months. The glacier velocity increases significantly when the surface is raised. Variations in basal water pressure have also been linked to glacier surges (see Section 3.5). For example, the surge of the Variegated glacier in Alaska, during 1982–1983 is believed to have been triggered by a change in the subglacial drainage system. Prior to the surge the glacier had a subglacial drainage system dominated by a few large tunnels. However, this appears to have changed to a system dominated by linked subglacial cavities in which the water pressure rose dramatically due to the lower rate of discharge possible from such a system. This rise in water pressure facilitated the rapid glacier flow of the surge. At the end of the surge the water that built up in this cavity system was released as a large flood and the subglacial system reverted to a large integrated tunnel system. The cause of this change in drainage system is unclear, but is believed to be central to the rapid glacier flow of this surge.

In summary, variation in basal water pressure has an important role to play in determining the flow dynamics of a glacier and is also important in the processes of glacial erosion (see Chapter 6).

4.4 DISCHARGE FLUCTUATIONS

The discharge of meltwater from glaciers varies dramatically both on a diurnal (daily) and on a seasonal basis. Diurnal discharge variations reflect atmospheric air temperatures and therefore the pattern of daily ablation on a glacier. Discharge is usually low in the early morning and rises in the late afternoon or evening (Figure 4.8). This diurnal fluctuation is suppressed in winter, but increases towards the late summer when the rate of daily ablation reaches its maximum. Seasonal fluctuations are equally dramatic (Figure 4.8). They reflect two factors: (1) the seasonal nature of ablation and (2) the seasonal development of the internal drainage network within warm-based glaciers. One can develop a simple model for the drainage pattern for glaciers in a strongly seasonal climate.

1. **Spring melt.** On the glacier, ablation of winter snow begins in the spring and, as a consequence, water pressure within the glacier begins to increase. In front of the glacier, ice in proglacial rivers (**Aufeis**) breaks up and melting of winter snow proceeds rapidly.
2. **Late spring melt.** On the glacier, ablation of winter snowfall is well advanced. Discharge in all channels, conduits and tunnels increases. The conduits grow in size and the internal drainage network within the glacier develops. The

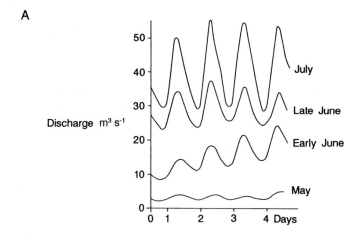

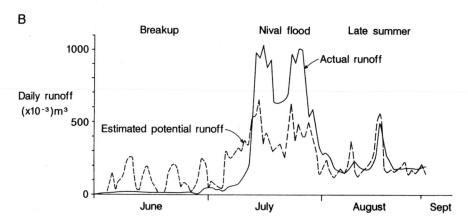

Figure 4.8   *Schematic discharge fluctuations for glacial meltwater streams. **A:** Diurnal fluctuations; these become more pronounced as the melt season proceeds. **B:** Seasonal fluctuations. The difference between potential runoff and actual runoff reflects the efficiency of the drainage network. At the start of the melt season the channel and tunnel network is poorly developed and therefore actual runoff is less than potential (i.e. water storage occurs in the glacier). However, as the melt season proceeds, actual and potential runoff become similar as the drainage system evolves into an efficient network*

amount of water available within the glacier exceeds that which can be discharged by the internal drainage network. Discharge from the glacier into the proglacial channel system steadily increases.

3. **Early summer.** With the development of a well-connected drainage network within the glacier, much of the stored water is released. The daily discharge exceeds the amount of daily melt on the glacier surface and in the course of a

few weeks the drainage system discharges the vast majority of its total annual discharge (**nival flood**).

4. **Late summer.** The drainage network within the glacier has reached its optimum efficiency. All the stored water within the glacier has been discharged. Daily discharge matches the amount of melt achieved each day. Water pressures within the glacier are usually at their minimum.

5. **Autumn.** Cessation of melting on the glacier causes a dramatic drop in discharge. The internal network of conduits and tunnels within the glacier begins to collapse as the water flowing within them declines and they close due to ice deformation. Only a few major drainage arteries contain sufficient flow for them to be maintained.

6. **Winter.** The degree to which the drainage network shuts down depends on the climate and severity of the winter. Most meltwater discharge, if there is any, will be derived from internal melting.

The degree to which the internal drainage network of conduits and tunnels within a glacier collapses each year is probably highly variable. Some shutdown will occur in all cases, but in large ice sheets the major discharge arteries are likely to be maintained since the rate of flow, due to melt generated by internal deformation and geothermal heating, is likely to be much larger. The smaller and thinner the glacier, the more likely it is that the drainage system evolves each year.

On some glaciers this pattern of diurnal and seasonal discharge fluctuation is interrupted by catastrophic events known as **jökulhlaups**. These are high-magnitude events, often several orders of magnitude greater than normal peak flows. Jökulhlaups may occur in one of two ways: (1) through subglacial volcanic activity or (2) through the drainage of ice-dammed lakes. Volcanic activity beneath glaciers is common today in Iceland. A spectacular example is Grímsvötn, beneath the Vatnajökull ice cap. Here water builds up subglacially above a volcano, which then drains catastrophically in the space of a few hours, through a 50 km long subglacial tunnel. The average volume of water discharged from Grímsvötn is between 3 and 3.5 km^3. In 1922 the jökulhlaup discharged approximately 7.1 km^3 of water with a maximum discharge of about 570 000 m^3 s^{-1}. This is more than the flow of the River Congo and about one-quarter of that of the River Amazon. The Grímsvötn jökulhlaup occurs every five to six years, when the water level above the volcano reaches a critical depth. Volcanically induced jökulhlaups may be regular events such as that at Grímsvötn or occasional events associated with irregular episodes of volcanism. It has been suggested that these events may occur with more frequency during deglaciation due to the unloading of the crust by the thinning ice (isostatic rebound). The reduction in crustal stress associated with unloading may cause an increase in the amount of volcanic activity.

Jökulhlaups due to the sudden drainage of ice-dammed lakes are much more widespread. Ice-dammed lakes occur wherever ice blocks the down-valley flow of water (Figure 4.9). The drainage of such lakes usually occurs at regular intervals, and when it does occur it is catastrophic. Why do these lakes drain catastrophically? There are several hypotheses: (1) vertical release of the ice dam by flotation caused by rising water levels in the lake; (2) overflowing of an ice dam, causing

Figure 4.9   *Ice-dammed lake of Vatnsdalur, Iceland, following a small jökulhlaup in 1990. **A:** View of the ice dam formed by Heinabergsjökull [Photograph: M. R. Bennett]. **B:** A view of the lake. The surface of the lake in the foreground is covered by a mass of small icebergs. Note the stranded icebergs along the sides of the lake due to lake lowering during the 1990 jökulhlaup. [Photograph: M. R. Bennett]*

rapid melting of the dam due to friction from the water flow; (3) destruction or fis-
suring of an ice dam by earthquakes; and (4) enlargement of pre-existing tunnels
beneath the dam by increased water flow and melt-enlargement due to frictional
heating. Of these mechanisms, the first is perhaps the most widely discussed,
although it is not without its problems. The idea is that the pressure of water
within the lake increases as the lake fills and the water level rises. If the pressure of
water within the lake exceeds the weight of the ice dam, the water will force its
way beneath the dam, effectively floating it. As water finds its way beneath the
dam, subglacial tunnels will develop due to frictional heating from the flowing
water, and the lake will drain. Once the lake is drained the ice dam will settle back
in place, any subglacial tunnels will close by ice deformation and the lake will
begin to fill again until the critical water depth or pressure is reached again. The
periodicity between jökulhlaups is determined by the rate at which the lake fills.
Thinning or thickening of the ice dam will also influence the periodicity of jökulhlaups and therefore the level of water necessary to trigger an event. This mechanism will only work where the lake can fill sufficiently to generate the necessary
pressure. In many situations the water level attainable within a lake may be limited by a spillway or channel via which the lake waters can drain, and in these
cases some other mechanism is required to trigger catastrophic lake drainage.
Moreover, many lakes appear to drain before the water depth necessary to cause
flotation is obtained. In this case one of the other three mechanisms listed above
must apply.

The shape of the jökulhlaup hydrograph will be determined by the nature of
the trigger or cause of the event and the volume of water available to be drained
(Figure 4.10). Jökulhlaups induced by volcanic activity tend to give floods of short
duration but of high magnitude, while jökulhlaups caused by the drainage of ice-
dammed lakes may produce floods with a greater duration but of lower magnitude (Figure 4.10). As one might expect, the size of the flood peak from a
jökulhlaup caused by the drainage of ice-dammed lakes is proportional to the volume of water within the lake (Box 4.3). The high magnitude nature of jökulhlaups
makes them of considerable geomorphological significance.

4.5 SUMMARY

Glacial meltwater is a very important component of the glacial system: it is the
main ablation product of most ice sheets; it is intimately involved in glacier sliding
and it is responsible for removing debris from the ice–rock interface and carrying it
beyond the confines of the glacier. The type of drainage network carrying glacial
meltwater within a glacier is dependent on its thermal regime. In cold glaciers
drainage is supraglacial, while in warm glaciers drainage is supraglacial, englacial
and subglacial. The direction of englacial and subglacial drainage within a glacier
follows the water pressure gradient, which is perpendicular to lines of equal water
pressure potential. These are controlled primarily by the surface slope of the glacier and to a lesser extent by subglacial topography. On hard substrates, subglacial

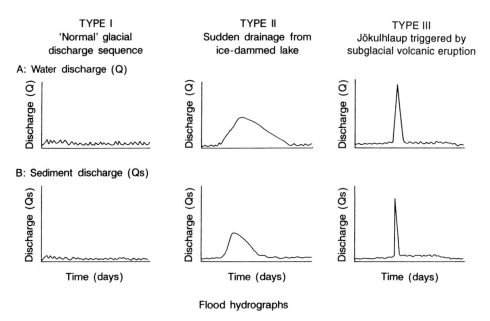

Figure 4.10   *Flood hydrographs for different types of jökulhlaup compared with non-catastrophic flow. [Modified from: Maizels & Russell (1992)* Quaternary Proceedings *2, Fig. 7, p. 142]*

drainage occurs through a linked system of cavities and/or via a few major subglacial conduits or tunnels. On soft substrates, subglacial drainage either occurs through the deforming sediment or via shallow channels cut in the surface of the sediment. Meltwater discharge varies diurnally and seasonally. Catastrophic events, known as jökulhlaups, may be superimposed on this discharge fluctuation. They are high-magnitude, low-frequency events caused either by subglacial volcanic activity or by the drainage of ice-dammed lakes.

4.6 SUGGESTED READING

The hydrology of glaciers is well covered in the book by Paterson (1994) and in the review papers by Röthlisberger & Land (1987) and Hooke (1989). Estimating the water pressure potential gradient within glaciers is covered by Shreve (1972) and its relationship to subglacial landforms is discussed in Shreve (1985a,b), Sugden *et al.* (1991) and Syverson *et al.* (1994). Sharp *et al.* (1993) use dye-tracing to reconstruct the subglacial hydrology of a small glacier in the European Alps. Holmlund (1988) discusses the internal geometry and evolution of moulins. The reorganisation of groundwater flow in aquifers by ice sheets is dealt with in Boulton *et al.* (1993).

The pattern of water discharge from glaciers has been observed in several papers: Meir & Tangborn (1961), Mathews (1963), and Elliston (1973). Björnsson

BOX 4.3: PREDICTING THE MAGNITUDE OF JÖKULHLAUPS

Attempts have been made to predict the peak flow magnitude of jökulhlaups, both for theoretical purposes and for use in the design of bridges or other structures that may have to withstand jökulhlaup flows. Developing theoretical relationships to predict discharge from ice-dammed lakes has proved difficult, not least because the exact mechanisms of drainage are poorly understood in many cases. However, Clague and Mathews (1973) developed an empirical equation based on data from the jökulhlaups for which the peak discharge could be estimated accurately. They obtained a regression relationship between peak discharge and the volume of the ice-dammed lake

$$Q_{max} = 75\left(\frac{V_o}{10^6}\right)^{0.67}$$

where Q_{max} = peak discharge (m^3 s^{-1}), and V_o = volume of the lake prior to drainage (m^3).

Although empirical, this equation gives surprisingly good results and has been used widely to predict jökulhlaup flood magnitudes.

Source: Clague, J. J. & Mathews, W. H. 1973. The magnitude of jökulhlaups. *Journal of Glaciology* 12, 501–504.

(1992) reviews the characteristics of jökulhlaups in Iceland, while a series of papers deal with the drainage of ice-dammed lakes (Russell 1989, 1990; Russell & de Jong 1989; Russell *et al.* 1990). Maizels & Russell (1992) review a variety of different aspects of jökulhlaups. Both Glen (1954) and Spring & Hutter (1981) discuss the mechanisms by which ice-dammed lakes drain.

Björnsson, H. 1992. Jökulhlaups in Iceland: prediction, characteristics and simulation. *Annals of Glaciology* 16, 95–106.

Boulton, G. S., Slott, T., Blessing, K., Glasbergen, P., Leijnse, T. & Van Gijssel, K. 1993. Deep circulation of groundwater in over-pressured subglacial aquifers and its geological consequences. *Quaternary Science Reviews* 12, 739–745.

Elliston, G. R. 1973. Water movement through the Gornergletscher. *International Association of Scientific Hydrology* 95, 79–84.

Glen, J. W. 1954. The stability of ice-dammed lakes and other water-filled holes in glaciers. *Journal of Glaciology* 2, 316–318.

Holmlund, P. 1988. The internal geometry and evolution of moulins, Storglaciären, Sweden. *Journal of Glaciology* 34, 242–248.

Hooke, R. LeB. 1989. Englacial and subglacial hydrology: a qualitative review. *Arctic and Alpine Research* 21, 221–233.

Maizels, J. & Russell, A. J. 1992. Quaternary perspectives on jökulhlaup prediction. *Quaternary Proceedings* 2, 133–152.

Mathews, W. H. 1963. Discharge of a glacial stream. *International Association of Scientific Hydrology* 63, 290–300.

Meier, M. F. & Tangborn, W. V. 1961. Distinctive characteristics of glacier runoff. *US Geological Survey Professional Paper* 424B, 14–16.

Paterson, W. S. B. 1994. *The Physics of Glaciers* (Third Edition). Pergamon, Oxford.

Röthlisberger, H. & Land, H. 1987. Glacial hydrology. In: Gurnell, A. M. & Clark, M. J. (Eds) *Glacio-fluvial Sediment Transfer*. John Wiley & Sons, Chichester, 207–284.

Russell, A. J. 1989. A comparison of two recent jökulhlaups from an ice-dammed lake, Søndre Strømfjord, West Greenland. *Journal of Glaciology* **35**, 157–162.

Russell, A. J. 1990. Extraordinary melt-water run-off near Søndre Strømfjord, West Greenland. *Journal of Glaciology* **36**, 353.

Russell, A. J. & De Jong, C. 1989. Lake drainage mechanisms for the ice-dammed Oberer Russellsee, Sondre Stromfjord, West Greenland. *Zeitschrift für Gletsherkunde und Glazialgeologie* **24**, 143–147.

Russell, A. J., Aitken, J. F. & De Jong, C. 1990. Observations on the drainage of an ice-dammed lake in West Greenland. *Journal of Glaciology* **36**, 72–74.

Sharp, M., Richards, K., Willis, I, Arnold, N., Nienow, P., Lawson, W. & Tison, J. 1993. Geometry, bed topography and drainage system structure of the Haut Glacier D'Arolla, Switzerland. *Earth Surface Processes and Landforms* **18**, 557–571.

Shreve, R. L. 1972. Movement of water in glaciers. *Journal of Glaciology* **11**, 205–214.

Shreve, R. L. 1985a. Late Wisconsin ice-surface profile calculated from esker paths and types, Katahdin esker system, Maine. *Quaternary Research* **23**, 27–37.

Shreve, R. L. 1985b. Esker characteristics in terms of glacier physics, Katahdin esker system, Maine. *Geological Society of America Bulletin* **96**, 639–646.

Spring, U. & Hutter, K. 1981. Numerical studies of Jökulhlaups. *Cold Region Science and Technology* **4**, 227–244.

Sugden, D. E., Denton, D. H. & Marchant, D. R. 1991. Subglacial meltwater channel system and ice sheet over riding of the Asgard range, Antarctica. *Geografiska Annaler* **73A**, 109–121.

Syverson, K. M., Gaffield, S. J. & Mickelson, D. M. 1994. Comparison of esker morphology and sedimentology with former ice-surface topography, Burroughs Glacier, Alaska. *Geological Society of America Bulletin* **106**, 1130–1142.

5
The Processes of Glacial Erosion

In this chapter we shall examine the processes of glacial erosion. As a glacier moves across the Earth's surface it detaches, picks up and transports rock and sediment: this is glacial erosion. Ice sheets are agents of erosion because they flow outward from a central point to their margins, moving bedrock within them. This reduces the elevation of the land under the ice sheet where erosion takes place and increases it by deposition at the ice sheet margin. The products of glacial erosion are crushed and transported in the basal layers of the glacier and are known as **basal debris**.

The processes and mechanics of glacial erosion are poorly understood. This is due to the fact that these processes occur beneath a glacier and cannot therefore be easily observed or monitored. Very few direct observations of the erosional processes beneath glaciers have been made. Instead, research has proceeded along theoretical lines and in many cases processes have been inferred solely from the analysis of the landforms produced by the glacial erosion revealed on deglaciation. Despite these problems, three principal mechanisms of glacial erosion are recognised: (1) **glacial abrasion**, the gradual wearing down of the bed by the passage of ice armed with debris; (2) **glacial plucking**, the removal of blocks of rock and sediment from the glacier bed; and (3) **glacial meltwater erosion**, caused by the flow of meltwater beneath a glacier. Each of these processes is examined within this chapter.

5.1 GLACIAL ABRASION

Glacial abrasion is the process by which rock particles transported at the base of a glacier are moved across a bedrock surface, scratching and wearing it away. This process is often likened to the action of sandpaper on a block of wood. It is the process that gives rise to striations and other rock-scoured features (see Section 6.1.1). Limited observations from beneath modern glaciers (Box 5.1) suggest that

BOX 5.1 DIRECT OBSERVATIONS AND MEASUREMENTS OF GLACIAL ABRASION

The direct observation of abrasion in action is extremely difficult since it involves digging tunnels through a glacier to access basal cavities. Some of the best observations are those of Boulton (1974), who described the movement of a basalt fragment over a large roche moutonnée 20 m below the surface of Breiðamerkurjökull in south-east Iceland. The basalt fragment was removed and the surface which had been in contact with the bed inspected. The fragment (clast) had been in contact with the bed at three points and between the points of contact crushed debris could be seen to have been ploughed up in front of them. Striations produced by the fragment could be traced for 3 m. The largest striation was seen to rapidly deepen to 3 mm but then to gradually shallow to 1 mm. Boulton (1974) related this decrease in depth to the build up of crushed debris which spreads the load at the interface over a wider area. The build up of a layer of crushed debris was thought to result in a change in the nature of motion from a jerky 'stick–slip' motion, to a relatively uniform sliding movement; the stick–slip motion produces a carpet of debris over which the particle subsequently slides. When this carpet is exhausted by commutation, the clast will again come into contact with the bed, thereby re-cutting the striation. This may explain the disappearance and reappearance of striae. These observations suggest that there are two abrasive processes: (1) the cutting of striae and (2) polishing of the bed by the fine debris which is ploughed up when a striation is cut. Boulton (1974) went on to measure the rate of abrasion by cementing rock and metal plates to bedrock surfaces adjacent to basal cavities beneath Breiðamerkurjökull and the Glacier d'Argentiere in the French Alps. These plates became quickly covered by basal ice and were later recovered for inspection.

Locality	Average abrasion rate		Ice thickness	Ice velocity
	Marble plate	Basalt plate		
Breiðamerkerjökull 1	3 mm yr^{-1}	1 mm yr^{-1}	40 m	9.6 m yr^{-1}
Breiðamerkerjökull 2	3.4 mm yr^{-1}	0.9 mm yr^{-1}	15 m	19.5 m yr^{-1}
Breiðamerkerjökull 3	3.75 mm yr^{-1}		32 m	15.4 m yr^{-1}
Glacier d'Argentiere	36 mm yr^{-1}		100 m	250 m yr^{-1}

Source: Boulton, G. S. 1974. Processes and patterns of glacial erosion. In Coates, D. R. (Ed.) *Glacial Geomorphology.* Proceedings of the Fifth Annual Geomorphology Symposia, Binghampton, Allen & Unwin, London, 41–87.

three main variables control the ability of a glacier to abrade its bed: (1) the basal contact pressure between the rock in the sole of the glacier and the bed; (2) the rate of basal sliding; and (3) the concentration and supply of rock fragments within the sole of the glacier. Each of these variables is examined below.

5.1.1 Basal Contact Pressure

This is probably the most important variable that determines the rate of glacial abrasion. Using the analogy of a block of wood being sanded with a piece of sandpaper, it follows that the harder you press down on the surface of the wood, the faster it is worn away. In a glacier this is the contact pressure between the clast in basal transport and the glacier bed beneath it: the greater the pressure, the more abrasion will occur. There are, however, two alternative views on what controls this contact pressure and the movement of basal clasts. The first view was developed by G. S. Boulton and the second was suggested later by B. Hallet.

1. **The Boulton model.** This model assumes that the contact pressure between a particle in basal transport which is in contact with the glacier bed is related to the **effective normal pressure**, which as we saw in Section 4.3 is a function of: (1) **normal pressure**, given by the weight of overlying ice; and (2) the **basal water pressure**, which acts in opposition to the weight of the overlying ice by buoying up the glacier like a hydraulic jack. Effective normal pressure will therefore be high when: (1) the ice is thick; and (2) basal water pressure is low. This last point is of some importance since in a cold-based glacier, where there is little or no meltwater present at the glacier bed, effective normal pressure will be much higher than for a warm-based glacier of similar thickness. Similarly, bedrock lithology beneath a warm-based glacier may also be important. Effective normal pressures will be much higher on porous rocks since this will reduce the basal water pressure (see Section 4.3). Where the bed is not horizontal (for example, where there is a bedrock obstacle), effective normal pressure is modified by an amount equal to the pressure of the ice flowing against the obstacle (Figure 4.6; see Section 4.3).

   In the Boulton model, effective normal pressure controls the rate of abrasion. As effective normal pressure increases, abrasion will also increase as the clast in the base of the ice is being pushed harder into the bed. However, as effective normal pressure increases, the friction between the clast and the bed is also increased and this friction will ultimately begin to slow the movement of the particle, and the basal ice which holds it will begin to flow around the clast. When this occurs, abrasion will start to decrease despite the fact that effective normal pressure is still increasing. Consequently, if all other variables are held constant (e.g. sliding velocity), abrasion will first increase with effective normal pressure and then decrease until the friction between the clast and the bed is such that it will stop moving and will lodge.

2. **The Hallet model.** This model assumes that the contact pressure between a clast in basal transport is independent of the effective normal pressure. This theory is based on the premise that clasts are completely surrounded by ice and can be considered to be essentially floating within it. This occurs because ice will deform around a clast by creep due to the weight of the ice above it and therefore basal clasts are effectively surrounded by ice at all times. In this case the contact pressure between a clast and the glacier bed is a function of the rate at which ice flows towards the bed, forcing the clast into contact with the bed. This depends on: (1) the rate of basal melting; and (2) the presence of extending glacier flow. In this model abrasion is independent of variations in effective normal pressure and is primarily a function of basal melting. As we saw in Section 3.4, basal melting is favoured by: (1) rapid ice flow which generates large amounts of frictional heat; (2) thick ice; (3) high ice surface temperatures; and (4) the advection of warm ice towards the glacier bed.

We shall return to these two different views of basal contact pressure below, since two very different models of glacial abrasion have been developed around them.

5.1.2 Basal Sliding

The rate of basal sliding controls the rate at which basal debris is moved across a bedrock surface. Using the sandpaper analogy, the faster you move the sandpaper back and forth, the faster the wood is worn away. The greater the rate of basal sliding, the greater the amount of abrasion. Basal thermal regime is important, since cold-based glaciers do not slide over their beds and will not have the ability to abrade them (see Section 3.4).

5.1.3 The Concentration and Supply of Rock Fragments

The concentration of debris within basal ice also controls the rate of abrasion. Ice on its own cannot cause significant abrasion without rock fragments within it. However, the rate of abrasion is not increased simply by increasing the concentration of basal debris. In fact, it has been suggested that abrasion is most effective where basal debris is relatively sparse. This is due to the fact that basal debris increases the frictional drag between the ice and its bed and therefore reduces the sliding velocity. Glaciers with relatively clean basal ice are able to slide faster than those with large amounts of basal debris. There is a certain threshold of debris concentration, above which the abrasion rate declines with increasing debris content because of its adverse effect on the rate of basal sliding (Figure 5.1).

The type and shape of basal debris is also important. Some rocks are more durable than others. A glacier armed with basal debris of a hard or resistant lithology will be more effective than a lithology which is quickly crushed. The most effective combination occurs where a glacier armed with debris entrained from a

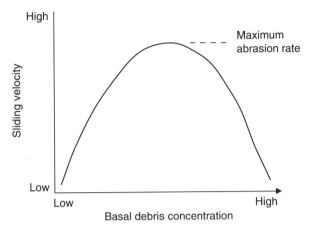

Figure 5.1   *Schematic representation of the relationship between basal sliding velocity and the concentration of debris within the base of the glacier*

hard substrate flows over a relatively soft lithology. If the debris is softer than the substrate, little abrasion will occur, since erosion would preferentially reduce the size of the basal clasts first. Equally, if the basal debris is of the same rock type as the substrate then both the debris and the glacier bed will be worn down at the same rate. The shape of basal clasts is also important. Sharp fragments are able to make deeper incisions into the underlying bedrock than those with blunter or more rounded points or edges. Laboratory observations have shown that clasts in contact with the bed rotate, which helps to improve their life span as erosive tools beneath the glacier.

The supply of basal debris is also of importance in controlling the rate of abrasion. Basal debris is quickly worn down and crushed. For abrasion to be effective it must therefore be replaced. This may occur either by: (1) the entrainment of fresh glacial debris at the glacier bed; or (2) basal melting which transfers debris from high level transport within a glacier to the bed (see Section 7.1). The entrainment of fresh basal debris is a result of glacial plucking, which is covered in the next section, while the controls on basal melting have already been discussed.

5.1.4 Abrasion Models

We have already seen that there are two alternative views concerning the nature and controls on the contact pressure between a basal clast and the bedrock beneath a glacier. Both Boulton and Hallet have developed numerical models with which to predict the patterns and amounts of glacial abrasion. The two models are very different.

Boulton's abrasion model assumes that the contact pressure on a rock particle at the base of a glacier is a function of the normal effective pressure. As a consequence, his model predicts that abrasion will be controlled by: (1) the effective normal pressure; and (2) the ice velocity. Effective normal pressure is controlled by ice thickness and basal water pressure (see Section 4.3). The relationship between abrasion and these two variables within Boulton's model is illustrated in Figure 5.2. This graph shows that for a given ice velocity, abrasion increases to a peak as effective normal pressure increases, and then falls rapidly to zero as the friction between debris and bed becomes sufficient to retard the movement of the particle. At effective normal pressures above a critical level no abrasion occurs, but instead debris hitherto transported is deposited. Erosion and deposition appear therefore to be two parts of a continuum.

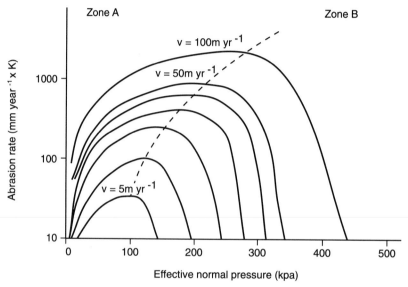

Figure 5.2   *Graphic representation of Boulton's abrasion model. The graph shows theoretical abrasion rates plotted against effective normal pressure for different ice velocities. In Zone A, abrasion rates increase with increasing pressure while in Zone B, abrasion rates fall with increasing pressure. Zone C, located to the right of the higher x-axis intercept for any one ice velocity is an area of no abrasion and basal debris is deposited as lodgement till. [Modified from: Boulton (1974) In: Coates (Ed.)* Glacial Geomorphology, *George Allen & Unwin, Fig. 7, p. 53]*

Boulton has used this equation to predict the evolution of bedrock bumps by glacial abrasion (Figure 5.3). In Section 4.3 we saw how effective normal pressure varied across an obstacle (Figure 4.6). Given this pattern of variation, Boulton used his abrasion model to predict how the shape of a two-dimensional obstacle would change with erosion. He assumed that the bump had a sinusoidal shape, that the ice velocity over the bump was 50 m per year and that the pressure fluctuation over the bump was 130 kpa (Figure 4.6). Given these values, he charted the evolution of two bedrock bumps under a glacier, one of which had an effective normal pressure of 70 kpa and one which experienced 240 kpa. The two patterns of evolution are quite different and are shown in Figure 5.3. The bump with an effective normal pressure of only 70 kpa evolved into a roche moutonnée shape (see Section

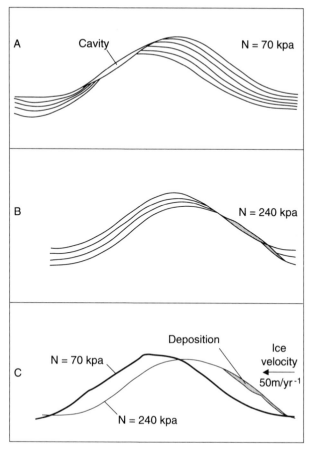

Figure 5.3 *Patterns of abrasion of a sinusoidal bedrock bump for a constant ice veloc-ity of 50 m per year for different values of normal effective pressure using Boulton's abrasion model. [Modified from: Boulton (1974) In Coates, D. R. (Ed.)* Glacial Geomorphology, *Allen & Unwin, London, Fig. 9, p. 56]*

6.2.2). The normal pressures all fall within the zone of rising abrasion and pressure (Zone A in Figure 5.2). The rate of abrasion was therefore highest on the up-glacier flank, low on the crest of the bump, zero on its lee flank where a cavity forms, and high at the foot of the lee flank. A roche moutonnée shape is therefore produced with time (Figure 5.3A). The morphology of the bump under 240 kpa of effective normal pressure evolves very differently. Here the normal pressures fall within zones of falling or zero abrasion with increasing pressure (Zones B and C in Figure 5.2). Lodgement occurs on the up-glacier flank, slight abrasion occurs on the crest of the bump, maximum abrasion occurs on the down-glacier flank, and slight abra-sion occurs at the foot of this flank. The effect is to produce a form that migrates in an up-glacier direction with a steep up-glacier flank similar to a crag and tail (Figure 5.3B). In summary, therefore, the key implications of Boulton's abrasion model are as follows:

A. Variations in ice thickness control abrasion and lodgement via effective normal pressure.
B. Variations in basal water pressure, controlled by such factors as bed permeability (geology) control abrasion via effective normal pressure.
C. Ice velocity controls the rate of abrasion.
D. Abrasion and lodgement form part of a continuum.

Hallet used an alternative approach to Boulton in formulating his abrasion model. Basal rock particles are envisaged as essentially floating hydrostatically in the ice and are therefore independent of effective normal pressure. In this model the rate of basal melting and ice velocity are the key controls on abrasion. As we will see in the next chapter, Hallet's model has been used widely in numerical models designed to study the evolution of large glacial landforms. The main implications of Hallet's model are as follows:

A. Abrasion is highest where basal melting is greatest.
B. Abrasion is independent of effective normal pressure and therefore of basal water pressure, although not of glacier thickness since this helps control the rate of basal melting.
C. Lodgement and abrasion are independent processes.

These two models contain very different predictions and at first sight these two theories seem to conflict. It is, however, possible that each model represents different but equally valid subglacial conditions. Boulton's model and predictions may apply where the basal ice is particularly dirty and debris-rich and therefore likely to behave as a solid slab. The rigid nature of this slab prevents the ice deforming around each clast. In contrast, Hallet's model and predictions may be more appropriate in areas where basal debris is thinly spread and the ice is consequently less rigid.

5.2 GLACIAL PLUCKING

Glacial plucking is the means by which a glacier is able to remove large chunks and fragments of rock from its bed. It is often referred to as **glacial quarrying**. It involves two separate processes: (1) the fracturing or crushing of bedrock beneath the glacier, and (2) the entrainment of this fractured or crushed rock.

5.2.1 Fracturing of Bedrock beneath a Glacier

The propagation of fractures within bedrock is essential for glacial plucking, but is poorly understood. Fractures may pre-date the advance of a glacier into an area and simply reflect the geological structure of the area, or may be generated by periglacial freeze–thaw weathering in advance of the glacier. However, this process does not occur beneath most glaciers and alternative processes are required to explain fracturing of bedrock as a glacier moves over it.

Pressure release as glacial erosion proceeds may generate fractures and joints parallel to the erosional surface. As rock surfaces are unloaded by erosion they may expand and fracture. This process has been used to explain the presence of large sheet joints parallel to eroded surfaces. Figure 5.4 shows sheet joints in granite developed on the side of a glacial valley in the Cairngorm Mountains, Scotland. These joints were produced by pressure release due to the unloading effect of glacial erosion along the line of the valley. Glacial erosion may therefore generate bedrock fracture and thereby accelerate the rate of erosion.

As a glacier moves over an irregular bedrock surface, complex patterns of basal ice pressure are generated, as illustrated in Figure 4.6. This pattern of pressure differences is transmitted to the underlying bedrock, causing stress fields to be set up within the bedrock. These stress patterns are often more pronounced if a cavity exists in the lee of a bedrock obstacle. These stress fields may be sufficient to cause the bedrock to fracture, although theoretical calculations suggest that this will only occur where pre-existing joints and weaknesses exist (Box 5.2). A rock mass with such weaknesses is referred to as a **discontinuous rock mass**. Most bedrock contains joints, bedding planes and other lines of weakness which may be exploited and expanded by the stress field generated by the flow of ice over them. The importance of these weaknesses is reflected in the fact that the morphology of many glacial erosional landforms is controlled by the pattern of discontinuities, joints and bedding planes within the parent rock mass (Figure 5.5). In general, the faster the rate of ice flow, the more pronounced are the variations in basal ice stress and therefore the stress field generated within the underlying bedrock. The pattern of effective normal pressure over obstacles may fluctuate with time, causing the stress fields within the bedrock obstacle to vary. Dramatic changes or repeated changes in the stress fields may be particularly important in propagating fractures along lines of weakness within the rock mass.

Temporal fluctuations in the pattern of ice pressure may also cause variation in the temperature of basal ice and in some cases may generate small cold-based patches within an otherwise warm-based glacier. This process is known as the **heat-pump effect**. At its simplest this process involves the melting of ice in areas of high basal ice pressure, for example on the upstream side of an obstacle. The high pressure reduces the freezing point, allowing melting to occur. Melting of ice consumes thermal energy (latent heat) and will cause the ice mass to cool. Some or all of the meltwater generated will move under the glacier to areas of lower basal ice pressure where it refreezes. If the basal ice pressure then falls over the original obstacle, refreezing of the available meltwater will occur around this obstacle. On freezing, latent heat is given off and will warm the basal ice. However, because some of the meltwater has now been lost, the temperature of the basal ice cannot regain its former level and a cold patch will form. In order to return the basal ice to its original temperature, the same amount of water would need to refreeze as was melted, but since meltwater has flowed away this cannot occur. Consequently, temporal pressure variations beneath an ice sheet, associated for example with diurnal fluctuations in ice velocity or meltwater discharge, may generate cold

Figure 5.4   *Sheet joints formed by pressure release along the sides of a glacial trough in the Cairngorm Mountains, Scotland. The surface of each sheet joint parallels the sides of the glacial trough. [Photographs: N. F. Glasser]*

BOX 5.2: CRUSHING OF BEDROCK OBSTACLES BENEATH A GLACIER

Mathematical modelling is a powerful tool in glacial geology. Numerical models may be used to make predictions which can then be tested against reality. Morland and Morris (1977) used this approach to study the potential for bedrock crushing by a glacier. Their aim was to see whether the stress field produced in bedrock by an overriding glacier was sufficient to cause failure in the bedrock. Using a theoretical and numerical model, they were able to calculate the likelihood of bedrock failure for different object shapes and different bedrock lithologies. The results of one experiment to predict the region of a bedrock bump where bedrock failure is most likely is shown below. The maximum stress generated by the glacier moving over this bump is located deep within the rock mass on the downstream flank of the bump. The value of this failure stress is less than the coherent strength of the bedrock itself, which led Morland and Morris (1977) to conclude that failure will not occur if the rock is coherent. In this situation the profile of a bedrock hummock, such as that illustrated below, will remain stable unless bedrock joints or other internal weaknesses are present within the rock. Morland and Morris (1977) conclude that it is the presence of these weaknesses that allows failure to occur and facilitates the development of a typical roche moutonnée profile.

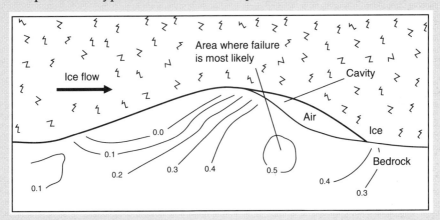

Source: Morland, L. W. & Morris, E. M. 1977. Stress in an elastic bedrock hump due to glacier flow. *Journal of Glaciology* 18, 67–75. [Diagram modified from: Morland & Morris (1977) *Journal of Glaciology* 18, Figure 7, p. 74]

patches on the glacier bed. Beneath these cold patches ice will be frozen to the bedrock, causing it to stick to the bed in these places. Lumps of rock beneath cold patches may therefore be entrained by freezing to the glacier as the ice flows forward (see Section 5.2.2).

Fluctuations in basal water pressure may also help to propagate bedrock fractures beneath a glacier (Figure 5.6). Basal water pressure may influence fracturing

Figure 5.5   *An ice margin of the Greenland Ice Sheet terminating on fractured bedrock. The blocky morphology of the bedrock surface is controlled by the joint spacing within the bedrock. [Photograph: N. F. Glasser]*

in two ways: (1) it affects the distribution and magnitude of the stress fields set up by ice in bedrock surfaces; and (2) its presence within fractures and microscopic cracks is important to the process of fracture propagation. As illustrated in Section 4.3, basal water pressure helps determine the presence or absence of basal cavities beneath a glacier. The presence or absence of basal cavities has an important influence on the distribution of stresses imposed on a bedrock obstacle by the glacier.

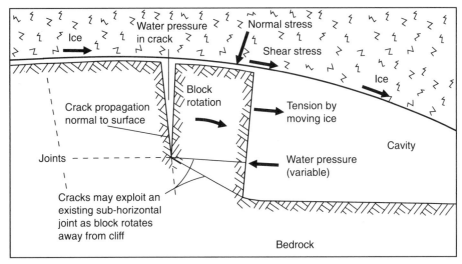

Figure 5.6 *Schematic diagram of glacial plucking, illustrating the role of water pressure in fracture propagation and the role of ice flow in rotating blocks out of the rock face*

Changes in the basal water pressure within lee-side cavities cause them to vary in size and may cause cavity closure and thereby alter the stress field within the bedrock obstacle. For example, the sudden drainage of meltwater from within a cavity may cause it to close even if the ice velocity remains unchanged, due to an increase in the effective normal pressure caused by the fall in the water pressure (see Section 4.3). Rapid or repeated changes of this sort may help to widen or propagate fractures within the bedrock mass (Figure 5.6). As a cavity forms due to an increase in water pressure, this hydrostatic pressure can effectively lift the base of the glacier up. If rock fragments are frozen to the glacier bed due to the heat-pump effect during this process, they will be lifted up and moved forward as the base of the glacier rises (Figure 5.7). This process is known as the **hydraulic-jack effect**.

Where fluctuations of basal water pressure are combined with the heat-pump effect a cycle of erosion may occur. As basal water pressure falls, the resulting increase in effective normal pressure over an obstacle may cause a period of fracture propagation. As basal water pressure rises, basal cavities may open and rock may be entrained by the hydraulic-jack effect and the heat-pump process. This would imply that glacial plucking will be most effective within a glacier where there are regular fluctuations in basal water pressure.

5.2.2 Rock Entrainment

Once a piece of bedrock has been detached from the glacier bed by the processes of rock fracture it must then be lifted and incorporated into the basal ice. This process is known as **entrainment**. The evacuation and entrainment of a rock fragment from the glacier bed is governed by the balance between the tractive force

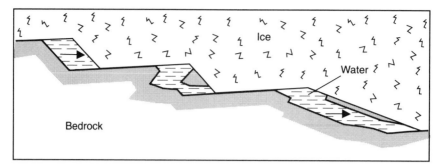

Figure 5.7   *Diagram to illustrate glacial plucking by a combination of the hydraulic-jack and heat-pump effects. Fragments of bedrock are frozen to the glacier bed by the heat-pump effect and then lifted from it as increasing water pressure opens basal cavities*

exerted on it by the overriding ice and the frictional forces which act to hold the rock in place. High basal water pressures may help to reduce the frictional forces holding debris in place. Entrainment may occur in the following ways:

A. The heat-pump effect may cause local patches of basal ice to freeze to the bed, detaching debris as the ice flows forward.
B. The drag between ice and bedrock may be sufficient to detach loose particles particularly if they become surrounded by ice.
C. Loose debris collected within basal cavities may become surrounded by ice and simply swept away if the cavity closes.
D. Freezing-on of material may occur in the lee of obstacles as meltwater generated on their upstream faces refreezes to form regelation ice in the low pressure zone in their lee. In warm-based glaciers the debris layer produced by the freezing of regelation ice is usually thin since debris is also released by melting on the upstream side of obstacles. However, if freezing-on dominates over melting, perhaps at the boundary between warm and cold ice, where there is a constant flux of meltwater freezing onto the glacier, a large thickness of debris-rich regelation ice may develop (Figure 5.8).
E. Beneath glaciers with a mixed thermal regime (polythermal), fluctuation of the thermal boundary between warm and cold ice may lead to the freezing-on of large rafts of sediment or rock. For example, if an area of previously warm basal ice was to turn cold, perhaps during deglaciation, large rafts of previously saturated sediment and bedrock may become frozen to the glacier bed and entrained as it flows forward.
F. Debris may also be incorporated into the ice along thrust planes. In glaciers with a mixed thermal regime in which there is a cold ice margin and a warmer interior, compressive flow is common since the cold ice moves less quickly than the warm ice within the glacier interior. This compression may lead to the development of thrusts within both the ice and underlying sediment along which debris may be incorporated (Figures 7.11 and 7.12; see Section 7.3).

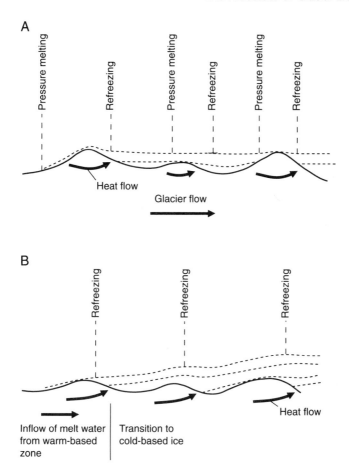

Figure 5.8 *Debris incorporated into regelation ice in the lee of rock bumps. **A:** In a warm-based glacier the layers tend to be destroyed by pressure melting associated with other bumps, and the debris layer remains thin. **B:** In a zone of thermal transition from warm-based ice to cold ice, the freezing-on of meltwater and debris may entrain a consid-erable thickness of basal debris. [Modified from: Boulton (1972) In: Price & Sugden (Eds) Polar Geomorphology, Institute of British Geographers, Special Publication 4, Fig. 5, p. 10]*

5.3 GLACIAL MELTWATER EROSION

Glacial meltwater erosion beneath glaciers may result from either mechanical or chemical processes. Traditionally the efficiency of chemical processes in glacial areas has been thought of as minimal. However, recent work has emphasised its importance within the meltwater system. The effectiveness of meltwater as an agent of erosion depends on: (1) the susceptibility of the bedrock involved, in par-ticular the presence of structural weaknesses or its susceptibility to chemical attack;

(2) the discharge regime, in particular the water velocity and the level of turbulent flow; and (3) the quantity of sediment in transport.

5.3.1 Mechanical Erosion

Mechanical erosion occurs through two process: (1) fluvial abrasion, and (2) fluvial cavitation.

Fluvial abrasion occurs by the transport of both suspended sediment and sediment in traction within the meltwater. This sediment impacts on the walls of rock channels or on the bed of subglacial tunnels, striating and grooving the rock surface. The rate of meltwater abrasion is controlled by the following:

1. **The properties of the sediment in transport.** Of particular importance is the hardness of the sediment relative to the bedrock surface over which the meltwater is flowing. The harder the sediment in transport is relative to the bedrock beneath the meltwater, the more erosion. The concentration of debris within the meltwater is also important. In general, rates of abrasion increase with increasing concentrations of debris over the range of concentration normal in most meltwater.
2. **Flow properties.** The rate of fluvial abrasion increases with the flow velocity. Similarly, the more turbulent the water flow, the greater the rate of abrasion since sediment particles are brought into contact with the bed and channel walls more frequently than where the level of turbulence is low.
3. **Properties of the channel.** The roughness and orientation of facets within a channel as well as its plan-form all affect the rate of fluvial abrasion. Erosion is greatest where sediment-charged water impacts at a near normal angle. Consequently, obstructions within the channel, such as large boulders, will be rapidly abraded. Similarly, where flow is directed towards the wall of a channel, for example at a bend, maximum erosion will be achieved (Box 5.3).

Fluvial abrasion may also occur without the transport of rock debris. Boulders which are too large or are wedged together may be vibrated by the passage of water. This vibration may cause abrasion or attrition of one boulder against the next or against the channel walls. Similarly, boulders or stones trapped within enclosed hollows may achieve considerable amounts of abrasion as they are swirled around within the hollow.

Fluvial cavitation occurs wherever the meltwater velocity exceeds about 12 m s^{-1}. It involves the creation of low-pressure areas within turbulent meltwater as it flows over a rough bedrock surface. These low-pressure areas form as the flow is accelerated around obstacles on the channel floor. If the pressure within the water drops sufficiently to allow the water to vaporise, bubbles of vapour (cavities) form. The cavitation bubbles grow and are moved along in the fluid until they reach a region of slightly higher local pressure where they will suddenly collapse. If cavity collapse occurs adjacent to a channel wall, localised but very high impact

BOX 5.3: FLUVIAL ABRASION AND THE ANGLE OF ATTACK

Much of our theoretical understanding of the processes of fluvial abrasion has been obtained from hydraulic engineers, who experience the problems of abrasion within pipes and on dam spillways. The erosion of pipes provides a suitable analogue for the erosion of subglacial N-channels. One of the important variables is the angle of incidence or attack between the sediment-rich water flow and the channel margins. Where the angle of attack is near normal, erosion is particularly pronounced. This is illustrated in the diagram below, which records the results obtained by Mills and Mason (1975) during an experiment involving a perspex pipe and water charged with aluminium solids. Erosion of the pipe on the bend commences when the angle of attack reached 21°. After erosion had proceeded for some time, the flow began to be deflected across the pipe causing erosion on the inside of the bend. It illustrates very neatly that any obstruction within a glacial meltwater channel or tunnel will be quickly abraded.

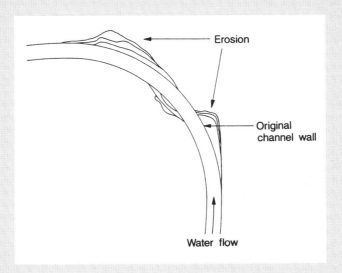

Source: Mills, D. & Mason, J. S. 1975. Learning to live with erosion of bends. *First International Conference on the Internal and External Protection of Pipes*, G1–G19. [Diagram modified from: Drewry (1986) *Glacial Geologic Processes*, Arnold, London, Figure 5.4, p. 67]

forces are produced against the rock. Repetition of these impact forces may lead to rock failure. In particular, the shock waves are often forced into microscopic cracks within a rock or between mineral grains, causing them to loosen and allowing their removal. Once a surface has become pitted or fretted from the loss of grains or small rock fragments the shock waves tend to become concentrated or guided into the pits, thereby accelerating the process of erosion. Large bowl-shaped depressions may be produced in this way.

5.3.2 Chemical Erosion

Glacial meltwater can also erode bedrock by the processes of chemical solution. Soluble components of rock and rock debris are dissolved and removed in solution. This process is particularly important on carbonate-rich lithologies (e.g. limestone and chalk), but is not restricted to them. Chemical denudation beneath glaciers is often neglected as a process of glacial erosion, although, in recent years its importance has been increasingly recognised. Chemical denudation is particularly effective beneath glaciers, despite the low temperatures and therefore reaction rates, for three main reasons:

1. **High flushing rates.** Meltwater passes through the glacial system rapidly and is rarely stored subglacially for long; its residence time is therefore short, and this ensures that it does not have time to become chemically saturated.
2. **Availability of rock flour.** Turbulent meltwater is able to transport large quantities of freshly ground rock particles in suspension which provide a very high surface area, or reaction surface, over which solution can occur.
3. **Enhanced solubility of carbon dioxide at low temperatures.** The solution of carbon dioxide by meltwater produces a weak acid. The solubility of carbon dioxide increases at low temperatures and consequently meltwater becomes more acidic and therefore more aggressive.

Chemical denudation is restricted to warm-based ice with abundant meltwater and may be particularly important in maritime areas where high rainfall adds significantly to the volume of water passing through the glacial system.

5.4 DETERMINING THE RATE OF GLACIAL EROSION

In the previous sections we examined the principal processes of glacial erosion. The efficiency of these processes coupled with the length of time over which they operate determines the amount of glacial erosion that can be achieved by a glacier. In this section we examine the methods available with which to calculate the depths and rates of glacial erosion. There are three ways in which estimates of the rates of glacial erosion can be obtained:

1. **Direct observation beneath modern glaciers.** Direct observations and experiments at the base of present-day glaciers can be conducted by drilling or tunnelling to the base of a glacier. This process is facilitated if access can be obtained via such tunnels to subglacial cavities. In these cavities and tunnels, measurements of glacial abrasion can be obtained by periodically measuring

either the micro-topography of bedrock surfaces or with reference to artificial plates fixed to the rock surface beneath the glacier. Measurements of glacial erosion observed beneath present-day glaciers in Iceland have given rates for glacial abrasion of between 0.5 and 1 mm per year (Box 5.1). No quantitative observations have yet been made of the process of glacial plucking beneath modern glaciers. This process operates periodically and is therefore difficult to monitor over a short time period.

2. **Geomorphological reconstructions of erosion in glaciated terrains.** Geomorphological observations and the distribution of preglacial deposits can be used to reconstruct the form of the preglacial landscape. This reconstructed surface is then compared to the present glacial surface to give a measure of the depth of glacial erosion (Box 5.4). This is a useful tool in areas where preglacial remnants are abundant, but is of limited use if the preglacial landscape has been largely removed. The more intense the glacial erosion, the more difficult it becomes to quantify the depth of erosion. In areas of multiple glaciation the erosion estimate reflects net erosion and also includes any fluvial erosion that may have taken place during interglacial periods. It is also important to note that this method relies on the extrapolation between preglacial fragments and

---

BOX 5.4: RATES OF GLACIAL EROSION FROM PREGLACIAL RECONSTRUCTIONS

The geomorphological reconstruction of preglacial landscapes provides a tool with which glacial geologists can estimate the depth and rate of glacial erosion. In the Sognefjord area of western Norway there is strong evidence that much of the landscape is preglacial in origin and has simply been dissected by a system of glacial troughs (fjords). Nesje *et al.* (1992) attempted to reconstruct the depths of erosion in this area by reconstructing the preglacial landscape. They assumed that the area was once dominated by a preglacial dendritic valley system, fragments of which survive on the interfluves between the glacial fjords. These fragments were used to reconstruct the preglacial landscape. From this reconstruction of the preglacial surface, they were then able to subtract the present-day topography to determine the depth of Quaternary erosion, assumed to have been primarily glacial. In the study the total volume of rock removed equalled 7610 km^3. This corresponds to an average erosion depth of 610 m over the area. Although during the Quaternary Ice Age the area was glaciated several times, it is possible to estimate the total length of time over which the area was glaciated from the sum of the duration of each period of glaciation. A figure of 600 000 years is obtained from these estimates. This corresponds to an average erosion rate of 1 mm per year.

Source: Nesje, A., Dahl, S. O., Valen, V. & Ovstedal, J. 1992. Quaternary erosion in the Sognefjord drainage basin, western Norway. *Geomorphology* 5, 511–520.

BOX 5.5: GLACIAL EROSION BENEATH MODERN GLACIERS INFERRED FROM SEDIMENT OUTPUT

There are two methods by which calculations of the rate of glacial erosion can be obtained from the sediment output of modern glaciers. The first (and most commonly used) method is to install sediment traps in proglacial streams leading from the glacier, and to measure the volume of sediment that accumulates within them over a given period of time. This volume is then averaged out over the catchment area of the glacier to give an average erosion rate. This method provides a minimum estimate since it does not include sediments deposited directly by the glacier.

A second method was used by Small *et al.* (1984) to examine the rate of glacial sedimentation directly. They surveyed the moraine embankments at the margins of the Glacier de Tsidjiore Nouve in Switzerland. This enabled them to calculate the growth of the moraines over a five-year period. From this they calculated the volume of sediment required to produced the changes in moraine volume and estimated that marginal sedimentation was of the order of 8727 to 11 230 tonnes per year. To this they added the volume of sediment moved each year as suspended load in the proglacial streams in front of the glacier. Averaged over the entire catchment area of the glacier, this is equivalent to an erosion rate of 1.55 to 2.29 mm per year. This figure is consistent with other estimates obtained for Alpine glaciers.

Source: Small, R. J., Beecroft, R. I. & Stirling, D. M. 1984. Rates of deposition on lateral moraine embankments, Glacier de Tsidjiore Nouve, Valais, Switzerland. *Journal of Glaciology* 30, 275–281.

is therefore open to error. As a consequence this method may only be used to provide order of magnitude estimates of glacial erosion.

3. **Sediment volume calculations.** A variety of methods have been used to relate sediment output from a glacial catchment to the rate of erosion within it. This method has been used to estimate erosion for small valley glaciers by measuring either the total sediment output from a glacial catchment, meltwater sediment discharge and ice-marginal accumulation, or simply just one component such as the volume of suspended sediment in meltwater. If the sediment output from the glaciated catchment and the total catchment area are known then the amount of erosion necessary to produce the sediment output can be calculated (Box 5.5). This method cannot be used for large ice sheets where catchments are difficult to define. However, a similar strategy has been developed to calculate the erosion beneath former ice sheets from the sediment deposited in offshore basins. If the volume of glacial sediment deposited in offshore sedimentary basins is calculated, it can be converted into an erosion depth over the source area. This calculation gives an indication of the amount of material eroded from the source area beneath the ice sheet (Box 5.6).

BOX 5.6: GLACIAL EROSION BENEATH FORMER ICE SHEETS INFERRED FROM SEDIMENT OUTPUT

Rates of glacial erosion can be estimated for small modern glaciers relatively easily by measuring the total amount of sediment output (Box 5.5). It is much more difficult to obtain estimates for former ice sheets. One method available is to examine the quantities of sediment found around the margins of former ice sheets. White (1972) attempted to do this from the record of marine sedimentation along the eastern margin of the former Laurentide Ice Sheet in North America. He calculated the total volume of glacially derived sediment in this area and estimated that the former Laurentide ice sheet would have had to erode its bed to an average depth of 120–200 m to produce the measured volume of marine sediment. This estimate is much higher than the rates of erosion obtained from modern glacier studies and may reflect the uncertainties present within White's method. These uncertainties include: (1) attributing all the offshore marine sediment to purely glacial erosion; (2) calculating the likely source area for this sediment; and (3) calculating the volume of bedrock which loose-grained marine sediment equates to. These problems seriously limit the reliability of the erosion estimates obtained by White's method.

Source: White, W. A. 1972. Deep erosion by continental ice sheets. *Geological Society of America Bulletin* **83**, 1037–1056.

5.5 PATTERNS OF GLACIAL EROSION

The primary or regional pattern of glacial erosion within a glacier is controlled by the basal thermal regime (Figure 5.9). Only when a glacier is warm-based is meltwater produced in large quantities, and only when meltwater is abundant can basal sliding and therefore glacial abrasion occur. Glacial plucking is also facilitated by the presence of meltwater. It is possible to predict where erosion is likely to occur within an ice sheet on the basis of its basal thermal regime (Figure 5.9). The following predictions can be made:

A. In areas where the ice is warm-based and basal melting is widespread, glacial erosion, in particular abrasion, will be most effective.
B. In areas where the ice is cold-based and there is no basal melting, glacial erosion will be limited.
C. Glacial erosion, in particular plucking, will be pronounced in areas where a zone of warm-based ice is replaced along a flow line by a cold-based zone. The meltwater moving from the warm-based zone into the cold zone will cause widespread freezing-on of bedrock and debris to the glacier sole, facilitating glacial plucking.
D. If the ice-sheet basal thermal regime evolves as a cold-based interior and a warm-based margin (Figure 3.13), then a belt of glacial erosion will advance

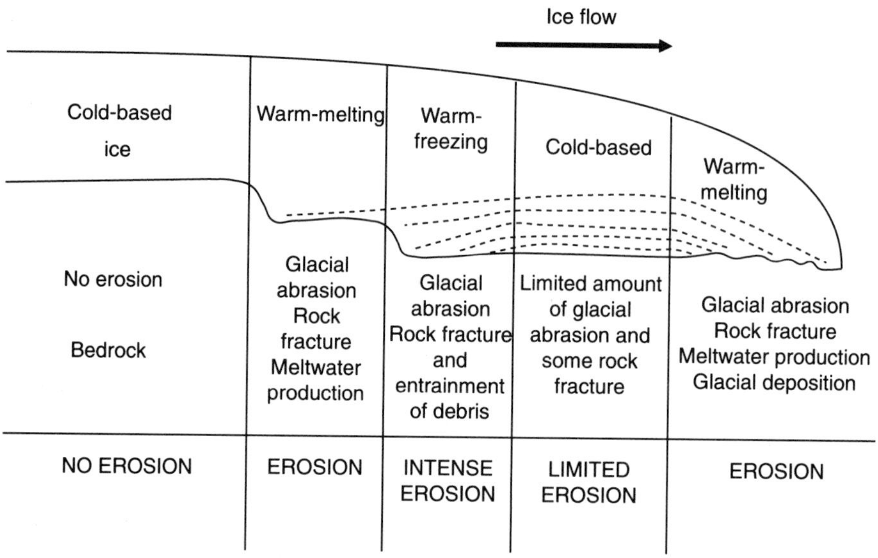

Figure 5.9   *Schematic cross-section through an ice sheet showing the influence of the basal thermal regime on the processes of glacial erosion*

across the landscape as the ice sheet grows. This belt of erosion will also retreat with the ice sheet as it decays.

The primary or regional pattern of glacial erosion is controlled by the basal thermal regime, but the secondary or local pattern is influenced by a complex range of variables, which include the following:

A. Bedrock geology: permeable lithologies may be associated with high values of effective normal pressure and therefore little glacial abrasion. Conversely, areas with well fractured and jointed bedrock will be easily plucked. Ice flow directions which pass from hard bedrock areas into softer ones will achieve more erosion.
B. Areas of fast-flowing ice will be zones of enhanced glacial abrasion. Consequently, ice sheet margins which terminate in maritime climates will be more effective agents of erosion than those which terminate in more continental climates.
C. Within warm-based zones of an ice sheet those areas which experience high levels of basal melting will experience greater levels of glacial abrasion than areas with lower rates of basal melting.
D. Areas subject to fluctuations in water pressure are likely to experience greater amounts of glacial plucking than areas with little or no fluctuation.
E. It has been suggested that an ice sheet which grows in a previously glaciated area with an ice divide located in the same place as in the past may achieve less erosion than an ice sheet growing for the first time within an area. Once an effi-

cient network of glacial troughs and discharge routes has been established there is less resistance to ice discharge and therefore less erosion. It follows therefore that in areas glaciated several times most of the glacial erosion may be completed during the first phase of glaciation.

It is these second-order variables which give rise to the complex and local variations we find within landscapes of glacial erosion.

5.6 SUMMARY

Glacial erosion is the removal of fragments of rock from a glacier bed. It may occur through three processes: (1) glacial abrasion; (2) glacial plucking; and (3) by the action of glacial meltwater. Glacial abrasion involves the scouring action of debris dragged over a bedrock surface by a glacier and is controlled primarily by effective normal pressure and ice velocity. Glacial plucking involves the fracturing and entrainment of bedrock beneath a glacier. Glacial meltwater erosion involves the breakdown and transport of bedrock through fluvial abrasion, fluvial cavitation and chemical solution. The nature of the glacial erosion undertaken by a particular glacier is dependent on its basal thermal regime since it controls the supply of meltwater at the base of the glacier and the presence of basal sliding. Unless the ice is warm there will be no basal sliding and processes of glacial erosion will be inhibited. Since the basal thermal regime changes over time, rates of glacial erosion may also vary with time. Rates of glacial erosion have been calculated from direct subglacial observations, geomorphological reconstructions and from the volume of sediment removed from a glaciated catchment. These studies confirm that fast-flowing, warm-based glaciers achieve significantly more erosion than slow-flowing, cold-based ones.

5.7 SUGGESTED READING

Subglacial observations of glacial erosional processes, usually in basal cavities, are recorded in several papers: Boulton & Vivian (1973), Boulton (1974), Vivian (1980), Anderson *et al.* (1982) and Rea & Whalley (1994). Boulton (1974, 1979) discusses the fundamental principles of glacial erosion and introduces his model of glacial abrasion. The alternative abrasion model is covered in Hallet (1979, 1981a,b). The supply of basal debris to the base of a glacier and its role in glacial abrasion is outlined in Rothlisberger (1968). The heat-pump effect is introduced and described by Robin (1976). This work is expanded by Rothlisberger & Iken (1981), while Iverson (1991) concentrates on the importance of fluctuations in subglacial water pressures and their role in glacial plucking. The fracturing of bedrock beneath glaciers is treated numerically by Morland & Morris (1977) and by Addison (1981), whilst Atkinson (1982, 1984) provides a good introduction to the role of bedrock conditions in fracture propagation within bedrock itself. Rea (1994) provides a good overview of rock fracture and glacial plucking. The factors which control rock resistance to erosion are outlined by Augustinus (1991). A readable discussion of the links between the glacier system, mass balance and glacial erosion is provided by Andrews (1972). A particularly good account of the role of glacial meltwater in both mechanical and chemical erosion is given by Drewry (1986) while Eyles *et al.* (1982), Fairchild *et al.* (1994) and Sharp *et al.* (1995) provide valuable case studies.

The debate between White (1972) and Sugden (1976) over the depths of erosion accomplished by the former Laurentide ice sheet makes interesting reading, while the more recent debate over rates of erosion in glacial and non-glacial areas is described in Hicks *et al.* (1990) and Harbor & Warburton (1992). The main points of these discussions are neatly summarised by Harbor & Warburton (1993). Good examples of specific studies attempting reconstructions of depths and rates of glacial erosion are those of Sugden (1978), Nesje *et al.* (1992) and Hall & Sugden (1987).

The linkage between patterns of glacial erosion and basal thermal regime are explored in papers by Sugden (1974, 1977, 1978), Dyke (1993) and Glasser (1995).

Addison, K. 1981. The contribution of discontinuous rock-mass failure to glacier erosion. *Annals of Glaciology* **2**, 3–10.

Anderson, R. S., Hallet, B., Walder, J. & Aubry, B. F. 1982. Observations in a cavity beneath Grinnell Glacier. *Earth Surface Processes and Landforms* **7**, 63–70.

Andrews, J. T. 1972. Glacier power, mass balances, velocities and erosion potential. *Zeitschrift für Geomorphologie* **13**, 1–17.

Atkinson, B. K. 1982. Subcritical crack propagation in rocks: theory, experimental results and applications. *Journal of Structural Geology* **4**, 41–56.

Atkinson, B. K. 1984. Subcritical crack growth in geological materials. *Journal of Geophysical Research* **7**, 63–70.

Augustinus, P. C. 1991. Rock resistance to erosion: some further considerations. *Earth Surface Processes and Landforms* **16**, 563–569.

Boulton, G. S. 1974. Processes and patterns of glacial erosion. In: Coates, D. R. (Ed.) *Glacial Geomorphology.* Proceedings of the Fifth Annual Geomorphology Symposia, Binghampton, Allen & Unwin, London, 41–87.

Boulton, G. S. 1979. Processes of glacier erosion on different substrata. *Journal of Glaciology* **23**, 15–37.

Boulton, G. S. & Vivian, R. 1973. Underneath glaciers. *Geographical Magazine* **45**, 311–319.

Drewry, D. 1986. *Glacial Geologic Processes.* Arnold, London.

Dyke, A. S. 1993. Landscapes of cold-centred Late Wisconsinian ice caps, Arctic Canada. *Progress in Physical Geography* **17**, 223–247.

Eyles, N., Sasseville, D. R., Slatt, R. M. & Rogerson, R. J. 1982. Geochemical denudation rates and solute transport mechanisms in a maritime temperate glacier basin. *Canadian Journal of Earth Science* **19**, 1570–1581.

Fairchild, I. J., Brady, L., Sharp, M. & Tison, J. 1994. Hydrochemistry of carbonate terrains in Alpine glacial settings. *Earth Surface Processes and Landforms* **19**, 33–54.

Glasser, N. F. 1995. Modelling the effect of topography on ice sheet erosion, Scotland. *Geografiska Annaler* **77A**, 67–82.

Hall, A. M. & Sugden, D. E. 1987. Limited modification of mid-latitude landscapes by ice sheets: the case of north-east Scotland. *Earth Surface Processes and Landforms* **12**, 531–542.

Hallet, B. 1979. A theoretical model of glacial abrasion. *Journal of Glaciology* **23**, 39–50.

Hallet, B. 1981a. Glacial abrasion and sliding: their dependence on the debris concentration in basal ice. *Annals of Glaciology* **2**, 23–28.

Hallet, B. 1981b General discussion. *Annals of Glaciology* **2**, 187–192.

Harbor, J. & Warburton, J. 1992. Glaciation and denudation rates. *Nature* **356**, 751.

Harbor, J. & Warburton, J. 1993. Relative rates of glacial and nonglacial erosion in Alpine environments. *Arctic and Alpine Research* **25**, 1–7.

Hicks, D. M., McSaveney, M. J. & Chinn, T. J. H. 1990. Sedimentation in proglacial Ivory Lake, southern Alps, New Zealand. *Arctic and Alpine Research* **22**, 26–42.

Iverson, N. R. 1991. Potential effects of subglacial water-pressure fluctuations on quarrying. *Journal of Glaciology* **37**, 27–36.

Morland, L. W. and Morris, E. M. 1977. Stress in an elastic bedrock hump due to glacier flow. *Journal of Glaciology* **18**, 67–75.

Nesje, A., Dahl, S. O., Valen, V. & Ovstedal, J. 1992. Quaternary erosion in the Sognefjord drainage basin, western Norway. *Geomorphology* **5**, 511–520.

Rea, B. R. 1994. Joint control in the formation of rock steps in the subglacial environment. In: Robinson, D. A. & Williams, R. B. G. (Eds) *Rock Weathering and Landform Evolution*. John Wiley, Chichester, 473–486.

Rea, B. R. & Whalley, B. 1994. Subglacial observations from Øksfjordjøkelen, North Norway. *Earth Surface Processes and Landforms* **19**, 659–673.

Robin, G. de Q. 1976. Is the basal ice of a temperate glacier at the pressure melting point? *Journal of Glaciology* **16**, 183–196.

Rothlisberger, H. 1968. Erosive processes which are likely to accentuate or reduce the bottom relief of valley glaciers. *International Association of Scientific Hydrology* **79**, 87–96.

Rothlisberger, H. & Iken, A. 1981. Plucking as an effect of water-pressure variations at the glacier bed. *Annals of Glaciology* **2**, 57–62.

Sharp, M., Tranter, M., Brown, G. B. & Skidmore, M. 1995. Rates of chemical denudation and CO_2 drawdown in a glacier-covered alpine catchment. *Geology* **23**, 61–64.

Sugden, D. E. 1974. Landscapes of glacial erosion in Greenland and their relationship to ice, topographic and bedrock conditions. In: Brown, E. H. & Waters, R. S. (Eds) *Progress in Geomorphology*. Institute of British Geographers Special Publication 7, 177–195.

Sugden, D. E. 1976. A case against deep erosion of shields by ice sheets. *Geology* **4**, 580–582.

Sugden, D. E. 1977. Reconstruction of the morphology, dynamics and thermal characteristics of the Laurentide ice-sheet at its maximum. *Arctic and Alpine Research* **9**, 21–47.

Sugden, D. E. 1978. Glacial erosion by the Laurentide ice-sheet. *Journal of Glaciology* **20**, 367–391.

Vivian, R. 1980. The nature of the ice–rock interface: the results of investigation on 20 000 m² of the rock bed of temperate glaciers. *Journal of Glaciology* **25**, 267–277.

White, W. A. 1972. Deep erosion by continental ice sheets. *Geological Society of America Bulletin* **83**, 1037–1056.

6
Landforms of Glacial Erosion

In the last chapter we examined the processes of glacial erosion. In this chapter we turn our attention to the landforms and landscapes created by these processes. The glaciers of the Cenozoic Ice Age have left distinct landscapes of glacial erosion composed of suites of individual and discrete landforms. In this chapter we examine some of the most commonly encountered landforms of glacial erosion which combine to make erosional landscapes (Figure 6.1) and attempt to explain their formation with reference to the processes of glacial erosion introduced in the previous chapter. For convenience, this chapter is divided into three sections based on spatial scale: micro-scale, meso-scale and macro-scale landforms. Figure 6.1 shows

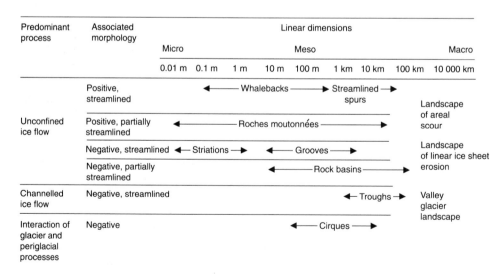

Figure 6.1 *Classification of landforms of glacial erosion. [Modified from: Summerfield (1991)* Global Geomorphology, *Longman, Table 11.4, p. 272]*

that some landforms may cross the boundaries between these three spatial categories; where this is the case, we consider the landforms at the scale at which they are most often encountered.

6.1 MICRO-SCALE FEATURES OF GLACIAL EROSION

Micro-scale landforms of glacial erosion are generally less than 1 m in size and are often found superimposed on larger landforms. We can recognise four types of micro-scale landform: (1) striations, (2) micro crag and tails, (3) friction cracks, and (4) p-forms and micro-channels.

6.1.1 Striations

Glacial abrasion produces lines or scratches on a rock surface as debris is dragged over it (Figure 6.2). These scratches are termed **striations**. They are useful because they are orientated parallel to local ice movement and can be used to make inferences about the pattern of ice flow both in space and time. It is important to note, however, that striations only provide the orientation and not the direction of ice flow. For example, a striation orientated north to south may have been cut by ice flowing either from the north or from the south. On their own, striae are therefore not sufficient to give the direction of ice flow. To determine this, reference must be made to other criteria, such as: (1) the morphology of the bedrock surface on which they are located, in particular the presence of stoss and lee forms (see Section 6.2.2); (2) the presence of micro crags and tails; (3) the presence of friction cracks; or (4) relationship to other larger landforms.

Striations are usually not more than a few millimetres in depth, but may be over several metres long. Their continuity is broken by small gaps or breaks where contact between the bed and basal clast was temporarily broken during their formation. This may occur due to the formation of small subglacial cavities or alternatively where a clast rides up over a cushion of debris (Box 5.1). The depth and continuity of striations is a balance between the effective normal pressure which keeps the base of the glacier in contact with its bed and changes in basal water pressure which allow small cavities to form (see Section 4.3). Striations formed by different ice flow directions may be superimposed in a cross-cut pattern (**cross-cut striations**; Figure 6.2B). This occurs when the ice flow direction changes either due to a readvance of ice over a deglaciated area or due to changes in ice flow direction within a glacier. Cross-cut striations record the fact that the second ice flow was unable to erode all the evidence of the earlier flow, either because of a lower efficiency of glacial abrasion or due to insufficient time. The occurrence of cross-cut striations can be used, therefore, to make inferences about former glacier dynamics (Box 6.1).

Given that striations are produced by abrasion, they require: (1) basal debris; (2) basal sliding; (3) moderate levels of effective normal pressure; and (4) transport of

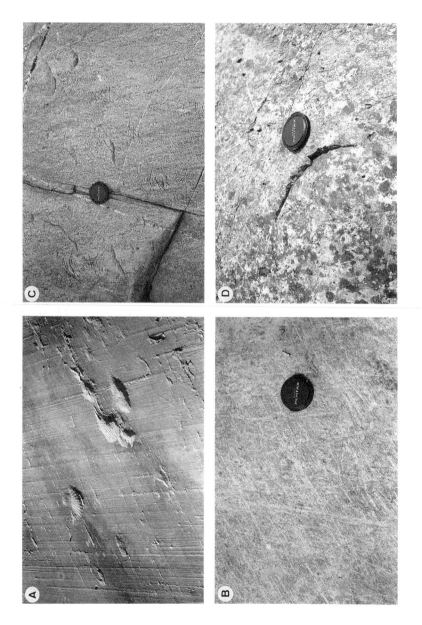

Figure 6.2 *Small-scale erosional landforms.* **A:** *Striations on an ice-smoothed slab in Patagonia [Photograph: N. F. Glasser].* **B:** *Cross-cut striations, North Wales [Photograph: J. M. Gray].* **C:** *Crescentic fractures, North Wales [Photograph: J. M. Gray].* **D:** *A crescentic gouge, North Wales [Photograph: J. M. Gray]*

BOX 6.1: USING LANDFORMS OF GLACIAL EROSION TO RECONSTRUCT GLACIER DYNAMICS

Landforms of glacial erosion are often used by glacial geologists to infer basic glaciological parameters such as ice movement direction and change over time, but are seldom used for anything more complex. Sharp *et al.* (1989) demonstrated for the first time how glacial erosional landforms might be used to reconstruct parameters which affect the operation of basal processes and glacier dynamics. Their field study area was Snowdon, North Wales—an area which has been subjected to multiple glaciations of different duration and intensity. During the height of the last glacial maximum the Snowdon area was overrun

by an ice sheet with a divide located in mid-Wales. At the close of the last glacial cycle, however, small cirque glaciers existed on Snowdon during a period known as the Younger Dyras. These two glacial episodes are marked by cross-cut striations. Detailed mapping of the glacial erosional landforms present, in particular of the size of striations and position of former lee-side cavities on bedrock surfaces, allowed Sharp *et al.* (1989) to suggest that the processes of glacial erosion were different during the two glacial events. Erosion beneath the ice sheet was dominated by lee-side fracturing of bedrock obstacles, by surface fracturing which created friction cracks, and by widespread abrasion. In contrast, erosion during the second phase, by the cirque glacier, was confined to glacial abrasion and there is little evidence for lee-side cavities. Many of the surface features eroded on the bedrock surface by the ice sheet were not removed by the later cirque glacier, suggesting that relatively little erosion took place during this final episode of glaciation. From this, Sharp *et al.* (1989) infer that the cirque glacier had low sliding velocities which prevented cavity formation. On the basis of these inferences and mass balance estimates, Sharp *et al.* (1989) were able to calculate the dynamics of these former ice bodies. The cirque glacier was shown to have a low sliding velocity (around 10 m per year), high contact pressure at the glacier base, low basal water pressures and therefore few lee-side cavities. In contrast, the earlier ice sheet was shown to have a much higher sliding velocity (>35 m per year), lower basal contact pressures, higher basal water pressures and therefore widespread cavity formation. This illustrates the detailed inferences that can be made using simple observations of glacial erosional landforms combined with numerical estimates.

Source: Sharp, M., Dowdeswell, J. A. & Gemmell, J. C. 1989. Reconstructing past glacier dynamics and erosion from glacial geomorphic evidence: Snowdon, North Wales. *Journal of Quaternary Science* **4**, 115–130. [Photograph: A. F. Bennett]

rock debris towards the bed by basal melting. If any of these prerequisites for glacial abrasion ceases to exist, then the formation of striations will also cease. In this situation, the striated surface will become fossilised and the last ice flow direction will be left imprinted on the bedrock surface. There are three situations where this may occur:

A. Where deglaciation occurs and the bedrock becomes re-exposed as the glacier margin retreats.
B. Where the basal thermal regime beneath a glacier changes from warm-based to cold-based. Basal sliding will stop and the basal ice will become frozen to the bedrock beneath. In this situation new striations cannot form and the existing striations are preserved.
C. Where a layer of basal till is deposited immediately on top of a striated bedrock surface. In this scenario, the last ice flow direction suggested by the striations will correspond approximately to the age of the overlying till. Striations may

therefore be placed in a stratigraphical framework if the relative ages of the till units within an area can be established.

Given large enough samples, striations may be used to reconstruct the local pattern of ice flow; however, their application to large-scale ice flow reconstruction is more problematic. This is because basal ice conditions, and especially basal thermal regimes, change markedly over both time and space and consequently the pattern of striations beneath a glacier may be asynchronous, that is of a variety of different ages. Striations located in close proximity on the bed of a former glacier may therefore date from different time periods or relate to different ice flows with radically different flow directions. Furthermore, the preservation of striations on the bed of a former ice sheet depends upon the basal boundary conditions during deglaciation: whether deglaciation occurs beneath a cold-based ice sheet (preservation of striations) or beneath a warm-based ice sheet (new striations forming constantly during deglaciation).

Striations formed during warm-based deglaciation will change direction as the orientation of the ice margin changes during retreat. The youngest striations on a bedrock outcrop will be orientated perpendicular to the ice margin, while older striations will not be related to older and therefore more distant ice-marginal positions. Consequently, the further a striation is from the current ice margin, the more difficult it becomes to relate its formation to a particular period of flow. This situation is complicated further if cold-based deglaciation occurs since this inhibits the formation of new striations and leads to the preservation of older ones. These complications mean that the interpretation of striation patterns over large areas such as an entire ice sheet bed is rarely a simple task.

6.1.2 Micro Crag and Tails

Micro crag and tails are small tails of rock which are preserved from glacial abrasion in the lee of resistant grains or mineral crystals on the surface of a rock. For example, in the slate rock of North Wales the presence of occasional pyrite crystals forms a point of resistance in an otherwise homogeneous rock. In the lee of these pyrite crystals small tails of rock are preserved. In many cases the pyrite weathers out on deglaciation (Figure 6.3). Micro crag and tails are important because they provide clear evidence of both the orientation and direction of ice flow.

6.1.3 Friction Cracks

Friction cracks are a family of small cracks, gouges, chattermarks and indentations created in bedrock as larger boulders or clasts beneath a glacier are forced into contact with the bed. They vary in form from crescentic-shaped gouges in which small chips of bedrock have been removed, to fracture lines or cracks. Three main types of feature can be recognised: (1) **crescentic fractures**, which are a series of small

Figure 6.3 *Micro crag and tails formed in the lee of pyrite crystals on a siltstone slab in the bed lake Lyn Peris, North Wales. In most cases the pyrite crystal has weathered out to leave a small hollow. Ice flow was from left to right.* [Photograph: J. M. Gray]

cracks often forming a distinct line which if crescentic in shape are usually convex up-ice (Figure 6.2C); (2) **crescentic gouges**, which occur where crescentic chips of rock have been removed and are normally concave up-glacier (Figure 6.2D); and (3) **chatter marks**, which are a series of irregular fractures. However, these features are not always consistently orientated in the direction of ice flow. For example, crescentic gouges are occasionally convex up-ice, when they are referred to as **reverse crescentic gouges**. In general, there is much morphological diversity to these features and a wide variety of different forms have been recorded. They tend to form preferentially on crystalline or homogeneous bedrock lithologies.

In general, friction cracks differ from striations because they are not produced by the continuous contact between a clast and the glacier bed, but are formed by intermittent contact. Local variations in effective normal pressure and bedrock topography are sufficient to make a clast 'bounce' or roll over a bedrock surface, creating small gouges or cracks when the clast comes periodically into contact with the bed. Friction cracks provide evidence of high effective normal pressures since considerable contact force between the clast and the bedrock is required to cause fracturing. The process of formation of these features has been successfully simulated in the laboratory (Box 6.2).

BOX 6.2: SIMULATING GLACIAL EROSION IN THE LABORATORY: THE PRODUCTION OF FRICTION CRACKS

In a series of papers, Prest (1921), MacClintock (1953) and Smith (1984) used a simple device consisting of a ball-bearing and a sheet of glass to replicate the formation of micro-erosional landforms. The ball-bearing is pressed against the glass under pressure, like a clast against the bedrock beneath a glacier. MacClintock (1953) was able to generate a series of friction cracks using this method. He noted that their orientation in relation to the direction in which the bearing was moving was dependent on whether it was rotating or not. A non-rolling bearing produced cracks that were concave in the direction movement, while a rolling bearing produced cracks that were convex in the direction of movement. In contrast, Smith (1984) found that both convex and concave fractures could be produced by a non-rolling bearing. However, Smith (1984) did note that convex forward cracks formed preferentially when a high level of pressure was applied. This would suggest that the orientation of friction cracks in relation to the direction of ice flow may be dependent on the effective normal pressure applied to the clast in contact with the bed. These experiments illustrate how simple laboratory simulations can provide insight into the processes operating beneath a glacier.

Sources: Prest, F. W. 1921. The structure of abrading glass surfaces. *Transactions of the Optical Society* **23**, 141–146. MacClintock, P. 1953. Crescentic crack, crescentic gouge, friction crack, and glacier movement. *Journal of Glaciology* **61**, 186. Smith, J. M. 1984. Experiments relating to the fracture of bedrock at the ice–rock interface. *Journal of Glaciology* **30**, 123–125.

6.1.4 P-forms and Micro-channel Networks

Smooth sinuous depressions and large grooves sculpted in bedrock are given the collective term **plastically moulded forms** or **p-forms**. The most commonly encountered types of p-forms are **sichelwannen, potholes or bowls** and **channels** (Figure 6.4). Sichelwannen are sickle-shaped bedrock depressions, usually occurring with an open end which points in the direction of ice flow. These open ends may be extended in the direction of ice flow as shallow runnels. Individual features are normally around 1 m in length but may occasionally exceed 10 m. Where they are particularly elongate in morphology they are referred to as **hairpin erosional marks** (Figure 6.4D). These features are found at a wide range of different sizes from a few millimetres to several metres. **Potholes** are more rounded and deeper depressions which often occur in conjunction with sichelwannen. They may be up to several metres in diameter and depth. Sinuous or linear channels cut into bedrock are also common (Figure 6.4). These channels are usually less than a metre in width and depth. All these features may occur either in isolation or in close association and may be found with striations and other features of glacial abrasion. Striations are sometimes found superimposed on p-forms.

The origin of p-forms is a source of debate. There are three main hypotheses: (1) formation by glacial abrasion; (2) formation due to abrasion by a till slurry; and (3) formation by meltwater. The presence of glacial striations on some p-forms has led many to argue for a mechanism involving glacial abrasion. The organisation of basal debris into distinct lines or streams at the base of a glacier (Figure 7.8) would tend to concentrate glacial abrasion in certain areas, allowing the process to sculpt grooves or channels. However, since the striations may have formed after the p-forms and since these features can also be generated by the transport of boulders in subglacial meltwater, their association with p-forms is not very instructive. The alternative mechanisms involve either a hyperconcentrated flow of till and water or simply a flow of normal meltwater. For example, the formation of hairpin erosional marks and sichelwannen can be explained by flow separation around a small obstacle on a bedrock surface (Figure 6.5). The size of the obstacle controls the size of the erosional mark produced. They may form around single crystals, grains or nodules which protrude up through a bedrock surface or alternatively around much larger bedrock knobs.

6.2 MESO-SCALE FEATURES OF GLACIAL EROSION

Meso-scale features of glacial erosion are those between 1 m and 1 km in size and comprise a family of landforms which includes: (1) streamlined bedrock features; (2) stoss and lee forms; (3) rock grooves and basins; and (4) meltwater channels.

6.2.1 Streamlined Bedrock Features

At a meso-scale the most common effect of glacial erosion is to streamline bedrock

Figure 6.4 P-forms. **A:** Large groove, Isle of Mull, Scotland [Photograph: J. M. Gray]. **B:** Sinuous channel, Isle of Mull, Scotland. Note the striated channel sides [Photograph: J. M. Gray]. **C:** Large channel, Canada [Photograph: J. Shaw]. **D:** Hairpin erosion marks, Canada. [Photograph: J. Shaw]

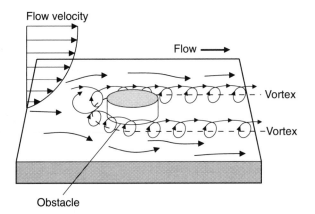

Figure 6.5   *Horseshoe vortices formed as meltwater is diverted around an obstacle. [Modified from: Shaw (1994) Sedimentary Geology **91**, Fig. 8, p. 276]*

protrusions to produce positive and upstanding landforms. The streamlined landforms, **whalebacks**, are referred to by a wide variety of different terms such as **rock drumlins**, **tadpole rocks**, and **streamlined hills** (Figures 6.6 and 6.7). Whalebacks are bedrock knolls that have been smoothed and rounded on all sides by a glacier. Individual whalebacks may be slightly elongated in the direction of ice flow although the structural attributes of the bedrock—joints, bedding planes and foliations—may dramatically affect their morphology. Whalebacks tend to have low height to length ratios, being relatively high (1–2 m) in comparison to their length (1.5–3 m). Striations and other small-scale features of glacial abrasion may be superimposed on any surface of a whaleback. Striations on whalebacks are often continuous along the entire length of the whaleback. From this it is possible to infer that there were no basal cavities around the whaleback during its formation and the ice was everywhere in contact with the landform. High effective normal pressures must therefore be present on both the proximal and distal faces of the whaleback in order to suppress the formation of basal cavities. Whalebacks are therefore landforms primarily produced by glacial abrasion.

6.2.2 Stoss and Lee Features

In contrast to whalebacks and other landforms dominated by glacial abrasion, stoss and lee features possess both abraded and plucked surfaces, and have therefore a pronounced asymmetry (Figures 6.6 and 6.7). They are defined as a bedrock knoll or small hill with a gently abraded slope on the up-ice side (stoss) and a steeper rougher plucked slope on the down-ice side (lee). The most common type of stoss and lee landform is a **roche moutonnée**. These landforms often occur in clusters or fields and may vary in size from several metres to tens or hundreds of metres. They form where high effective normal pressures occur on the stoss side of a bedrock hummock, but the pressure is sufficiently low on the down-ice side to

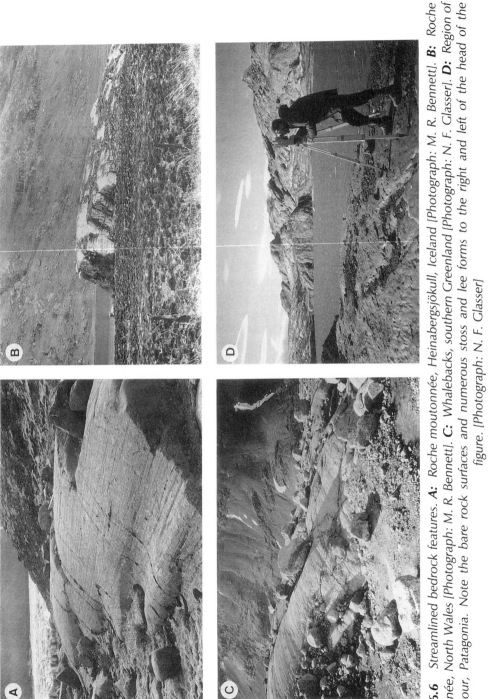

Figure 6.6 Streamlined bedrock features. **A:** Roche moutonnée, Heinabergsjökull, Iceland [Photograph: M. R. Bennett]. **B:** Roche moutonnée, North Wales [Photograph: M. R. Bennett]. **C:** Whalebacks, southern Greenland [Photograph: N. F. Glasser]. **D:** Region of areal scour, Patagonia. Note the bare rock surfaces and numerous stoss and lee forms to the right and left of the head of the figure. [Photograph: N. F. Glasser]

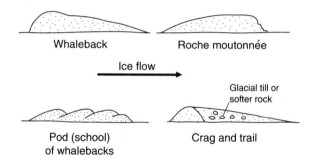

Figure 6.7   *Main types of streamlined erosional landforms*

allow a cavity to form (see Section 4.3). Consequently, the up-ice side experiences glacial abrasion while the down-ice side is glacially plucked. The presence of a lee-side cavity is prerequisite for the formation of a roche moutonnée and consequently they are restricted to areas where the ice velocity is high enough and the effective normal pressure sufficiently low to allow cavities to open. They form preferentially therefore in areas of thin and fast-flowing ice. Since glacial plucking is facilitated by variations in basal water pressure, their formation is also helped by the presence of abundant meltwater.

Once a lee-side cavity has opened and glacial plucking is initiated, the properties of the parent bedrock determine the detailed morphology of the resulting roche moutonnée (Figure 6.8). Bedrock jointing is particularly important since joint depth and spacing determine the size of the blocks that can be plucked from the lee of the original bedrock hummock. It is possible to predict the evolution of the plucked surface of a roche moutonnée (Figure 6.9). Block removal will begin at the furthest point down-ice in the cavity and as successive blocks are removed the plucked surface will migrate further up-ice. This results in a plucked lee-side face that resembles a staircase (Figure 6.9). The spacing of the horizontal and vertical joints within the bedrock will determine the dimensions of each step within the staircase. In bedrock that is not heavily jointed the glacier may create its own fractures due to basal pressure fluctuations; in this case the staircase formed by the removal of blocks may be more varied.

Roches moutonnées usually play host to a variety of other micro-scale erosional landforms, the typical distribution of which is illustrated in Figure 6.10. Although roches moutonnées are characterised by abraded stoss slopes and plucked lee sides, the orientation of these two surfaces is not always a reliable indicator of ice movement direction since their detailed morphology is controlled in part by the pattern of joints or foliations within the rock mass (Figure 6.8). This reduces their reliability as palaeo-ice flow indicators.

6.2.3 Rock Grooves and Rock Basins

Glacial grooves can be formed by either glacial abrasion or by meltwater erosion. Grooves are similar in morphology to striations, except for their greater size and

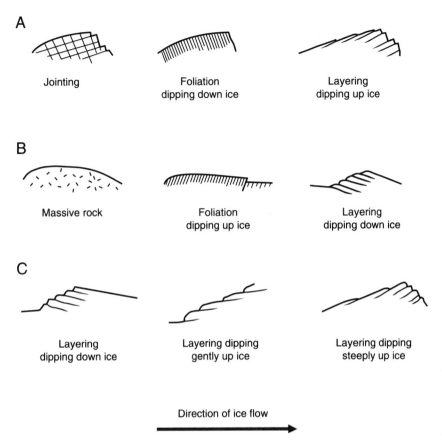

Figure 6.8 *Schematic representation of some relationships between roche mouton-née morphology and geological structure.* **A:** *Plucked lee-side slopes.* **B:** *Abraded lee side slopes.* **C:** *Abraded and plucked stoss slopes.* [*Modified from: Chorley* et al. *(1984) Geomorphology, Methuen, Fig. 17.18, p. 449]*

greater depth. They range from tens of metres to hundreds of metres in length and may be up to several metres wide and a metre deep. They are probably the product of glacial abrasion although the flow of meltwater may be important in the expansion of the groove once formed. The location and orientation of individual grooves may also be influenced by the presence of structural weaknesses within the rock mass.

Rock basins are individual depressions carved in bedrock. They are often found in association with roches moutonnées and may fill with water on deglaciation. Rock basins range in size from several metres to hundreds of metres in diameter. The development of these basins is controlled by the distribution of structural weaknesses within a rock mass which can be exploited by glacial plucking. The size and density of the basin is therefore usually a function of the spacing of joints, or other lines of weakness, within the rock mass.

The formation of rock basins beneath a glacier provides a good illustration of

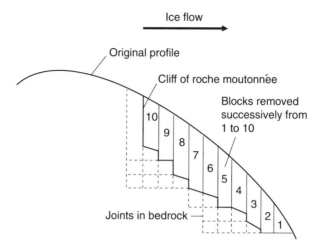

Figure 6.9   *A theoretical model of the evolution of a roche moutonnée. Blocks are removed successively from 1 to 10. [Modified from: Sugden et al. (1992) Geografiska Annaler **74A**, Fig. 4, p. 256]*

the role of **positive feedback** within landform development. The situation is as follows: as a glacier flows over an irregular bed it develops zones of compressional and extensional flow (Figure 3.8). Zones of compressional flow will tend to transport material away from the bed, while extensional flow will cause ice to move towards the bed. Erosion by glacial plucking will be limited in areas of extensional flow since bedrock blocks cannot be transported away from the bed easily, although abrasion may be facilitated by the increased contact pressure between basal clasts and the bed. Net erosion by plucking is favoured in areas dominated by compressional flow since the eroded blocks can be transported away from the bed. Consequently, as a glacier flows over a slight depression it will first experience extensional flow on the up-ice side and then compressive flow on the down-ice side of the depression. The extensional flow component will increase basal pressure and abrasion of the basin floor while the compressional phase will facilitate block removal and plucking. The form of the basin will therefore be accentuated by erosion. This will, in turn, increase the degree of extensional and compressional flow experienced by the glacier as it flows over the basin, which will in turn accelerate its erosion. In this way the basin will grow as a consequence of the positive feedback between basin erosion and compressional/extensional flow.

6.2.4 Meltwater Channels

The final group of meso-scale landforms of glacial erosion are meltwater channels (Figures 6.11 and 6.12). Meltwater channels may form in three environments: (1) subglacially, (2) along the ice margin, and (3) in proglacial locations associated with the flow of water away from the glacier or out of ice-contact lakes. These different channels are often difficult to identify from one another.

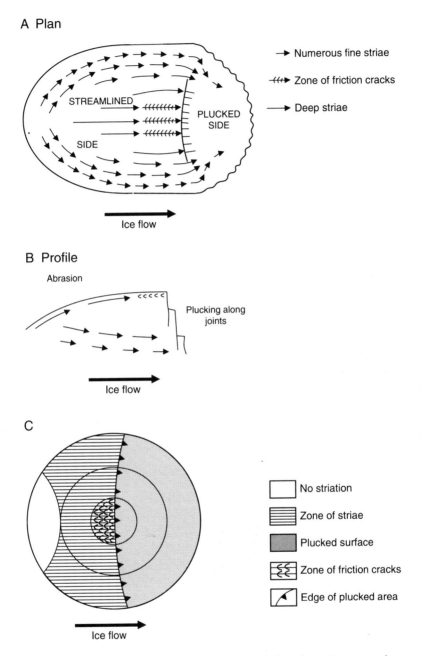

Figure 6.10   *The distribution of micro-scale features of glacial erosion on roches moutonnées.* **A** *and* **B:** *Plan and profile views of a roche moutonnée.* **C:** *A stereographic model of the distribution of micro-scale features on a roche moutonnée. [Reproduced with permission from: Chorley et al. (1984)* Geomorphology, *Methuen, Fig. 17.17, p. 448]*

124

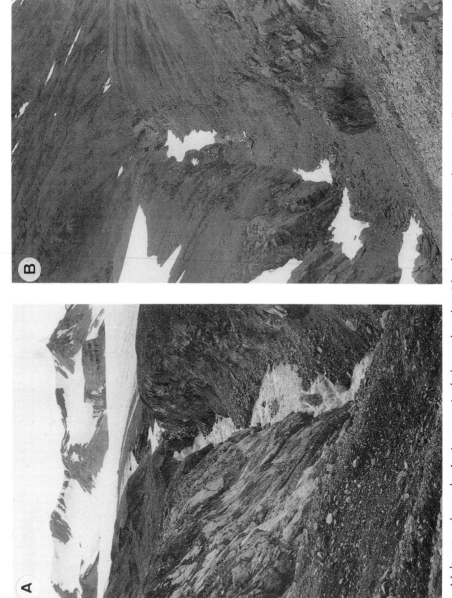

Figure 6.11 Meltwater channels. **A:** Ice-marginal channel at the side of austra Brøggerbreen, Svalbard [Photograph: M. R. Bennett]. **B:** Former overflow channel from the Vatnsdalur ice-dammed lake in Iceland (see Figure 4.9). [Photograph: M. R. Bennett]

Figure 6.12 *Subglacial meltwater channel in the Cheviot Hills, northern England. The channel is cut into thick glacial till and has a cross-section typical of many subglacial meltwater channels. [Photograph: M. R. Bennett]*

Subglacial meltwater channels are sometimes known as ice-directed channels and their orientation is believed to be controlled by the hydraulic potential gradient within the glacier (see Section 4.2). This gradient is determined primarily by the glacier slope and secondarily by the topography beneath the glacier. As a consequence, subglacial meltwater channels may cut across or be orientated transverse to the surface contours and drainage patterns of the present-day topography. Subglacial meltwater is also able to flow uphill if driven by the hydraulic gradient within the glacier. Consequently, the long profile of a subglacial meltwater channel may not have a constant gradient but have an 'up and down' profile.

In contrast, ice-marginal channels run parallel to the glacier front and are most commonly found in the ablation area of glaciers, where rates of surface ablation and meltwater production are highest. The morphology of these channels is more variable than that of subglacial ones. They may also stop suddenly as meltwater is diverted into the glacier via a crevasse or moulin. This may be recorded by a **chute** or sudden right-angle bend in the channel. Ice-marginal channels may sometimes start in large bowl-shaped depressions. These have been interpreted as plunge pools formed by water falling from the glacier into the channel in the same way that plunge pools form on rivers beneath waterfalls. Some channels have a distinct cross-sectional form, while others may simply have a cross-section that resembles a bench or step cut into the hillside. The longitudinal gradient of ice-marginal channels approximates that of the glacier surface at the time of their formation.

Where a glacier terminates on a bedrock surface, proglacial streams may cut distinct channels or gorges as the water moves away from the ice margin. Equally, ice-dammed lakes may drain over cols or ridges via **overspill channels**.

6.3 MACRO-SCALE FEATURES OF GLACIAL EROSION

Macro-scale features of glacial erosion are those features which are 1 km or greater in dimension. They form significant landscape items and may contain within them many of the smaller landforms already considered within this chapter. Five main landforms are recognised at this scale: (1) regions of areal scour, (2) troughs, (3) cirques, (4) giant stoss and lee forms, and (5) tunnel valleys.

6.3.1 Areal Scouring

The most commonly encountered landscape of glacial erosion is one of **areal scour**. It consists of an area of scoured bedrock composed of an assemblage of whalebacks, roches moutonnées and rock basins which all merge into one another (Figure 6.6D). The detailed morphology of the individual landforms within the region of areal scour is primarily controlled by the orientation, spacing and density of joints, foliations and other lines of weakness within the bedrock. Every part of the landscape is affected by glacial erosion. This type of landscape develops under extensive areas of warm-based ice. In Britain this type of terrain is sometimes referred to as **knock and lochan topography**. This term describes the upstanding rounded bedrock lumps (knocks) and the water-filled depressions (lochans) which separate them.

6.3.2 Glacial Troughs

Glacial troughs are deep linear features carved into bedrock (Figure 6.13). These troughs represent the effects of glacial erosion where ice flow is confined by topography and is therefore channelled along the trough or valley. Troughs may be cut beneath ice sheets as well as by valley glaciers and larger outlet glaciers. Glacial troughs are formed by a combination of both glacial abrasion and plucking. Both these processes are required to produce steep-sided troughs, although the effects of glacial abrasion are more obvious due to the smoothed and polished trough walls. The processes of glacial plucking are most evident on the valley floors, where plucked bedrock landforms such as roches moutonnées and rock basins are common.

The cross-sectional form of individual glacial troughs is often described as being 'U-shaped', but their true shape is more accurately described by an empirical power-law or quadratic equation. If a slope profile is surveyed from the centre of a trough up one of its sides then this profile can be described by mathematical equations. In this way the profile of a trough can be compared to that of another to examine, for example, the effects of rock type on trough morphology. The description of a landform through numbers and equations is known as **morphometry**.

Figure 6.13   *A glacial trough occupied by an outlet glacier of the Greenland Ice Sheet, southern Greenland. [Photograph: N. F. Glasser]*

The simplest equation that can be used to describe the cross-sectional morphology of a trough is a power-law equation such as

$$Y = aX^b \tag{6.1}$$

where Y = vertical distance from the valley floor, X = horizontal distance from the centre of the valley, a = constant, and b = a measure of the profile curvature. Most glacial troughs have a value for b of between 1.5 and 2.5. A parabola would have a value of 2. Alternatively, the cross-sectional shape of a glacial trough can be described by a quadratic equation, such as

$$Y = a + bX + cX^2 \tag{6.2}$$

where Y = vertical distance from the valley floor, X = horizontal distance from the centre of the valley, and a, b, c = coefficents determined statistically for each trough.

The choice of equation used in studies of troughs depends on the aim of the study. If the aim is to compare the variation of trough profiles from a standard shape such as the parabola then the power-law equation is most applicable. However, if the aim is to compare the shape of individual troughs with one another then the quadratic equation is most appropriate.

The morphometric description of troughs is a powerful tool, since it allows the variation in trough form to be examined objectively. For example, one might expect trough morphology to vary with the lithology, or the strength of the rock mass, into which it is cut. By comparing mathematically the shape of troughs cut in one type of bedrock with those cut in another, such hypotheses may be tested (Box 6.3).

One of the most important characteristics of glacial troughs is the uniformity in their overall form. This uniformity has led some researchers to suggest that glacial troughs may represent an equilibrium landform. Once the initial morphology of the glacial trough is established, it changes very little. This suggests that troughs represent the adaptation, by erosion, of preglacial valleys, in which ice flow is difficult, into forms which are able to accommodate ice flow comfortably and efficiently. The amount of glacial erosion needed to create a glacial trough is therefore equal to the amount of adaptation needed to modify the preglacial valley in order to discharge the available ice efficiently. Once the shape of a glacial trough is established its size should be simply a function of the amount of ice that it has to discharge. This has been confirmed by studies of outlet glaciers draining ice caps in Greenland and of valley glaciers in New Zealand where a strong relationship

BOX 6.3: ROCK MASS STRENGTH AND TROUGH MORPHOMETRY

It is possible to hypothesise that the morphology of a glacial trough will be affected by the strength of the rock mass or bedrock into which it is cut. Augustinus (1992) examined this hypothesis in the Southern Alps of New Zealand. Determining the strength or resistance of bedrock to erosion is difficult. It is a function not only of the intact strength of the rock but also the density, spacing and orientation of the joints or other lines of weakness within the rock mass. Augustinus (1992) used a method developed in slope studies, known as the **rock mass strength**, to estimate the geomorphological strength of rock in order to compare it with trough morphology. The rock mass strength classification involves scoring a rock mass in the field against a series of eight properties which collectively determine its strength. They are: (1) intact rock strength measured with a Schmidt hammer; (2) the degree of weathering; (3) the spacing between joints or partings; (4) the width of joints or partings; (5) the continuity of the joints or partings; (6) the orientation of the joints or partings in relation to the slope; (7) the presence or absence of infills along joints or partings; and (8) the presence of the outflow of water from joints or partings. At a series of sample sites Augustinus (1992) determined trough morphometry and estimated the rock mass strength of the rock mass into which each trough was cut. The result revealed a strong relationship between rock mass strength and trough form. Glacial troughs appear to become narrower and the sides become steeper as the rock mass strength increases. The number of troughs in a given area also appears to increase with rock mass strength. The work illustrates how the morphology of troughs is partly controlled by the strength of the bedrock into which it is eroded.

Source: Augustinus, P.C. 1992. The influence of rock mass strength on glacial valley cross-profile morphometry: a case study from the Southern Alps, New Zealand. *Earth Surface Processes and Landforms* **17**, 39–51.

between the drainage or accumulation area which supplies the glacier and the size of its trough has been found. The relatively simple morphology of glacial troughs and their close relationship to ice discharge has also enabled their formation and evolution to be modelled effectively (Box 6.4). This type of analysis suggests that over time glacial valleys become adjusted to the glacial system that forms them. This might imply that most erosion will occur in an area when it is first glaciated, and once an efficient system of glacial troughs and other ice discharge routes has been established little modification may occur.

The longitudinal profile of glacial troughs usually contains a series of enclosed basins within it. These are excavated as outlined in Section 6.2.3 through the oper-

BOX 6.4: A NUMERICAL MODEL OF THE EVOLUTION OF A GLACIAL TROUGH

Understanding the formation of glacial troughs is difficult due to the lack of access beneath current glaciers and the length of time involved in their formation. Consequently, several researchers have tried to model the evolution of glacial troughs using computer models. Of particular note is the work of Harbor *et al.* (1988) who used a computer model to simulate the development of a glacial valley from an initial V-shaped fluvial valley. They first modelled the pattern of glacier flow within a valley with V-shaped cross-section and then used this to calculate the pattern of erosion within the valley as the ice flowed through it. Within the model the rate of erosion is assumed to be proportional to the square of the sliding velocity experienced by the glacier. After a period of time the new cross-sectional shape produced by the predicted erosion was calculated. The ice flow through this new cross-section was then modelled and the pattern of erosion predicted again. In this way the evolution of the valley's cross-sectional shape was modelled through a series of time steps and Harbor *et al.* (1988) were able to examine the evolution of a U-shaped glacial valley from an original V-shaped valley cross-section. Their results are summarised in the diagram below and three stages of valley evolution were recognised.

Time 1: initial V-shaped valley cross-section. During this period maximum erosion occurs at two points mid-way along the valley walls, producing a curved cross-section.

Time 2: intermediate stage. As the channel shape changes, the velocity and erosion patterns under the glacier are also changed. The two peaks of erosion reduce in magnitude and begin to shift towards the valley centre.

Time 3: final U-shaped cross-section. At this stage erosion is concentrated at the base of the valley and the valley morphology remains relatively constant as the valley is incised into the landscape.

To reach this final steady-state requires the glacier to excavate a valley almost double the original valley depth. Harbor *et al.* (1988) calculated that this would take around 100 000 years, given a rate of glacial erosion of around 1 mm per year and a final valley depth of 100 m.

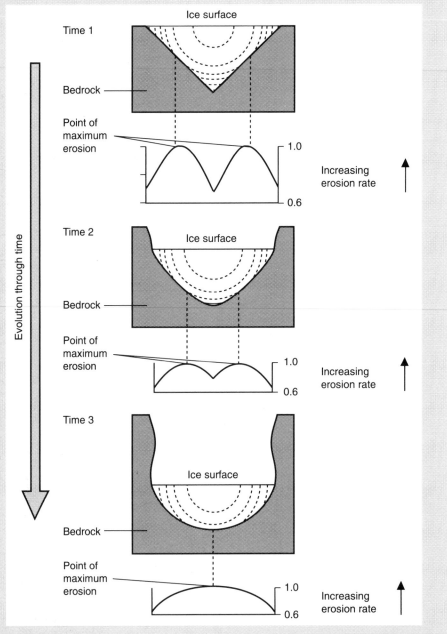

Source: Harbor, J. M., Hallet, B. & Raymond, C. F. 1988. A numerical model of landform development by glacial erosion. *Nature* **333**, 347–349. [Diagram modified from: Harbor *et al.* (1988) *Nature* **333**, Figure 1, p. 348]

ation of positive feedback between compressional flow and basin development. **Hanging valleys** are formed where two glacial troughs have been eroded into the landscape at different rates. If one trough is cut at a faster rate, perhaps because of a greater drainage area than the tributary trough then the floor of the tributary will become perched at a higher altitude than that of the first trough. There may therefore be a substantial height difference between the altitude of the two valley floors, and the highest is said to be left 'hanging' above the lowest.

The planimetric pattern of troughs evolves with the intensity or length of glaciation. When an area is first glaciated the ice occupies preglacial valleys. These glaciers will modify the cross-sectional morphology of these valleys to produce troughs. Remnants of the preglacial valleys may survive as truncated spurs along the margins of the trough. The flow of ice between two valleys across the interfluve, perhaps via a low col, will modify the preglacial valley pattern further. This process is known as **diffluence** if it involves transfer of ice between just two valleys, or **transfluence** if ice is transferred between several valleys. Via this process the preglacial watershed between valleys may be eliminated—a process known as **divide elimination**—and new troughs may be formed. With time, the preglacial valley pattern will be modified and adjusted until it is able to discharge all the available ice efficiently. Modification of the valley pattern therefore provides a crude guide to the intensity of glacial erosion. If the valley pattern is predominantly dendritic it is therefore only slightly modified from its preglacial fluvial form and the intensity of erosion or duration of glacial erosion must have been low. On the other hand, if the trough pattern is highly modified then the intensity of glacial erosion has probably been much greater. This has been used to estimate the intensity of glacial erosion within Scotland (Box 6.5).

BOX 6.5: THE MODIFICATION OF VALLEY PATTERNS BY GLACIAL EROSION

Although glacial geologists are able to identify individual landforms of erosion, it is much more difficult to quantify the extent to which a landscape has been modified or affected by erosion. Haynes (1977) tried to quantify the effects of glacial erosion on the Scottish landscape by examining the degree of valley connectivity in Highland areas. Fluvial landscapes tend to produce dendritic drainage patterns with low valley connectivity, whereas erosion by ice sheets will modify drainage patterns, breach watersheds and cut new troughs into the landscape. This will tend to increase the interconnectivity of the valley system. By using topological measures of connectivity, originally developed to study the connectivity of transport networks, Haynes (1977) compiled maps to show the valley connectivity across Scotland (α-index: a high value indicates high connectivity). Diagram A shows the percentage of the terrain occupied by glacial valleys, while Diagram B shows the connectivity of these valleys as indicated by the α-index. The results show that valley connectivity, and therefore ice sheet erosion, is highest in north and west Scotland and lowest in the east and south. This pattern is to be expected since the high rates of accumulation and ablation of the more maritime west coast would give steep mass-balance gradients and therefore faster flowing ice which would cause intense erosion, while the lower mass balance gradients of the more continental eastern and southern part of Scotland would give lower rates of ice flow and therefore less intense erosion. This work provides a good example of a simple quantitative method by which glacial geologists may determine the intensity of glacial erosion.

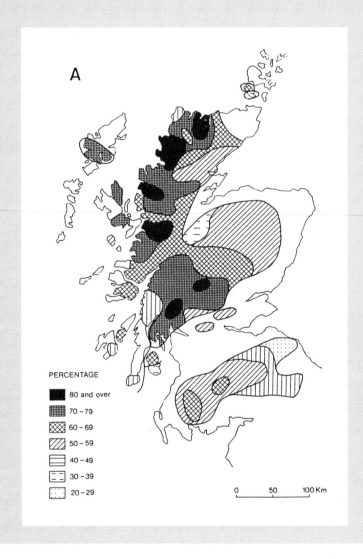

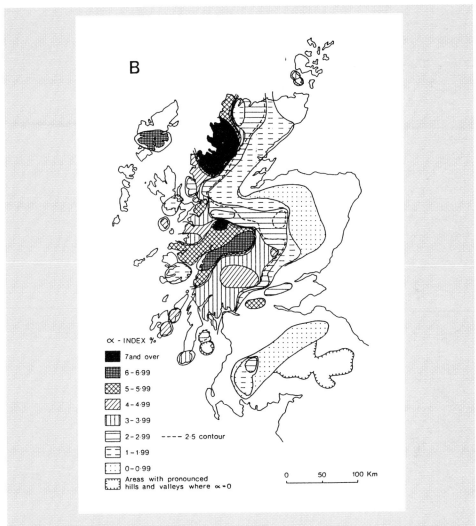

B

α - INDEX %

- 7 and over
- 6 - 6·99
- 5 - 5·99
- 4 - 4·99
- 3 - 3·99
- 2 - 2·99   ---- 2·5 contour
- 1 - 1·99
- 0 - 0·99
- Areas with pronounced hills and valleys where α = 0

0   50   100 Km

Source: Haynes, V. M. 1977. The modification of valley patterns by ice-sheet activity. *Geografiska Annaler* **59A**, 195–207. [Reproduced with permission from: Haynes (1977) *Geografiska Annaler* **59A**, Figs 3 and 4, p. 109]

6.3.3 Cirques

Cirques are hollows which open downslope and are bounded upslope by a cliff or steep slope known as a headwall. The headwall is usually arcuate in plan and is much steeper than the cirque floor. The floor of the cirque may contain an enclosed rock basin and show evidence of glacial erosion, while the headwall is predominantly formed by glacial plucking and periglacial freeze–thaw weathering. Cirques are eroded by discrete patches of glacier ice, although they may drain into valley glaciers.

Glacial cirques are found in mountainous terrain subject to local glaciation. They may also occur as part of landscapes of ice sheet erosion, like those in Britain (Figure 6.14). Here phases of local glaciation allowed cirques to develop although their morphology has been modified during periods of more intense ice sheet glaciation. Most cirques are the product of cumulative erosion during several phases of glaciation.

The precise definition of a cirque varies; for example, in three separate studies of Scottish cirques the total number identified in the landscape varied between 347 and 876. Part of the reason for this is that cirques tend to grow in clusters, and individual features may become amalgamated over time to become composite features. For example, one large cirque may actually be a composite of several smaller feeder cirques. Similarly, such cirques may cluster around the heads of glacial troughs, for which they act as the accumulation area. This variety of form makes the identification of individual cirques highly subjective. However, some of this subjectivity may be removed by using mathematical formulae to describe and classify cirque morphology.

In longitudinal profile, the headwall of a cirque is normally much steeper than

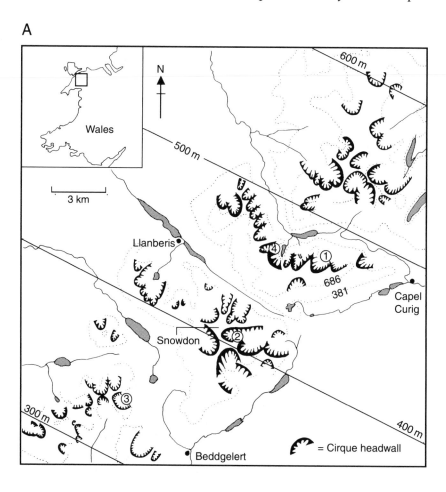

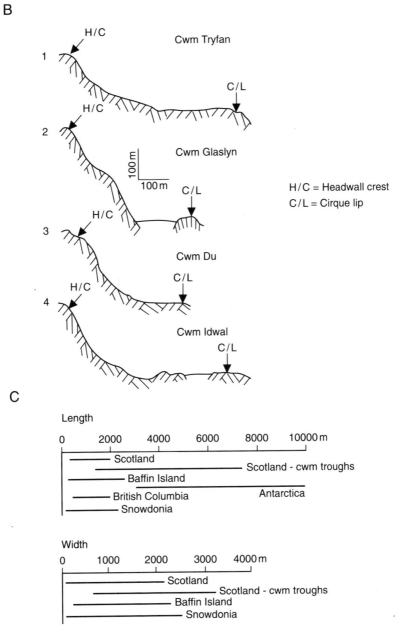

Figure 6.14 *Cirques in Snowdonia, North Wales.* **A:** *The distribution of cirques in northern Snowdonia. Contours on a linear trend surface of cirque floor altitude are also shown at 300, 400, 500 and 600 m. This records a rise in cirque elevation from the south-west to the north-east.* **B:** *Typical long profiles through selected cirques (numbers refer to the location of each cirque on (A).* **C:** *Comparison of the size of cirques in Snowdonia with those found elsewhere. [Modified from: Bennett (1990) Queen Mary and Westfield College Research Papers in Geography* **2**, *Figures 4, 10 and 13, pp. 4, 16 and 20]*

its floor. The basic long profile shape of a cirque can therefore be described by a logarithmic curve

$$Y = k(1 - X)e^{-X} \qquad (6.3)$$

where Y = vertical distance from the floor of the cirque, X = horizontal distance from the front of the cirque, e = constant, and k = shape contant. This type of curve is known as a k-curve and the value of k is known as the k-number. For most cirques the k-number is between 0.5 and 2. The greater the value of the k-number, the steeper the headwall of the cirque (Figure 6.15). A cirque fitted by a k-curve with a k-number of 2 will have a steep headwall and will be over-deepened to such an extent that its floor will mostly likely contain a lake. The type of long profile that a cirque possesses is a function of its bedrock lithology and of the structural weakness within it. For example, in areas where bedrock dips into a cirque then the floor of the cirque will also tend to dip inwards and the cirque may contain a small lake. Where bedrock dips out of the cirque it will have a floor which slopes outwards and will be best described by a low k-number (Figure 6.15). Jointing within the bedrock also determines the nature of the headwall. Closely spaced joints tend to produce blocky headwalls, while widely spaced joints produce smoother headwall profiles.

The evolution of cirque morphology through time can be examined using

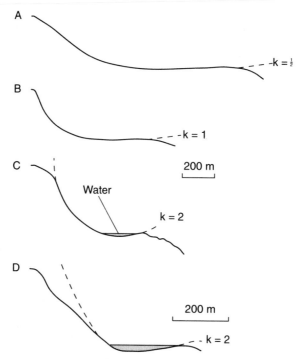

Figure 6.15   *Empirical curves, k-curves, fitted to the longitudinal profiles of four Scottish cirques. The higher the value of k, the more enclosed the cirque basin.* [Modified from: Haynes (1968) Geografiska Annaler **50A**, Fig. 3, p. 223]

ergodic reasoning. Ergodic reasoning suggests that under certain circumstances sampling in space can be equivalent to sampling through time; and that space–time transformations are permissible as a working tool. This works on the assumption that within a population of evolving landforms there will be a range of different examples at different stages of development. By comparing each example within a given area (i.e. in space), a model of landform evolution through time may be established. For example, within a mountainous area there may be a range of cirques each of which may have been initiated at a different time or developed at a different rate. Consequently, the spatial variation in cirque morphometry within the area will provide insight into how cirques evolve through time.

This type of methodology has been used to establish the following models for the evolution of cirques in different areas. As cirques increase in size they appear to become more enclosed both in plan and in profile. Figure 6.16 shows a simple

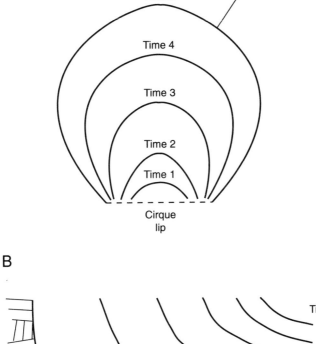

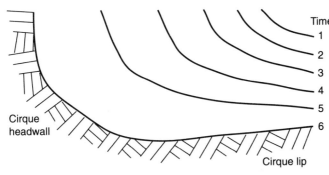

Figure 6.16 *A model of the evolution of a cirque, based on observation within the Scottish Highlands. [Modified from: Gordon (1977) Geografiska Annaler **59A**, Fig. 7, p. 192]*

model of cirque evolution based on morphometric observation in the Scottish Highlands in which a well-defined cirque basin develops with time. In practice this type of model may only hold in areas of relatively uniform bedrock geology in which the structural weaknesses are present in all directions. In most rock masses this is not case and structural or lithological weaknesses tend to be orientated in one direction. In these situations the morphology of the cirque and its evolution may be strongly guided by orientations of these structural and lithological weaknesses. For example, the data in Table 6.1 show that, in the mountains of North Wales, cirques cut along the strike of the outcropping geology tend to be more elongated than those cut across strike. Work on these cirques also suggests that cirques which are orientated parallel to geological structures may evolve in a different fashion from those which are orientated at right angles to the geological structure (Figure 6.17). Those cirques orientated along the geological structure appear to have experienced faster rates of headwall retreat than downcutting and consequently have evolved an elongated, flat-floored morphology. In contrast, those which have developed transverse to the geological and lithological trend appear to have experienced similar rates of downcutting and headwall retreat. Consequently, they have evolved into more compact and enclosed basins. Once a basin has developed in the floor of a cirque it will continue to grow due to the same positive feedback mechanism outlined for rock basins (see Section 6.2.3).

Table 6.1   *The difference in cirque morphometry between cirques cut along and across geological strike in Snowdonia, North Wales. Average data is shown for a total sample of 81 cirques. Plan closure is a measure of the degree of planimetric cirque development; a circle would have a value of 360°. Profile closure is a measure of the vertical development of a cirque; the higher the value, the closer a cirque is to an enclosed basin. The length/width ratio is a measure of cirque elongation; a high value indicates greater elongation. The width/amplitude ratio measures the degree of cirque incision; low values indicate a deeply incised cirque. Concavity is a measure of the overall cirque development; the higher the value, the greater the basinal development within the cirque (a classic armchair-shaped cirque would have a high value). [Modified from: Bennett (1990) Queen Mary and Westfield College Research Papers in Geography* **2***, Table 14, p. 21]*

	Headwall height	Cirque area	Plan closure	Profile closure	Length/ width ratio	Width/headwall height ratio	Concavity
Along strike	131 m	565 122 m²	150°	25.7	1.3	2.4	253°
Across strike	166 m	630 679 m²	162°	35.8	1.1	3.2	305°

The elevation of cirques and their aspect can be used to provide general palaeoclimatic information. The altitude at which an abandoned cirque lies may be used as a measure of the former regional snowline or equilibrium line altitude within the area on the assumption that cirques are formed by discrete glaciers (Figure 6.14). In most cases this reflects a composite snowline averaged over numerous periods of glaciation during which the cirques were occupied. The closer the altitude of a cirque to sea-level, the lower the snowline. Cirques close to sea-level pro-

A Cirque evolution along-strike or parallel to the structural trend

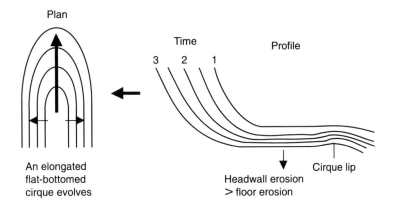

An elongated flat-bottomed cirque evolves

Headwall erosion > floor erosion

B Cirque evolution across-strike or perpendicular to the structural trend

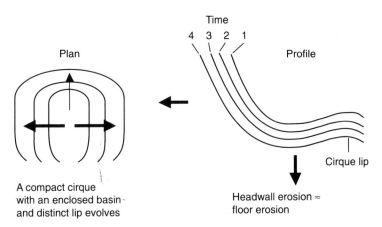

A compact cirque with an enclosed basin and distinct lip evolves

Headwall erosion ≈ floor erosion

Figure 6.17 *An alternative model of the evolution of a cirque for a rock mass with a strong structural trend. [Modified from: Bennett (1990) Queen Mary and Westfield College Research Papers in Geography 2, Fig. 16, p. 25]*

vide evidence of relatively harsh climates, while those at high altitudes may form when the regional snowline is much higher. For example, cirques within the Snowdon mountains of North Wales rise in elevation from the south-west to the north-east, a trend which has been interpreted as reflecting the direction of prevailing snow winds during the later part of the Cenozoic Ice Age (Figure 6.14). The aspect of a cirque or the direction in which it opens is also indicative of palaeoclimate, since the location of the cirque glacier which cut the cirque is controlled by the direction of snow-bearing winds and the direction of incoming solar radiation. In the northern hemisphere most cirques face towards the north-east, although this aspect may be modified by strong snow-winds such that they form in the lee of mountain slopes crossed by the winds.

Cirques provide another good example of a positive feedback system since the deeper the cirque becomes, the more efficient its shape is for trapping snow and shading the accumulation areas. The cirque glacier will consequently be bigger. A bigger glacier is likely to achieve more efficient erosion and therefore to continue to excavate the cirque; a cycle which will continue while the glacier survives.

6.3.4 Giant Stoss and Lee Features

Preglacial valley spurs and other hills may be eroded by ice into giant stoss and lee forms. These are given the collective term of **giant roches moutonnées** and range in size from hundreds of metres to several kilometres across. These features are carved in bedrock and may appear as either plucked valley spurs or as free-standing and isolated bedrock protrusions. Their morphology is similar to that of smaller roches moutonnées, with ice-smoothed proximal surfaces and plucked distal surfaces. This suggests that they are formed by a similar process but simply on a much larger scale. Smaller roches moutonnées and other erosional landforms are usually found superimposed on these asymmetrical spurs and hills.

Good examples of giant stoss and lee features are to be found on Deeside, Scotland, where large streamlined hills with lee-side cliffs, up to 160 m high, are to be found (Figure 6.18). These large-scale landforms formed under thin ice near the end of the last glaciation, during a time of high ice velocity and abundant melt-

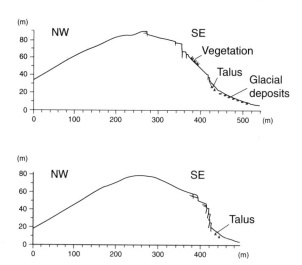

Figure 6.18 *Cross-sections through two giant stoss and lee forms in Glen Dee, Scotland. [Modified from: Sugden et al. (1992) Geografiska Annaler **74A**, Fig. 7, p. 259]*

water. Other large stoss and lee features are to be found in south Greenland. These large composite landforms are ice polished on their upstream side, while their downstream face is plucked. The upstream sides of these landforms contain smaller whalebacks superimposed upon the main landform.

6.3.5 Tunnel Valleys

Tunnel valleys are linear valleys or incisions found beneath land or sea areas with flat or low relief. At their largest they may be 2 km wide, over 100 m deep and extend for between 6 and 30 km in length. In general, however, they tend to be smaller, perhaps only 600–800 m wide and less than 60 m deep. In long profile they contain enclosed hollows or isolated, often elongated, basins. They are normally infilled to varying depths with a variety of different types of sediment. Tunnel valleys are found extensively on the continental shelf of northern Europe and North America, and buried examples have also been recorded on land. Although small tunnel valleys have been recorded on land for some time, their widespread occurrence on the continental shelf has only come to light in recent years. As a consequence, the morphological variation present within tunnel valleys and its spatial variation is poorly understood at present.

The origin of these valleys is also unclear. Four main hypotheses have been proposed: (1) over-deepening of preglacial valleys by meltwater flowing away from an ice margin, (2) glacial erosion, (3) erosion by subglacial meltwater, and (4) erosion by catastrophic glacial floods, jökulhlaups, associated with the drainage of subglacial lakes. At present there are insufficient data on the morphology and distribution of tunnel valleys to distinguish between these various hypotheses.

6.4 SUMMARY

A range of glacial landforms are produced by the processes of glacial erosion. These can be organised on the basis of scale. Common micro-scale landforms are striations, micro crag and tails, friction cracks and p-forms. At a meso-scale one can recognise streamlined bedrock landforms, stoss and lee forms, grooves and basins and meltwater channels. Macro-scale landforms include areas of areal scour, glacial troughs, cirques, giant stoss and lee forms and tunnel valleys. Each of these landforms provides a component of the landscape produced by glacial erosion and a clue about the dynamics and character of the glacier which formed them. In fact, each landform tells a story about the glacier that formed it. Figure 6.19 shows the distribution of these landforms beneath a hypothetical cross-section through an ice sheet. This conceptual model assumes that the ice is warm-based throughout (isothermal) and that its velocity profile is simply a function of mass balance (see Section 3.5). Roches moutonnées occur where the ice is fast-flowing and relatively thin, conditions which are ideal for cavity formation, while whalebacks occur under thicker ice. Valley modification is greatest in the zone of fast-flowing ice close to the equilibrium line.

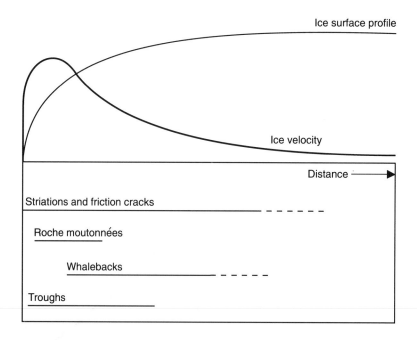

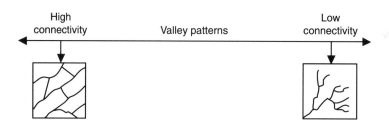

Figure 6.19   *Schematic cross-section through a warm-based ice sheet showing the distribution of different erosional landforms*

Table 6.2 summarises the inferences that can be made about the dynamics and glacial conditions under which each erosional landform is normally produced. It provides a key to the clues left about the former glaciers which shaped the erosional landscape visible in areas once covered by Cenozoic glaciers.

Table 6.2   *Summary table of the principal landforms of glacial erosion and their significance in the reconstruction of former glaciers*

Striations	Morphology: small grooves scratched into bedrock surfaces, usually continuous over several metres. Indicative of: warm-based ice, orientated in the direction of local ice flow.
Friction cracks	Morphology: crescentic-shaped gouges cut into bedrock surfaces. Indicative of: warm-based ice, may be used to indicate the direction of local ice flow.
Micro crag and tails	Morphology: small tails of rock formed in the lee of resistant crystals, grains or nodules. Indicative of: warm-based ice and not only the orientation but also the sense of direction of local ice flow.
p-forms	Morphology: smooth-walled 'sculpted' depressions and channels cut into bedrock. Indicative of: warm-based ice, abundant meltwater, low effective normal pressures, typical of thin ice.
Whalebacks	Morphology: upstanding streamlined bedrock lumps with abraded surfaces. Indicative of: warm-based ice, high effective normal pressures, the absence of basal cavities. Such conditions may occur beneath thick, slow-moving ice with little available basal meltwater.
Roches moutonnées	Morphology: upstanding bedrock lumps with both abraded and plucked faces. Indicative of: warm-based ice, low effective normal pressure, presence of basal cavities, abundant meltwater with regular fluctuations in basal water pressure. These conditions arise under thin, fast-flowing ice.
Subglacial meltwater channels	Morphology: steep-sided channels cut into bedrock or till which may have an orientation that is discordant to the local topography. These channels may have an irregular 'up and down' long profile. Indicative of: warm-based ice and of the pattern of subglacial drainage.
Ice-marginal meltwater channels	Morphology: these may have a complete channel cross-section or alternatively may consist of a channel floor and one wall, the other wall having been formed by the glacier (half channel). Their long profile gradient may parallel that of the glacier margin. They start and end abruptly. These channels are commonly associated with other ice-marginal landforms. Indicative of: the location of the former ice margin.
Regions of areal scour	Morphology: areas of low relief smoothed into streamlined bumps and basins. Numerous roches moutonnées and whalebacks. All surfaces usually contain abundant striations and friction cracks. Indicative of: warm-based ice, basal cavities and fluctuating

	basal meltwater pressure. Common under relatively thin ice where low effective normal pressures and high ice velocities exist.
Glacial troughs	Morphology: over-deepened valleys with smoothed, polished, steep walls and flat floors. Indicative of: warm-based ice, abundant meltwater and high ice velocities. The cross-sectional area of a trough may be related to ice discharge; the larger a trough, the greater the ice discharge.
Cirques	Morphology: large armchair-shaped depressions cut into bedrock with steep walls and sides. They may contain a small enclosed basin which is often lake-filled. Indicative of: warm-based ice and abundant meltwater. Climate can only support local glaciers. Elevation of cirques may provide information about average regional snowlines within a deglaciated areas.
Giant stoss and lee forms	Morphology: large upstanding bedrock hills or spurs with abraded up-ice and plucked down-ice faces. Indicative of: warm-based ice, low effective normal pressure, presence of basal cavities, abundant meltwater with regular fluctuations in basal water pressure. These conditions arise under thin, fast-flowing ice. They may indicate relatively low levels of glacial erosion, since they may be associated with preglacial valley spurs or bedrock hills.
Tunnel valleys	Morphology: large, linear, steep-sided valleys or depressions which may contain enclosed basins in their floor. Tunnel valleys are usually infilled with sediment and occur both on the continental shelf and in lowland areas. Indicative of: origin unclear at present, although may be formed by subglacial meltwater.

6.5 SUGGESTED READING

The study of glacial erosion is traditionally a neglected part of glacial geology. As a consequence, much of the literature is old and obscure. Although they are now rather old, the classic papers by Matthes (1930) and Linton (1963) provide excellent reviews of the main features of glacial erosion. Also of note is the work by Linton on landscapes of glacial erosion that was published by Clayton (1974). The paper by Laverdiere *et al.* (1979) presents a classification of features of glacial erosion, while Dionne (1987) shows how streamlined glacial features may also be classified.

Abrasional bedrock forms such as glacial striations, micro crag and tails and friction cracks are described in papers by Gilbert (1906), Harris (1943), Dreimanis (1953), Lamarche (1971), Gray (1982), Shaw (1988) and Thorp (1981). Kleman (1990) provides a methodological discussion of the use of striations in reconstructing former ice flow patterns, and further examples can be found in Gray & Lowe (1982), Gemmell *et al.* (1986), Thorp (1987) and Lindstrom (1991). Sharp *et al.* (1989a) show how sophisticated inferences may be made from micro-erosional landforms.

McCarroll *et al.* (1989) illustrate how striations may be formed by glaciofluvial processes.

Classic papers on the origins of p-forms and glacial grooves are provided by Dahl (1965) and Gjessing (1965). Gray (1981) describes some excellent examples in Scotland. Boulton (1974) argues for their formation by glacial abrasion while Shaw (1994) discusses their origin as fluvial landforms. Sharp *et al.* (1989b) discuss the significance and formation of micro-channels cut by subglacial meltwater.

Streamlined bedrock features are discussed by Dionne (1987), while a qualitative theory for the development of stoss and lee forms is presented by Carol (1947). The influence of bedrock structure on roche moutonnée morphology is covered by Rastas & Seppala (1981) and Gordon (1981). Lindstrom (1988) presents an interesting argument that the morphology of most roches moutonnées is inherited from the preglacial landscape. The morphology of medium- and large-scale stoss and lee features is recorded in Rudberg (1973) and their formation is discussed by both Glasser & Warren (1990) and Sugden *et al.* (1992).

The morphology and origin of meltwater channels are described in a range of papers, including Sissons (1960, 1961), Price (1963), Clapperton (1968), Rodhe (1988) and Sugden *et al.* (1991).

There is a large range of literature on glacial troughs which ranges from the qualitative descriptions of Linton (1963, 1967) to the more recent quantitative descriptions of Boulton (1974), Hirano & Aniya (1988), Harbor *et al.* (1988), Harbor (1990, 1992), Harbor & Wheeler (1992) and Augustinus (1992a). The role of ice discharge in creating glacial troughs is developed in papers by Haynes (1972) and Augustinus (1992b). The role of ice sheets in changing large-scale drainage patterns is discussed by Linton (1963, 1967) and Haynes (1977).

An introduction to the study of glacial cirques and problems associated with their definition is provided by Evans & Cox (1974), while the global distribution of cirque forms is discussed in Evans (1977). Part of volume 59A of the journal *Geografiska Annaler* is devoted to cirques and to glacial erosion in general. The volume edited by McCall & Lewis (1960) also contains useful information. The detailed morphology of cirques and the control exercised on it by bedrock lithology and jointing is covered in Haynes (1968, 1995) and Gordon (1977). The morphology, distribution and possible origin of tunnel valleys is discussed by Wingfield (1990).

Augustinus, P. C. 1992a. The influence of rock mass strength on glacial valley cross-profile morphometry: a case study from the Southern Alps, New Zealand. *Earth Surface Processes and Landforms* **17**, 39–51.

Augustinus, P. C. 1992b. Outlet glacier trough size–drainage area relationships, Fiordland, New Zealand. *Geomorphology* **4**, 347–361.

Boulton, G. S. 1974. Processes and patterns of glacial erosion. In: Coates, D. R. (Ed.) *Glacial Geomorphology*. Proceedings of the Fifth Annual Geomorphology Symposia, Binghampton, Allen & Unwin, London, 41–87.

Carol, H. 1947. The formation of roches moutonnées. *Journal of Glaciology* **1**, 57–59.

Clapperton, C. M. 1968. Channels formed by the superimposition of glacial meltwater streams, with special reference to the east Cheviot Hills, northeast England. *Geografiska Annaler* **50A**, 207–220.

Clayton, K. M. 1974. Zones of glacial erosion. In: Brown, E. H. & Waters, R. S. (Eds) *Progress in Geomorphology.* Institute of British Geographers Special Publication 7, 163–176.

Dahl, R. 1965. Plastically sculptured detail forms on rock surfaces in northern Nordland, Norway. *Geografiska Annaler* **47A**, 83–140.

Dionne, J. C. 1987. Tadpole rock (rockdrumlin): a glacial streamline moulded form. In: Menzies, J. and Rose, J. (Eds) *Drumlin Symposium.* Balkema, Rotterdam, 149–159.

Dreimanis, A. 1953. Studies of friction cracks along the shore of Cirrus Lake and Kasakokwag Lake, Ontario. *American Journal of Science* **251**, 769–783.

Evans, I. S. 1977. World-wide variations in the direction and concentration of cirque and glacier aspects. *Geografiska Annaler* **59A**, 151–175.

Evans, I. S. & Cox, N. 1974. Geomorphology and the operational definition of cirques. *Area* **6**, 150–153.

Gemmell, J., Smart, D. & Sugden, D. 1986. Striae and former ice-flow directions in Snowdonia, North Wales. *Geographical Journal* **152**, 19–29.

Gilbert, G. K. 1906. Crescentic gouges on glaciated surfaces. *Geological Society of America Bulletin* **17**, 303–313.

Gjessing, J. 1965. On plastic scouring and subglacial erosion. *Norsk Geografisk Tidsskrift* **20**, 1–37.

Glasser, N. F. & Warren, C. R. 1990. Medium scale landforms of glacial erosion in South Greenland: process and form. *Geografiska Annaler* **72A**, 211–215.

Gordon, J. E. 1977. Morphometry of cirques in the Kintail–Affric–Cannich area of northwest Scotland. *Geografiska Annaler* **59A**, 177–194.

Gordon, J. E. 1981. Ice-scoured topography and its relationships to bedrock structure and ice movement in parts of northern Scotland and West Greenland. *Geografiska Annaler* **63A**, 55–65.

Gray, J. M. 1981. p-forms from the Isle of Mull. *Scottish Journal of Geology* **17**, 39–47.

Gray, J. M. 1982. Un-weathered, glaciated bedrock on an exposed lake bed in Wales. *Journal of Glaciology* **28**, 483–497.

Gray, J. M. & Lowe, J. J. 1982. Problems in the interpretation of small-scale erosional forms on glaciated bedrock surfaces: examples from Snowdonia, North Wales. *Proceedings of the Geologists' Association* **93**, 403–414.

Harbor, J. M. 1990. A discussion of Hirano and Aniyaís (1988,1989) explanation of glacial-valley cross profile development. *Earth Surface Processes and Landforms* **15**, 369–377.

Harbor, J. M. 1992. Numerical modelling of the development of U-shaped valleys by glacial erosion. *Geological Society of America Bulletin* **104**, 1364–1375.

Harbor, J. M. & Wheeler, D. A. 1992. On the mathematical description of glaciated valley cross sections. *Earth Surface Processes and Landforms* **17**, 477–485.

Harbor, J. M., Hallet, B. & Raymond, C. F. 1988. A numerical model of landform development by glacial erosion. *Nature* **333**, 347–349.

Harris, S. E. 1943. Friction cracks and the direction of glacier movement. *Journal of Glaciology* **51**, 244–258.

Haynes, V. M. 1968. The influence of glacial erosion and rock structure on corries in Scotland. *Geografiska Annaler* **50A**, 221–234.

Haynes, V. M. 1972. The relationship between the drainage areas and sizes of outlet troughs of the Sukkertoppen Ice Cap, West Greenland. *Geografiska Annaler* **54A**, 67–75.

Haynes, V. M. 1977. The modification of valley patterns by ice-sheet activity. *Geografiska Annaler* **59A**, 195–207.

Haynes, V. M. 1995. Alpine valley heads on the Antarctic Peninsula. *Boreas* **24**, 81–94.

Hirano, M. & Aniya, M. 1988. A rational explanation of cross-profile morphology for glacial valleys and of glacial valley development. *Earth Surface Processes and Landforms* **13**, 707–716.

Kleman, J. 1990. On the use of glacial striae for reconstruction of palaeo-ice sheet flow patterns. *Geografiska Annaler* **72A**, 217–236.

Lamarche, R. Y. 1971. Northward-moving ice in the Theford mines area of southern Quebec. *American Journal of Science* **271**, 383–388.

Laverdiere, C., Guimont, P. & Pharand, M. 1979. Marks and forms on glacier beds: formation and classification. *Journal of Glaciology* **23**, 414–416.

Lindstrom, E. 1988. Are roches moutonnées mainly preglacial forms? *Geografiska Annaler* **70A**, 323–331.

Lindstrom, E. 1991. Glacial ice-flows on the islands of Bornholm and Christianso, Denmark. *Geografiska Annaler* **73A**, 17–35.

Linton, D. L. 1963. The forms of glacial erosion. *Transactions of the Institute of British Geographers* **33**, 1–28.

Linton, D. L. 1967. Divide elimination by glacial erosion. In: Wright, H. E. & Osburn, W. H. (Eds) *Arctic and Alpine Environments*. Indiana University Press, Indiana, 241–248.

Matthes, F. E. 1930. Geologic history of the Yosemite Valley. *US Geological Survey Professional Paper* **160**, 54–103.

McCall, J. G. & Lewis, W. V. 1960. Investigation on Norwegian cirque glaciers. *Royal Geographical Society Research Series IV*, 39–62.

McCarroll, D., Matthews, J. A. & Shakesby, R. A. 1989. Striations produced by catastrophic subglacial drainage of a glacier-dammed lake, Mjolkedalsbreen, Southern Norway. *Journal of Glaciology* **35**, 193–196.

Price, R. J. 1963. The glaciation of part of Peebleshire, Scotland. *Transactions of the Edinburgh Geological Society* **19**, 326–348.

Rastas, J. & Seppala, M. 1981. Rock jointing and abrasion forms on roches moutonnées, SW Finland. *Annals of Glaciology* **2**, 159–163.

Rodhe, L. 1988. Glaciofluvial channels formed prior to the last deglaciation: examples from Swedish Lapland. *Boreas* **17**, 511–516.

Rudberg, S. 1973. Glacial erosion forms of medium size – a discussion based on four Swedish case studies. *Zeitschrift für Geomorphologie* **17**, 33–48.

Sharp, M., Dowdeswell, J. A. & Gemmell, J. C. 1989a. Reconstructing past glacier dynamics and erosion from glacial geomorphic evidence: Snowdon, North Wales. *Journal of Quaternary Science* **4**, 115–130.

Sharp, M., Gemmell, J. C. & Tison, J. 1989b. Structure and stability of the former subglacial drainage system of the Glacier de Tsanfleuron, Switzerland. *Earth Surface Processes and Landforms* **14**, 119–134.

Shaw, J. 1988. Subglacial erosional marks, Wilton Creek, Ontario. *Canadian Journal of Earth Science* **25**, 1256–1267.

Shaw, J. 1994. Hairpin erosional marks, horseshoe vortices and subglacial erosion. *Sedimentary Geology* **91**, 269–283.

Sissons, J. B. 1960. Some aspects of glacial drainage channels in Britain, Part I. *Scottish Geographical Magazine* **76**, 131–146.

Sissons, J. B. 1961. Some aspects of glacial drainage channels in Britain, Part II. *Scottish Geographical Magazine* **77**, 15–36.

Sugden, D. E., Denton, D. H. & Marchant, D. R. 1991. Subglacial meltwater channel system and ice sheet over riding of the Asgard range, Antarctica. *Geografiska Annaler* **73A**, 109–121.

Sugden, D. E., Glasser, N. F. & Clapperton, C. M. 1992. Evolution of large roches moutonnées. *Geografiska Annaler* **74A**, 253–264.

Thorp, P. W. 1981. An analysis of the spatial variability of glacial striae and friction cracks in part of the Western Grampians of Scotland. *Quaternary Studies* **1**, 71–94.

Thorp, P. W. 1987. Late Devensian ice sheet in the western Grampians, Scotland. *Journal of Quaternary Science* **2**, 103–112.

Wingfield, R. 1990. The origin of major incisions within the Pleistocene deposits of the North Sea. *Marine Geology* **91**, 31–52.

7
Glacial Debris Transport

In this chapter we examine how debris is transported by glaciers. Debris may be derived either from above a glacier (**supraglacial sources**) or from its bed (**subglacial sources**). Supraglacial sources consist of debris which falls or flows onto the surface of the glacier from valley sides or mountains which extend above its surface (**nunataks**). Debris may also be deposited on a glacier surface by windfall (e.g. volcanic ash, dust, sea salt), avalanches or by human action. Subglacial debris is derived through glacial erosion and basal debris entrainment.

Within a glacier, debris is transported at two levels (Figure 7.1): high-level transport and low-level transport. **High-level transport** involves the transport of debris without it coming into contact with the glacier bed. If debris falls onto the glacier surface above the equilibrium line it will be buried beneath successive layers of snow accumulation. If the debris falls onto the glacier surface below the equilibrium line, it will stay on the glacier surface until it reaches the snout. **Low-level transport,** or basal transport, involves the movement of debris at the base of the glacier. The material may be derived from the bed by subglacial erosion or indirectly from debris which falls onto the glacier surface and makes its way to the bed via crevasses or by the downward flow of ice associated with extending flow or basal melting (Figure 7.1). This debris may remain at the base of the glacier until it is deposited or reaches the glacier snout. Alternatively, it may be elevated onto the surface of the glacier by upward flow or along thrusts within the ice formed by compressive flow, a phenomena common at the glacier snout. The imprint left on glacial debris by transport depends on the pathway it follows through the glacier. The debris structure within a given glacier is important in determining the geomorphological impact it has.

7.1 HIGH-LEVEL DEBRIS TRANSPORT

In the accumulation area of a glacier, debris falling onto the surface from adjacent

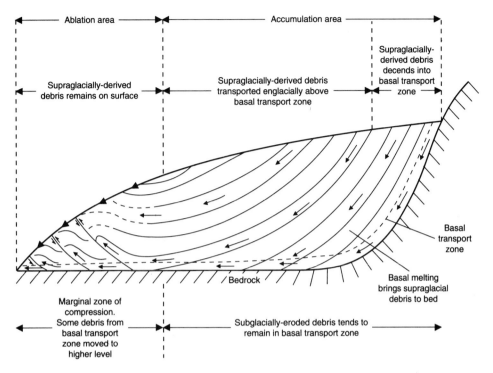

Figure 7.1 *The transport pathways of debris through a valley glacier. Two transport pathways can be identified: (1) a high-level pathway in which the debris does not come into contact with the glacier bed, and (2) a low-level pathway in which the debris is in contact with the glacier bed. [Diagram reproduced with permission from: Boulton (1993) In: Duff (Ed.)* Holmes' Principles of Physical Geology, *Chapman & Hall, Fig. 20.35, p. 425]*

rock walls or nunatacks will become buried by fresh accumulations of snow. The debris will become concentrated in layers along the primary **ice stratification** formed by successive annual layers of accumulation (Box 7.1). The debris therefore becomes englacial and will emerge from the glacier in the ablation zone to form a supraglacial cover as the surface is lowered by melting. Any rock debris which accumulates on the glacier surface in the ablation zone will not be permanently buried by ice, unless it falls into a crevasse, and will form a supraglacial cover.

Debris on the surface of a glacier may either be concentrated into down-glacier ridges known as **medial moraines** or will form an irregular layer over the glacier surface (Figure 7.2). There are two broad categories of medial moraine: **ice stream interaction medial moraines** and **ablation-dominant medial moraines**. Ice stream interaction medial moraines are formed by the confluence of two lateral moraines at the junction of two glaciers (Figure 7.3). The medial moraine consists of a debris-covered ice ridge which extends down the trunk of the glacier and marks the line of suture between the two glaciers. It is important to note that the moraine is the

BOX 7.1: THE STRUCTURE OF GLACIERS

Glacier ice contains a number of distinct structures. These can be divided into: (1) **primary structures**, which result from accumulation; and (2) **secondary structures**, which develop due to ice deformation during flow.

The principal primary structure is **ice stratification**, which results from the accumulation of snow each year. Summer surfaces are usually indicated by a refrozen melt-layer of bluish ice and by a concentration of debris which has fallen onto the glacier surface. Another type of primary structure is regelation layers, which result from the freezing of regelation ice to the base of the glacier (see Section 3.3.2). These primary structures become deformed during ice flow by internal deformation to produce secondary structures. The most important secondary structure is **ice foliation**. Foliation is a layered fold structure of different sizes of ice crystals which develops by the deformation of primary ice structures during ice flow. It may develop either parallel to or transverse to the direction of glacier flow. Longitudinal foliation normally occurs at the ice margin and parallel to the direction of glacier flow. Most accumulation basins are wider than the channel which drains them; as a consequence of this, primary ice stratification is compressed and folded to form longitudinal folds or foliation. Transverse or arcuate foliation develops in regions with transverse crevasses or **ice falls**. Ice falls occur where a glacier descends a steep slope and consists of a densely crevassed region on the surface of the glaciers. Crevasses open as ice enters the ice fall, due to the extensional stress caused by flow acceleration in the ice fall. These crevasses then close at the base of the ice fall as extensional stresses change to compression. Former crevasses can be traced by the presence of different types of ice crystal which reflect light in different ways. These crevasse traces are deformed into an arcuate pattern by flow, extending further downstream in the centre of the glacier than at the sides. On some glaciers these crevasse traces become depicted by alternating bands of light and dark snow, known as **ogives**. It has been suggested that each pair of dark and light ice bands represents a year's movement through the ice fall. The darker bands result from the concentration of dust and debris into crevasses during the summer months.

Source: Hambrey, M. J. & Müller, F. 1978. Ice deformation and structures in the White Glacier, Axel Heiberg Island. *Journal of Glaciology* 25, 215–228.

surface expression of a vertical debris septa within the glacier which may extend to the glacier bed (Figure 7.3).

Ablation-dominant medial moraines form where ridges of englacial debris are revealed down-glacier by surface melting. Such moraines appear to grow out of the glacier surface in the ablation zone. This type of medial moraine may form in two ways. First, the supply of debris, from a rockwall or nunatak, onto a small area of a glacier may form a linear debris plume down-glacier of that point. In the accumulation zone, this debris will slowly descend through the glacier in the direction

Figure 7.2   *Medial moraines on Pedersenbreen, Svalbard [Photograph: M. R. Bennett]*

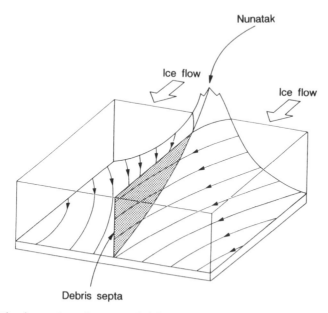

Figure 7.3   *The formation of a vertical debris septa and medial moraine by the conflu-ence of two glaciers*

of flow as it becomes buried by snowfall. In the ablation zone, melting will reveal this debris plume as a medial moraine on the glacier surface.

The second method by which ablation-dominant medial moraines may form is through the folding of the debris-rich ice stratification. In the accumulation zone, debris may collect as a diffuse layer over much of the glacier's surface, where it will become buried by snowfall to form debris strata. These debris strata may become folded across the glacier during ice flow, particularly if the ice flows into a restricted channel (Figures 7.4 and 7.5). The axes of these folds are usually parallel to the direction of ice flow. When the debris-rich ice reaches the ablation zone, surface melting will lower the glacier surface to reveal the anticlinal crests of the longitudinal debris-rich folds. If the folding is relatively open then a series of small medial moraines may each define the axis of a fold. However, if the folding is intense then the debris in the individual folds may merge to form a single medial ridge (Figure 7.4). The intensity of folding is determined by the amount of transverse compression within the glacier. For example, if the glacier flows from a wide accumulation basin through a narrow trough then the level of compression and therefore the

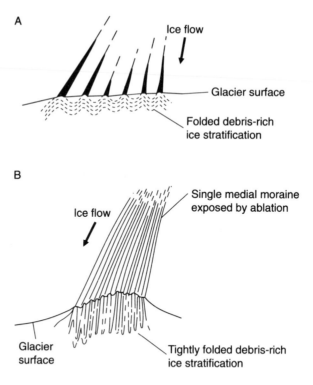

Figure 7.4   *The formation of medial moraines by the folding of debris-rich ice stratification. A diffuse debris layer is folded and thereby concentrated during glacier flow to form a medial moraine. The fold axes (ice foliation) are parallel to the flow direction and the folding results from transverse compression within the glacier usually caused by ice flowing from a broad accumulation basin into a narrow outlet. This model of medial moraine formation is based upon the unpublished work of M. J. Hambrey*

Figure 7.5 *Medial moraines on austra Lovénbreen, Svalbard, formed by the folding of debris-rich ice stratification. The folded structure can be seen in the foreground. The medial moraines give rise to thin bands of lithologically distinct debris on the glacier forefield, which are visible in the middle distance. [Photograph: M. R. Bennett]*

intensity of folding will be high. In practice it is important to recognise that the englacial and supraglacial debris structure of any glacier may be very complex and is a product of the deformation history of the ice. Debris may be deposited on the surface of a glacier in the accumulation zone and may be redistributed and concentrated by creep, folding and thrusting of the glacier ice during flow.

The debris structure on the surface of a glacier is also affected by the rate of debris supply relative to the glacier's flow velocity. If the rate of supply is high and the glacier's velocity low, a thick layer of debris will result. However, if the flow velocity, and therefore the rate of down-glacier transport is high, then the debris cover will be thinner. The presence of debris on a glacier surface has an important influence on its ablation dynamics. Supraglacial debris acts as insulation and will retard surface melting on a glacier. Thick patches or ridges of supraglacial debris may be associated with large ice-cored mounds and ridges on the glacier surface. Melting of the debris-free ice occurs either side of the debris, but is retarded beneath the cover of supraglacial debris. This process is particularly important in the development of ice-cored moraines, where blocks of glacier ice become buried beneath debris and are left as the glacier retreats (see Section 9.1.3).

Debris transported at a high level within the glacier retains its primary characteristics and is unaltered by glacial transport. Consequently, it is typically angular, coarse and contains little in the way of fine material (Figures 7.6 and 7.7). In fact, the characteristics of debris transported at a high level within a glacier are similar to those of talus or scree, which reflects the common origin of both as rockfall debris.

7.2 LOW-LEVEL DEBRIS TRANSPORT

Low-level debris transport involves the transport of debris at the base of the glacier. The material may be derived from the bed by subglacial entrainment or indi-

Figure 7.6   *Supraglacial debris on the surface of a valley glacier in the European Alps. Note the coarse angular nature of the debris. [Photograph: A. F. Bennett]*

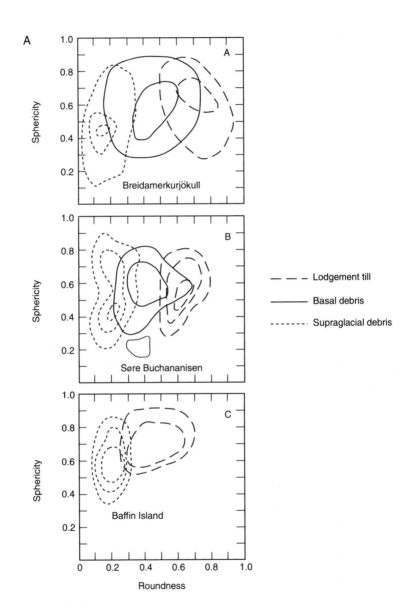

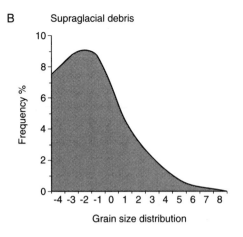

Supraglacial debris

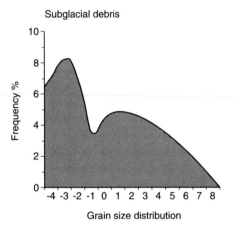

Subglacial debris

Figure 7.7 *Transport pathways and debris characteristics. **A:** Particle shape—a series of plots of Krumbein particle roundness versus Krumbein particle spherictity for debris transported in different ways within a glacier. Note the difference between supraglacial, subglacial and lodged debris. The lodged debris is more rounded than the subglacial debris from which it was deposited because of continued post-depositional clast modification. **B:** Particle size—typical grain-size distributions for supraglacial, high-level transported debris, and subglacial, low-level transported debris. The grain-size units are phi units where zero equals 1 mm, positive values indicate smaller particles and negative values larger ones. [Modified from: **A:** Hambrey (1994) Glacial Environments, UCL, Fig. 1.4, p. 21. **B:** Boulton (1993) In: Duff (Ed.) Holmes' Principles of Physical Geology, Chapman & Hall, Fig. 20.36, p. 425]*

rectly from debris which falls onto the glacier surface and makes its way to the bed via crevasses and the downward movement associated with extending flow or basal melting (Figure 7.1). This debris may remain at the base of the glacier until it is deposited or reaches the glacier snout. Alternatively, it may be elevated to the surface by upward flow or by thrusting associated with glacier compression, a phenomenon typical of some snout regions. Debris transported in this way is altered by glacial transport. The process of entrainment of basal debris has been discussed in Section 5.2.2.

Once entrained, basal debris is concentrated both laterally and vertically. The vertical extent of the basal debris may be increased by folding of debris-rich basal ice. Laterally debris may become concentrated into streams around bedrock obstacles. As a uniform debris-rich basal ice layer approaches a bedrock obstacle, its flow accelerates due to enhanced creep around the sides of the obstacle, such that little debris-rich ice is carried over the top of the obstacle (Figure 7.8). It is possible, therefore, by ice streaming around a field of such obstacles, to develop a complex and laterally concentrated basal sediment layer.

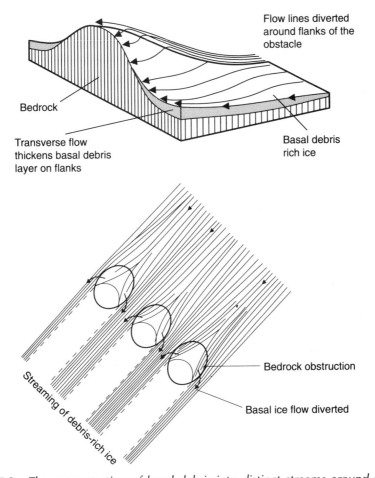

Figure 7.8   *The concentration of basal debris into distinct streams around subglacial bedrock obstacles. [Modified from: Boulton (1974) In: Coates (Ed.)* Glacial Geomorphology, *George Allen & Unwin, Fig. 12, p. 62]*

The basal transport zone can be divided into two subzones: (1) a **zone of traction** in which debris is moved along the bed of the glacier; and (2) a **zone of suspension** in which debris is transported immediately above the glacier base. The transfer of debris between these two zones is controlled primarily by pressure melting and regelation. Pressure melting on the up-ice side of an obstacle causes particles to move downwards, while the freezing-on of ice and debris on the down-ice side causes clasts to move upwards. Folding, particularly against bedrock steps, may also cause particles to move up from the zone of traction.

Particles in transport within the zone of traction experience considerable modification through the processes of crushing and abrasion (**commutation**). They are, therefore, typically more spherical, rounded and usually have a bimodal or multimodal grain-size distribution (Figure 7.7B). The grain-size distribution is normally composed of three separate populations: (1) large rock particles, **lithic fragments**; (2) mineral grains produced by crushing of the rock fragments; and (3) sub-mineral-sized particles produced by the abrasion of mineral grains (Figure 7.9). The principal size modes are composed of the lithic fragments and the mineral grains. In an ideal environment in which debris is not removed or added then the relative

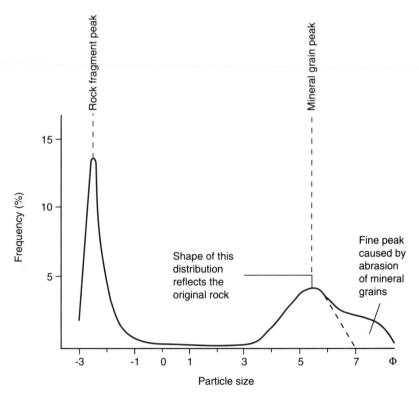

Figure 7.9   *The grain-size distribution of subglacially transported debris is composed of three distinct populations: (1) lithic or rock fragments, (2) mineral grains produced by crushing of the rock fragments, and (3) fines produced by the abrasion of individual mineral grains*

importance of these two modes should change as the transport distance increases: the fine mineral mode should grow at the expense of the coarse mode (Figure 7.9) (Box 7.2). In practice this is rarely observed due to the constant addition of new material.

Roundness should also increase with transport distance as the corners of a particle are blunted and smoothed off. Particle abrasion is responsible for increasing roundness. However, observations suggest that roundness does not increase indefinitely, but reaches a **terminal roundness**. This reflects the fact that rock particles are constantly being crushed, which increases their angularity. The degree of roundness a particle can achieve will be controlled by the length of time between crushing events, which will depend on its strength and the force applied. The stronger or more resistant a particle is to crushing, the more rounded it may become. Particles transported at the base of a glacier are also characterised by faceted and striated surfaces which often give the clast a bullet-shaped appearance (Figure 7.10).

Particles in transport within basal ice also develop a strong **particle fabric**. Elongated particles become aligned with the direction of ice flow. This is true of any elongated particles immersed within a fluid. First a preferred orientation develops in the direction of flow, and subsequently particles may also become orientated in a direction perpendicular to flow. This reflects the fact that the most effi-

Figure 7.10 *A basal clast showing faceting and striations. Note the bullet shape.*
[Photograph: M. R. Bennett]

BOX: 7.2 GRAIN-SIZE DISTRIBUTIONS AND TRANSPORT DISTANCES

Dreimanis and Vagners (1971) argued on the basis of crushing experiments that the relative magnitude of the two grain-size peaks of subglacial sediment should change with transport distance. As the distance increases, the finer mode, composed predominantly of mineral grains, should grow at the expense of the coarser mode, composed of predominantly lithic fragments. They demonstrated this by examining the grain-size distributions of till samples in

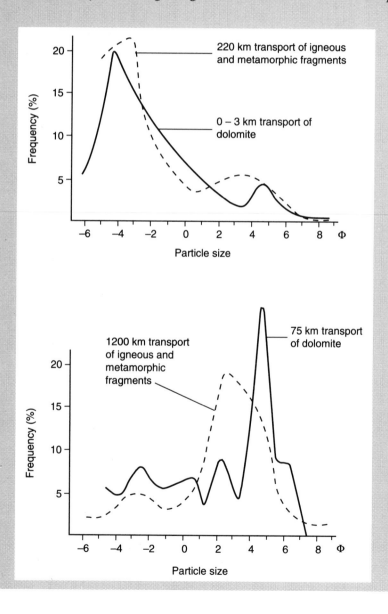

the Hamilton Niagara region of south-east Canada. The grain size of only one clast lithology was analysed in order to keep the results relatively simple. The results show that with increased transport distance the finer mode increases in importance, although the amount of transport necessary depends on the lithology.

In practice, when multiple lithologies are considered the patterns are much more complex. Moreover, the introduction of new material may complicate this picture such that in very few cases can this type of pattern be recorded.

Source: Dreimanis, A. & Vagners, U. J. 1971. Bimodal distribution of rock and mineral fragments in basal tills. In: Goldthwait, R. P. (Ed.) *Till: A Symposium*, Ohio State University Press, Ohio, 237–250. [Diagram modified from: Dreimanis & Vagners (1971) In: Goldthwait, R. P. (Ed.) *Till: A Symposium*. Ohio State University Press, Ohio, Fig. 4, p. 243]

cient orientation (the one giving the least resistance) is one in which the long axis is parallel to the direction of flow. This property is of particular importance since, as we will see in the next chapter, glacial sediments may inherit this particle fabric.

7.3 DEBRIS TRANSFER BETWEEN LOW- AND HIGH-LEVEL TRANSPORT

The transfer of basal debris into an englacial or supraglacial position is of particular importance in determining the depositional processes which operate within a glacier (see Section 8.1.8). Debris transfer occurs to some extent at most glacier snouts due to compressive flow in the ablation area which facilitates the upward flow of debris-rich ice. However, this process is facilitated by the development of **glacial thrusts** within some glaciers which may transfer basal debris into an englacial or supraglacial location. Thrusts, sometimes referred to as **shear planes**, develop in three types of situation: (1) in glaciers with a mixed or polythermal basal thermal regime; (2) where glaciers flow against large subglacial bedrock scarps; and (3) in surging glaciers.

They are most common in glaciers with a mixed or polythermal thermal regime. The zone of thermal transition from warm- to cold-based ice is associated with strong compressional stress. The warm-based ice slides over its bed, but the cold-based ice does not. This causes a deceleration in the rate of basal ice flow, which generates intense compressional stress usually beyond that which ice creep can accommodate. As a consequence of this, thrusts develop. These thrusts form at the base of the glacier and allow basal ice to move up and over the colder ice in front (Figures 7.11 and 7.12). Thrusts may penetrate throughout the thickness of a glacier or terminate englacially as **blind thrusts**. Thrust planes often exploit structural weaknesses within the ice such as the line or trace of former crevasses. They may occur singularly and extend considerable distances across an ice margin (Figure

Figure 7.11   *Debris-rich thrusts in the surge-type glacier Kongsvegen, Svalbard. Glacier flow is from left to right. The thrust is blind, that is it does not reach the glacier surface and was formed during a surge in 1948. The debris within the thrust varies from a few millimetres to over 0.5 m thick. The melt-out of debris from the thrust has stained the ice beneath. [Photograph: M. R. Bennett]*

7.12), or alternatively occur in anastomosing networks in which each thrust has a short arcuate outcrop on the surface of the glacier. The angle between the glacier bed and the thrust plane is generally low (30°–40°), although high-angle thrusts

Figure 7.12   *Large linear thrust outcropping on the right-hand side of the glacier margin. Outlet glacier of the Greenland Ice Sheet, southern Greenland. [Photograph: N. F. Glasser]*

occur within surge-type glaciers (40°–80°). Thrusting is a particularly important mechanism within surging glaciers.

Debris is incorporated into thrusts in a variety of different ways and may vary from thin layers of debris-rich ice to large rafts of frozen sediment, which were frozen to the glacier bed prior to its movement up a thrust plane (Figure 7.13).

The presence or absence of thrusts within a glacier is therefore important to its debris structure. Thick englacial and supraglacial debris concentrations result from thrusting and consequently have important implications for the release of debris from the glacier. In warm-based glaciers thrusts are comparatively rare and debris transport is dominated by high-level transport. Thick basal debris layers are unusual, since basal melting prevents their development. Only small quantities of debris are incorporated englacially and most of the supraglacial debris at an ice margin has the characteristics of high-level transport. In contrast, mixed regime or polythermal glaciers often have well-developed thrust planes. Thick englacial debris concentrations result and the supraglacial debris at the glacier snout contains a mixture of debris transported both at a high and low level within the glacier. The subglacial debris layer is also often thick due to freezing-on of debris in zones of thermal transition between warm and cold ice. Glaciers which are completely cold-based throughout are dominated by high-level debris transport. Subglacial debris layers are thin if the glacier is completely frozen to its bed.

Figure 7.13 *Debris within englacial thrusts and fractures.* **A:** *Raft of frozen fine-grained gravel with a thrust at the margin of Pedersenbreen, Svalbard. Melt-out of the debris has mantled the glacier surface to form an ice-cored mound centred on the debris-rich thrust [Photograph: M. R. Bennett].* **B** *and* **C:** *Dykes of frozen sediment within englacial fractures on the surface of Kongsvegen, Svalbard. [Photograph: M. R. Bennett]*

7.4 SUMMARY

There are two transport pathways for debris within a glacier: a high-level transport route (supraglacial and englacial) and a low-level or basal pathway (subglacial). The transport pathway followed determines the properties of the glacial debris that results. High-level transport does not involve modification of the debris, while debris moving at the base of the glacier is strongly modified. Basal debris is typically more rounded, spherical and finer than debris which does not come into contact with the glacier bed. Basal debris often has a multimodal or bimodal grain-size distribution, composed of a coarse mode of lithic fragments and a finer mode of mineral-sized grains. The long axis of debris in basal transport tends to become preferentially orientated in the direction of ice flow. The characteristics of debris transported either at a high or low level within a glacier are sufficiently different for them to be used by glacial geologists to infer the transport history of glacial debris. This is particularly important in the analysis of ancient glacial deposits (Box 7.3).

BOX: 7.3: DETERMINING THE TRANSPORT HISTORY OF ANCIENT GLACIAL DEBRIS

In north-eastern Svalbard there is a well-preserved sequence of tillites (lithified till) of Late Precambrian age. These tillites are of uncertain origin. Dowdeswell *et al.* (1985) used the typical sediment characteristics of modern glacial sediments to interpret these ancient sediments. In particular they used particle fabric (orientation), but also made reference to particle shape. On the basis of particle fabric they were able to suggest that most of the tillites were either flow tills or glaciomarine diamicton. From the particle shape analysis they were able to determine that most of the clasts within the tillites had undergone some form of basal transport, being typically well rounded, striated and spherical. The relative absence of clasts with supraglacial characteristics led them to suggest that the supply of supraglacial sediment to the glacier that deposited the tillites in the Late Precambrian may have been very small. This would suggest that the glacier which deposited them had the dimensions of an ice cap or ice sheet. One would expect a much higher proportion of supraglacial clasts if the tillites had been deposited in an ice-field or by a valley glacier overlooked by large mountains providing a large supraglacial debris source area. The work illustrates how simple sedimentological observations of particle shape can be used to make important inferences about the transport history of glacial debris and therefore about ancient ice sheets.

Source: Dowdeswell, J. A., Hambrey, M. J. & Wu, R. 1985. A comparison of clast fabric and shape in Late Precambrian and modern glacigenic sediments. *Journal of Sedimentary Petrology* **55**, 691–704.

7.5 SUGGESTED READING

The key paper on this subject is that of Boulton (1978), which quantified for the first time the difference between glacial sediments that have and have not been in contact with the glacier bed. This work has been reproduced several times since (e.g. Ballantyne 1982) and has been used to determine the transport history of both ancient and modern glacial sediments (e.g. Dowdeswell *et al.* 1985; Benn & Ballantyne 1994). The development of particle fabric within ice is discussed in the classic paper by Glen *et al.* (1957). Changes in basal particle characteristics with transport are discussed in papers by Dreimanis & Vagners (1971), Drake (1972) and Humlum (1985). The work of Eyles & Rodgerson (1978) and Small (1987) provide useful information about the formation of medial moraines. The tectonic structure of glaciers is covered by Hambrey (1975), Hambrey & Milnes (1977) and Hambrey & Müller (1978).

Ballantyne, C. K. 1982. Aggregate clast form characteristics of deposits near the margins of four glaciers in the Jotunheimen Massif, Norway. *Norsk Geografiska Tidsskrift* **36**, 103–113.

Benn, D. I. & Ballantyne, C. K. 1994. Reconstructing the transport history of glacigenic sediments: a new approach based on the co-variance of clast form indices. *Sedimentary Geology* **91**, 215–227.

Boulton, G. S. 1978. Boulder shapes and grain-size distributions of debris as indicators of transport paths through a glacier and till genesis. *Sedimentology* **25**, 773–799.

Dowdeswell, J. A., Hambrey, M. J. & Wu, R. 1985. A comparison of clast fabric and shape in Late Pre-Cambrian and Modern glaciogenic sediments. *Journal of Sedimentary Petrology* **55**, 691–704.

Drake, L. D. 1972. Mechanisms of clast attrition in basal till. *Geological Society of America Bulletin* **83**, 2159–2166.

Dreimanis, A. & Vagners, U. J. 1971. Bimodal distribution of rock and mineral fragments in basal tills. In: Goldthwait, R. P. (Ed.) *Till: A Symposium*, Ohio State University Press, Ohio, 237–250.

Eyles, N. & Rogerson, R. J. 1978. A framework for the investigation of medial moraine formation: Austerdalsbreen, Norway, and Berendon Glacier, British Columbia, Canada. *Journal of Glaciology* **20**, 99–113.

Glen, J. W., Donner, J. J. & West, R. G. 1957. On the mechanism by which stones in till become orientated. *American Journal of Science* **255**, 194–205.

Hambrey, M. J. 1975. The origin of foliation in glaciers: evidence from some Norwegian examples. *Journal of Glaciology* **14**, 181–185.

Hambrey, M. J. & Milnes, A. G. 1977. Structural geology of an Alpine glacier (Griesgletscher, Valais, Switzerland). *Eclogae Geologicae Helvetiae* **70**, 667–684.

Hambrey, M. J. & Müller, F. 1978. Ice deformation and structures in the White Glacier, Axel Heiberg Island. *Journal of Glaciology* **25**, 215–228.

Humlum, O. 1985. Changes in texture and fabric of particles in glacial traction with distance from source, Myrdalsjökull, Iceland. *Journal of Glaciology* **31**, 150–156.

Small, R. J. 1987. Englacial and supraglacial sediment: transport and deposition. In: Gurnell, A. M. & Clark, M. J. (Eds) *Glacio-fluvial Sediment Transfer*. John Wiley & Sons, Chichester, 111–145.

8
Glacial Sedimentation on Land

Glaciers that terminate on land may deposit sediment in two ways: either directly or via meltwater. Direct glacier deposition is restricted to the immediate vicinity of the glacier while meltwater may carry sediment beyond the glacier. Traditionally, glacial sediments are divided into those which are non-sorted, formed by direct glacier sedimentation, and those which are sorted or stratified, deposited by meltwater. In practice this is a rather artificial distinction since meltwater deposits are not always stratified. In this chapter we shall first consider the products of direct glacier sedimentation and the deposition by meltwater.

8.1 DIRECT GLACIAL SEDIMENTATION

Debris deposited directly by a glacier is known as **till**. A till may be defined as a sediment deposited by glacier ice, but one that has not been disaggregated, although it may have suffered glacially induced flow either in the subglacial or supraglacial environment. Till is a **diamicton**, which is a non-genetic term for a non-sorted or poorly sorted unconsolidated sediment that contains a wide range of particle sizes (Figure 8.1). It normally consists of large pebbles, cobbles or boulders, referred to generally as **clasts**, set within a fine-grained matrix of silt and clay. Its characteristics are, however, highly variable.

There are four primary processes by which debris in transport within a glacier may be deposited:

1. **Lodgement**, which occurs when the frictional resistance between a clast in transport at the base of a glacier and the glacier bed exceeds the drag imposed by the overlying ice such that the clast ceases to move.
2. **Melt-out**, i.e. the direct release of debris by melting.
3. **Sublimation**, vaporisation of ice causing the direct release of debris.

Figure 8.1 *A typical glacial till. **A:** A coastal section at Glanllynnau, North Wales, where a flow till unit occurs above a lodgement till. In places these two till units are separated by sands and gravels to form a tripartite sequence. **B:** Close-up of a lodgement till. [Photographs: M. R. Bennett]*

4. **Subglacial deformation**, which involves the assimilation of sediment into a deforming layer beneath a glacier (Box 3.3).

The process of sublimation is currently only well documented from Antarctica and occurs due to the extreme cold and aridity of this environment.

Together these processes combine to give rise to six types of till: (1) **lodgement till**, (2) **subglacial melt-out till**, (3) **deformation till**, (4) **supraglacial melt-out (moraine) till**, (5) **flow till**, and (6) **sublimation till**. The formation and characteristics of each of these deposits is discussed below.

8.1.1 Lodgement Till

Lodgement till results from the subglacial lodgement of debris in basal traction beneath a glacier. It can occur in one of three ways: (1) direct lodgement of debris in traction over the glacier bed; (2) basal melting and debris release; or (3) deposition in basal cavities.

1. **Direct lodgement.** In order to understand this process it is best to first consider a single clast in transport at the base of a glacier. A clast in transport at the base of a glacier need not move forward at the same speed as the basal ice: the ice may flow around the particle as it transports it. The particle will lodge (i.e. stop moving) when its forward velocity is reduced to zero. This will occur whenever the friction between the particle and the bed exceeds the drag on the particle imposed by the ice flowing over it. At this point the ice will simply flow around the particle without moving it forward. A particle may, therefore, lodge beneath flowing ice. Figure 8.2 shows several ways in which the velocity of an individual particle or mass of particles may be reduced to zero.

    According to Boulton's model of glacial abrasion, lodgement forms part of a continuum with erosion (see Section 5.1.1). Lodgement occurs when the effective normal pressure is sufficiently high to prevent a clast from moving forward (Figure 5.2). Effective normal pressure increases the friction between the clast and the bed until it reduces the forward velocity of the clast to zero. In this model lodgement is favoured by: (1) an increase in ice thickness, which will increase the effective normal pressure; (2) a fall in basal water pressure; and (3) a fall in ice velocity. The significant aspect of this model is that lodgement can occur under fast-flowing ice.

    The alternative view advocated by Hallet, in which basal clasts are envisaged as essentially floating within the ice, considers abrasion and lodgement to be two distinct processes. According to Hallet, lodgement is independent of effective normal pressure and lodgement will only occur at: (1) low ice velocities; and (2) where the rate of basal melting is high. In this model lodgement cannot occur under fast-flowing ice. As suggested in Section 5.1.4, both models may represent equally valid subglacial conditions, although it is important to note that many of the structures present within lodgement till are indicative of deposition under active ice.

2. **Subglacial melting.** Sediment is supplied to the sole of the glacier and released by basal melting. It may then be transported subglacially before it finally lodges. Subglacial melting is principally driven by geothermal heat, although heat generated by the internal deformation of ice may also be important. Geothermal heat tends to be concentrated towards depressions and hollows in the glacier bed. Consequently, debris will be released preferentially into the hollows within the glacier bed. The rate of basal melting is determined by: (1) the geothermal heat flux; (2) the amount of frictional heating, which increases with ice velocity towards the equilibrium line; (3) ice thickness, increasing the thickness of a glacier may increase basal ice temperature; (4) the rate of advection of cold or warm snow; and (5) ice surface temperatures. This is discussed in more detail in Section 3.4.

3. **Cavity deposition.** Subglacial cavities are known to occur where glaciers flow over irregular bedrock surfaces. If the sliding velocity of a glacier is high, its thickness small and the bedrock obstacles acute, then large cavities may form in their lee. Sediment can accumulate within these cavities in the following ways:

A. Roof melting: melting of ice from the roof of the cavity may release debris (Figure 8.3A).
B. Till slurry: basal debris may enter as a slurry squeezed from between the ice bed interface (Figure 8.3B).
C. Clast expulsion: a large clast moving along the ice–bedrock interface will experience downward pressure, pushing the clast against the bed. If the clast moves over a cavity this pressure will be unopposed due the absence of the bed and may cause the clast to be pushed out of the ice roof (Figure 8.3D).
D. Till-curls: as debris-rich ice moves over the roof of a basal cavity, pressure release may cause it to peel away from the cleaner ice above to form till-curls which fall onto the cavity floor (Figure 8.3E).

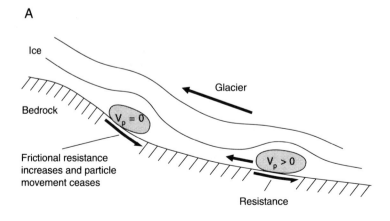

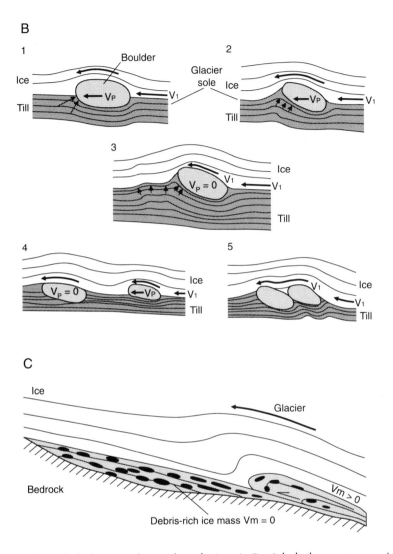

Figure 8.2   *Particle lodgement beneath a glacier. **A:** Particle lodgement on a rigid substratum. The clast in transport stops moving when the frictional resistance between it and the bed exceeds the drag imposed on the clast by the flowing ice. **B:** Particle lodgement on a soft substratum. Clasts plough through soft sediment and will be stopped when the sediment ploughed up in front of them provides sufficient resistance to retard forward movement. Subsequently other clasts may jam against the first, and in this way boulder pavements or concentrations may form. **C:** Lodgement of a debris-rich ice mass. A debris-rich ice body may lodge beneath a glacier when the frictional resistance between it and the bed exceeds that of the ice above, which shears over the debris-rich ice mass*

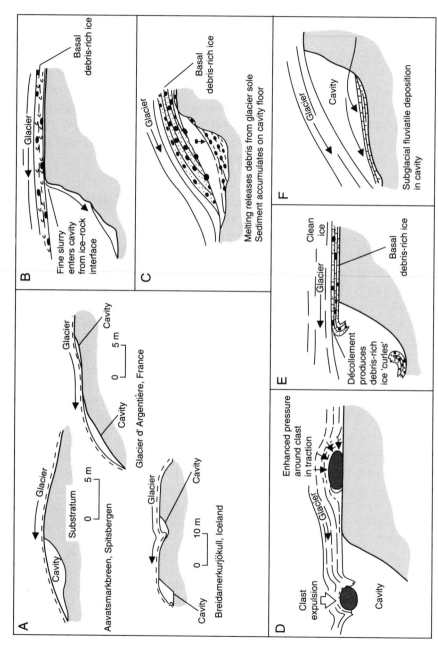

Figure 8.3 *Subglacial cavity occurrence (A) and observed mechanisms (B–F) by which debris accumulates on the floor of subglacial cavities. [Modified from: Boulton (1982) In: Davidson-Arnot, R. et al. (Eds) Research in Glacial, Glacio-fluvial and Glacio-lacustrine Systems, Geo Books, Fig. 1, p. 4]*

E. Fluvial deposition: meltwater flow may deposit sediment within basal cavities. Subsequent cavity closure may remould this sediment to give a till (Figure 8.3F).

All these processes may lead to deposition within basal cavities. The properties of sediment deposited in such cavities may vary considerably from those of sediment deposited by direct lodgement.

In summary, lodgement till may form by: (1) direct lodgement; (2) basal melting; and (3) deposition in subglacial cavities. These processes leave an imprint on lodgement tills which gives them distinct characteristics.

The character and properties of lodgement till reflect the nature of its deposition beneath active flowing ice. The clasts within lodgement tills are typical of basal transport with rounded edges, spherical form, and striated and faceted faces (see Section 7.2). These characteristics are usually more pronounced than for basal debris within a glacier, which reflects the continued post-depositional modification of clasts (Figure 7.7A). Bullet-shaped clasts are common (Figure 7.10). The particle-size distribution is also typical of basal debris transport, being either bimodal or multimodal. The sediment is dominated by the presence of a fine-grained matrix into which larger clasts are set. These clasts have a strong particle orientation or fabric, in which elongated particles are aligned closely with the direction of local ice flow. The clasts are usually dominated by local rock types; many subglacial clasts are deposited almost as quickly as they are entrained. Lodgement tills are typically dense and well-consolidated sediments. They are also usually massive or structureless and may contain well-developed shear planes or foliations. Sheared or brecciated clasts, **smudges**, may be present if the clast lithology is weak. Till is often squeezed into bedding planes and joints at the till–rock interface, known as the **rock-head**. The lower contact of the till units may also infill grooves and striations cut by erosion in the rock-head prior to the deposition of the till. Boulder clusters or pavements may occur within the sediment along with evidence of ploughing of clasts (Figure 8.2). These characteristics are summarised in Table 8.1.

8.1.2 Subglacial Melt-out Till

Subglacial melt-out till forms by the direct release of debris from a body of stagnant debris-rich ice by melting of the interstitial ice. Till formation by subglacial melt-out has been widely postulated in the past as an important till-forming process, but recent work has suggested that it may be only of local importance. This process is driven primarily by geothermal heat, since the ice is not deforming and no heat is generated by friction. Bodies of stagnant ice may occur in a variety of situations. For example, they may occur when ice buried by surface debris becomes isolated and detached from the main body of the glacier, or they may occur in the vicinity of bedrock obstacles where fresh active ice is thrust over stagnant ice trapped

Table 8.1 Summary table of the main sedimentary characteristics of the main types of till

A	Lodgement till	Subglacial melt-out till	Deformation till
Particle shape	Clasts show characteristics typical of basal debris transport: rounded edges, spherical form, and striated and faceted faces. Large clasts may have a bullet-shaped appearance	Clasts show characteristics typical of basal transport, being rounded, spherical, striated and faceted. These characteristics are less pronounced than those of lodgement till	Dominated by the sedimentary characteristics of the sediment which is being deformed, although basal debris may also be present
Particle size	The particle size distribution is typical of basal debris transport, being either bimodal or multimodal	The particle size distribution is typical of basal debris transport, being either bimodal or multimodal. Sediment sorting associated with dewatering and sediment flow may be present	Diverse range of particle sizes reflecting that found in the original sediment. Rafts of the original sediment may be present, causing marked spatial variability
Particle fabric	Lodgement tills have strong particle fabrics in which elongated particles are aligned closely with the direction of local ice flow	Fabric may be strong in the direction of ice flow, although it may show a greater range of orientations than that typical of lodgement till	Strong particle fabric in the direction of shear; this may not always be parallel to the ice flow direction. High-angle clasts and chaotic patterns of clast orientation are also common
Particle packing	Typically dense and well-consolidated sediments	The sediment may be well packed and consolidated, although this is usually less marked than in a lodgement till	Densely packed and consolidated
Particle lithology	Clast lithology is dominated by local rock types	Clast lithology may show an inverse superposition	Diverse range of lithologies reflecting that present within the original sediments
Structure	Massive structureless sediments, with well-developed shear planes and foliations. Sheared or brecciated clasts (smudges) may be present. Boulder clusters or pavements may occur within the sediment, along with evidence for ploughing of clasts	Usually massive but if it has been subject to flow it may contain folds and flow structures. Crude stratification is sometimes present. The sediment does not show evidence of shearing and overriding during formation	Fold, thrust and fault structures may be present if the level of shear homogenisation is low. Rafts of undeformed sediment may be included. Smudges (brecciated clasts) may also be present

Table 8.1 *Continued*

B	Supraglacial melt-out (moraine) till	Flow till	Sublimation till
Particle shape	Usually dominated by sediment typical of high-level transport, but subglacially transported particles may also be present. The majority of clasts are not normally striated or faceted	Broad range of characteristics, but dominated by particles which are angular and have a non-spherical form. The majority of clasts are not striated or faceted	Clasts typical of basal transport, being rounded, spherical, striated and faceted
Particle size	The size distribution is typically coarse and unimodal. Some size sorting may occur locally where meltwater reworking has occurred	The size distribution is normally coarse and unimodal, although locally individual flow packages may be well sorted	The particle size distribution is typical of basal debris transport, being either bimodal or multimodal
Particle fabric	Clast fabric is unrelated to ice flow, is generally poorly developed and is spatially highly variable	Variable particle fabric; individual flow packages may have a strong fabric, reflecting the former palaeo-slope down which flow occurred	Strong in the direction of ice flow, although it may show a greater range in orientation than a typical lodgement till
Particle packing	Poorly consolidated, with a low bulk density	Poorly consolidated with a low bulk density	Typically has a low bulk density and is loose and friable
Particle lithology	Clast lithology is usually very variable, and may include far-travelled erractics	Variable, but may include far-travelled erractics	Clast lithology may show an inverse superposition
Structure	Crude bedding may occur but generally it is massive and structureless	Individual flow packages may sometimes be visible. Crude sorting, basal layers of tractional clasts, may be visible in some flow packages. Sorted sand and silt layers may be common, associated with reworking by meltwater. Individual flow packages may have erosional bases. Small folds may also be present	The deposit is usually stratified and may preserve englacial fold structures

around obstacles close to the ice margin. Large areas of stagnant ice may also develop in the low-angle ice lobes produced during glacier surges.

Subglacial melt-out may involve the lowering of debris onto the bed with little modification. Alternatively, if saturated, it may flow beneath the weight of ice into low-pressure areas such as basal crevasses and abandoned subglacial tunnels. The degree of flow and disturbance during melting will determine the degree to which the till retains the particle orientation (fabric) of the debris-rich ice from which it melted.

The clasts within subglacial melt-out till are typical of basal transport and are rounded, spherical, striated and faceted. These clast characteristics are similar to those of basal debris within ice, but are normally less pronounced than those in lodgement till since deposition occurs under stagnant ice and little post-depositional modification of clasts therefore takes place. The particle size distribution is also typical of basal debris transport, being either bimodal or multimodal. Sediment sorting associated with the release of water during melt-out, **dewatering**, and sediment flow may be present. Provided that subglacial flow or free fall of debris did not occur during melt-out of the basal ice, it may be strongly orientated in the direction of ice flow, although there is usually a greater range or spread of orientations around the mean than for a typical lodgement till due to the disturbance of fabric during melt-out. If the sediment has, however, undergone subglacial flow its fabric will not reflect the direction of ice flow. Subglacial melt-out tills may be well packed and consolidated, although this is usually less marked than in a lodgement till. Clast lithology usually shows an inverse superposition with the youngest, least-travelled clast occurring at the base of a deposit and the oldest, furthest-travelled clasts at the top. Subglacial melt-out till is usually massive but if it has been subject to flow it may contain folds and flow structures formed at the time of deposition. Crude stratification inherited from the basal ice is sometimes present if melt-out has occurred in an undisturbed fashion. The sediment does not show evidence of shearing and overriding during formation, although it may have been deformed subsequently. These characteristics are summarised in Table 8.1.

8.1.3 Deformation Till

We have already seen in Section 3.3.3 how sediments may form a mobile deforming layer beneath glaciers flowing over soft substrates (Box 3.3). This layer may consist of soft preglacial sediment which has been overrun and deformed, or alternatively it may consist of glacial sediment, lodgements tills and glaciofluvial sediments, which may either pre-date or date from the current phase of glaciation.

Glaciotectonic deformation takes place whenever the stress imposed by the glacier exceeds the strength of the material beneath or in front of it. The material may be subject to both **brittle deformation** (faults, thrusts) and **ductile deformation** (folds). Deformation of this sediment proceeds in stages, depending upon the amount of deformation or shear the sediment is subjected to by the glacier above.

Figure 8.4 *Sediment deformation within fine-grained sands at the head of Loch Morar in the Scottish Highlands. The photograph shows a small fold which has been sheared. This type of structure is typical of relatively low levels of deformation. [Photograph: M. R. Bennett]*

At low levels of shear the sediment is simply folded and faulted (Figure 8.4). As the level of shear increases, these structures slowly become attenuated and the nose of folds may be detached from their core, **de-rooted**, to create **boudins**. Boudins are sausage-shaped blocks of less-ductile material surrounded by a more-ductile medium. In time they may become attenuated and drawn out at high levels of shear to form **tectonic laminations** (Figure 8.5). Sediments which experience very high levels of shear become completely mixed and homogenised. The product, therefore, of intense deformation is a homogeneous diamicton in which all the original sedimentary structures of the deposit have been destroyed. Diamictons produced by such deformation are referred to as **deformation, soft bed** or **assimilation tills**. Since the end product of subglacial deformation is a homogeneous till, the recognition of these deposits is particularly difficult, especially where the whole sequence has been homogenised or mixed to a similar level (Box 8.1).

Deposits of deformation till may accumulate by either: (1) the transport and accumulation of sediment within the subglacial deforming layer; or (2) down-cutting and the assimilation of new material into the deforming layer.

Sediment may be transported within the deforming layer beneath glaciers as they flow forward; a process sometimes referred to as **till advection**. Sediment flow occurs whenever the stress field imposed by the glacier exceeds the strength of the sediment: the greater the stress field relative to the material's strength, the

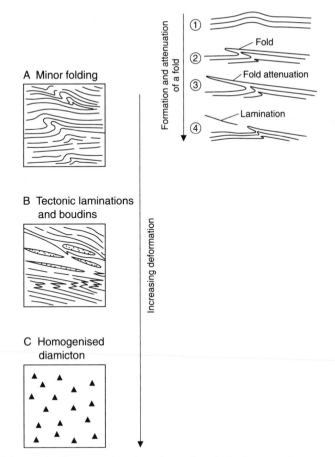

Figure 8.5 *Schematic diagram showing the style of glaciotectonic structures associated with different levels of subglacial deformation. The insert shows how a fold may be attenuated to form a tectonic lamination. [Modified from: Hart & Boulton (1991) Quaternary Science Reviews **10**, Fig. 14, p. 347]*

greater the rate of sediment flow within the deforming layer. Consequently, the rate of sediment transport in a deforming layer varies spatially in response to changes in: (1) basal shear stress; (2) effective normal pressure; and (3) variation in sediment properties. For example, the rate of transport will decrease if the sediment becomes more rigid, due to either the introduction of fresh basal debris, or an increase in the effective normal pressure.

   If the flow of deformable sediment into a given area equals the outflow of sediment from that area then deposition will not occur unless the geometry or dynamics of the glacier change. However, if more sediment flows into an area than out, for example from an area of rapid transport to one of little transport, then sediment accumulation will occur. If one considers a glacier flowing over a deforming horizon of uniform character zones, with rapid extending ice flow due to increasing basal shear stress down-ice, a zone of erosion due to transport by subglacial

BOX 8.1: AN EXAMPLE OF A DEFORMATION TILL

Eyles *et al.* (1994) describe the tills of the Holderness coast in Yorkshire, England. These tills have been traditionally interpreted as lodgement tills formed by the lodgement of individual particles beneath a glacier. The dark grey and brown till units exposed on the coast consist of a stony mud which contains up to 30% silt. It contains chalk clasts and 'erratic' inclusions of marine mud, containing fossils, and glauconitic sands. These erractics range from small pellets to large rafts that have been streaked-out or attenuated by shearing and deformation. Folds within the deposit are picked out by lines of chalk clasts, many of which have been brecciated, forming smudges. Some of the chalk clasts have steep dips and random orientations. The deposit also contains bullet shaped, striated and faceted clasts typical of basal debris.

Eyles *et al.* (1994) argue that these characteristics are not consistent with a lodgement till in which particles are lodged against a hard substrate, but instead reflect the reworking of marine muds. The sediment's character results from the partial mixing of marine sediment with basal debris, giving sheared and folded rafts of marine sediment as well as high-angle clasts. They argue therefore that these sediments provide a good example of a deformation till and provide evidence of widespread subglacial deformation. Eyles *et al.* (1994) suggest that this reworking or assimilation of marine sediment occurred beneath a surging lobe of the last main British Ice Sheet.

Source: Eyles, N., McCabe, A. M. & Bowen, D. Q. 1994. The stratigraphic and sedimentological significance of Late Devensian ice sheet surging in Holderness, Yorkshire, UK. *Quaternary Science Reviews* **13**, 727–759.

deformation will occur (output > input). In areas of compressional flow due to decreasing basal shear stress down-ice, a zone in which transport by subglacial deformation decreases and sediment accumulates will occur (output < input). For an ice sheet flowing over a deforming bed of uniform character this would result in a pattern of erosion and deposition by subglacial deformation like that shown in Figure 8.6, which corresponds to the large-scale pattern of extending and compressional flow within an ice sheet.

The thickness of the deforming layer may also be increased by down-cutting into the sediment pile beneath the deforming layer. In this way, new undeformed sediment is assimilated into the deforming layer.

The homogeneous character of many deformation tills and their formation beneath active ice has in the past caused them to be mistakenly interpreted as lodgement tills (Box 8.1). They are important because they introduce a range of non-glacial sediment into glacial deposits and may account for much of the diversity and variation in subglacial tills deposited by the former mid-latitude ice sheets.

The characteristics of deformation till are summarised in Table 8.1 and are described below. The clast shape in deformation till is dominated by the sedimentary characteristics of the sediment that is being deformed. Some of the debris may

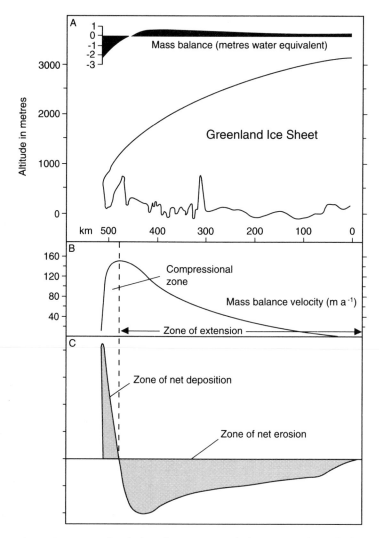

Figure 8.6 *Patterns of subglacial erosion and deposition by subglacial deformation within the Greenland ice sheet. [Diagram reproduced with permission from: Boulton (1987) In: Menzies, J. & Rose, J. (Eds) Drumlin Symposium, Balkema, Fig. 5, p. 37]*

not show any signs of glacial transport, although basal debris is usually incorporated and mixed into the deformation till. A diverse range of particle sizes reflecting that found in the original sediment will also be present and rafts of sediment may cause marked spatial variation in these properties through the deposit. Strong particle fabrics in the direction of shear are common in deformation till. This may not, however, always be parallel to the ice flow direction and will vary spatially through the deposit. Clasts with high dips and locally chaotic patterns of clast orientation are also diagnostic of deformation till. This reflects the complex stress fields and patterns of deformation within the deforming layer. Deformation tills are commonly well consolidated and contain a diverse range of clast lithologies reflecting the nature of the source material. Fold, thrust and fault structures may be present if the level of shear homogenisation is low. Boudins, de-rooted folds, and tectonic lamination may occur at high levels of shear. Rafts of undeformed sediment may be included within the till unit, along with deformed or 'kneaded' rafts. Smudges (brecciated clasts) may also be present.

8.1.4 Supraglacial Melt-out (Moraine) Till

Supraglacial melt-out till forms by the direct release of debris from debris-rich ice by melting of the surrounding or interstitial ice. This process is principally driven by solar radiation. A large thickness of debris may accumulate on the surface of a glacier. This debris may be confined to debris transported at a high-level within the glacier or alternatively may incorporate debris transported in basal ice if englacial thrusting occurs (see Section 7.3).

As the debris accumulates on the glacier surface it first accelerates melting, since dark surfaces absorb more heat than reflective ones, but then insulates the surface from further melting as the debris thickness increases. Variation in the thickness of debris causes variation in insulation and therefore variation in the surface ice topography. The debris becomes concentrated on ridges and mounds of buried ice. This debris is unstable and prone to slumping and surface redistribution. Due to the almost constant movement of debris on the glacier surface it rarely retains any of the characteristics of the debris-rich ice from which it is derived. Scree-like characteristics, crude bedding and downslope clast fabrics may develop due to the flow or fall of material down ice-cored slopes. Several different facies of supraglacial moraine till may be identified, depending on the thickness of the original supraglacial debris cover and upon the level of fluvial reworking (Box 8.2). The debris on a glacier surface is commonly concentrated into ice-cored ridges which may trace the outcrop of thrust planes or other debris-rich structures on the glacier surface (Figures 7.12 and 7.13). The characteristics of a supraglacial melt-out till are summarised in Table 8.1. The clast content is usually dominated by sediment typical of high-level transport (see Section 7.1), but subglacially transported particles may also be present where englacial thrusting occurs. Consequently, clasts show a broad range of characteristics, but are dominated by particles that are angular and have a non-spherical form (Figure 7.7A). Clasts are not universally striated or faceted. The size distribution is typically coarse and unimodal. Some size sorting may occur locally where meltwater reworking has occurred. Clast fabric is unrelated to ice flow, is generally poorly developed and is spatially highly variable. However, locally strong fabrics may develop where sediment has fallen or slumped down ice-cored slopes. Supraglacial melt-out till is poorly consolidated and has a low bulk density. Clast lithology is usually very variable, and far-travelled erractics, from distant nunataks, are common. These may have been transported on the surface of the glacier for considerable distances. Crude bedding, like that found in scree slopes, may develop locally within supraglacial till units, but

BOX 8.2: SUPRAGLACIAL MORAINE FACIES

Supraglacial debris transported by warm-based valley glaciers is deposited as supraglacial melt-out or moraine till. Eyles (1979) identifies three facies associated with the deposition of supraglacial melt-out till.

Facies 1 occurs where the supraglacial debris on the glacier surface slows the rate of ice melt such that the till is slowly superimposed onto the subglacial landsurface by the melt-out of buried ice. This gives the till surface an irregular relief known as **hummocky moraine** (see Section 9.1.3). This process is illustrated in the diagram below.

Facies 2 occurs where the supraglacial debris cover is much thinner or too coarse to retard ice melt and the till is deposited as a thin dispersed bouldery veneer as the ice margin retreats.

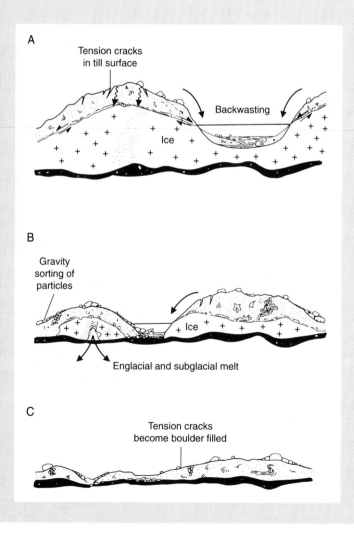

Facies 3 occurs where supraglacial debris is found interbedded with proximal outwash sediments. The characteristics of proximal outwash are such that it is often difficult to distinguish from supraglacial melt-out till.

Estimates for Icelandic glaciers suggest that the total volume of supraglacial melt-out till may be of the order of 200–2000 m^3 a year, compared with a maximum discharge of 26 000 m^3 of lodgement till.

Source: Eyles, N. 1979. Facies of supraglacial sedimentation on Icelandic and Alpine temperate glaciers. *Canadian Journal of Earth Science* **16**, 1341–1361. [Diagram modified from: Eyles (1979) *Canadian Journal of Earth Science* **16**, Fig. 5, p. 1348]

generally they are massive and structureless, although some meltwater reworking may be present.

8.1.5 Flow Till

When debris on the ice surface becomes saturated it may begin to creep, slide and flow. The degree of fluid flow depends on the water content, the debris character, the surface gradient and whether the surface over which the debris is moving is composed of ice or debris. In general, the greater the water content, the more fluid the debris flow (Figure 8.7). There are three main types of flow recognised at modern glacier margins:

1. **Mobile flows**: thin, highly fluid, rapid flows which are erosive and show crude size sorting, with coarse particles tending to settle to the base of the flow. The particles are usually strongly orientated in the direction of flow.
2. **Semi-plastic flows**: thick, slow-moving tongues of debris, which are erosive. They may show size sorting, with coarse particles settling to the base of the flow, and their upper surfaces may be resorted by the flow of meltwater. Fold structures and a weak particle orientation may develop.
3. **Creep**: slow downslope movement of debris, not visible to the naked eye. This may occur either as a general non-channelised lobe of debris or as a more or less continuous sheet of creeping mass. Particles are rarely orientated in the direction of flow.

In practice, although it is possible to recognise different types of flow on modern glaciers, ancient flow till deposits simply consist of numerous **flow packages** or individual flows, of varying type stacked one on top of the other. Consequently, flow tills are characteristically very diverse in nature. The characteristics of a unit of flow till consisting of several flow packages is summarised in Table 8.1. The clasts within flow tills show a broad range of characteristics, but are dominated by particles which are angular and have a non-spherical form. Clasts are commonly

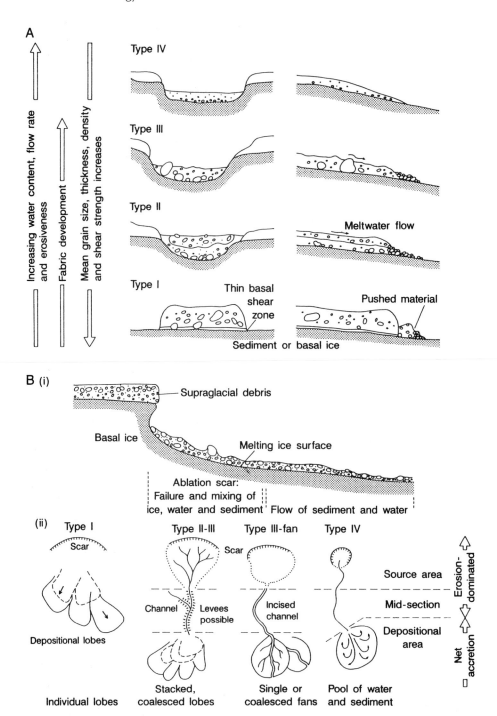

Figure 8.7   *The characteristics of flow tills. **A:** Variation of flow type with water content. **B:** Morphology of the source and depositional area of different types of flow. [Modified from: Lawson (1982) Journal of Geology **90**, Figs 3, 5 and 13, pp. 282, 287 and 296]*

not striated or faceted. The deposit is usually dominated by sediment typical of high-level transport, but subglacially transported particles may be present. In general, the size distribution is coarse and unimodal, although locally individual flow packages may be well sorted. Flow tills have a variable fabric, although individual flow packages can have strong fabrics, reflecting the former slope down which flow occurred. This type of sediment is poorly consolidated with a low bulk density, although occasionally flow packages may be closely packed. Individual flow packages may sometimes be visible within a flow till unit, and crude sorting within some flow packages may be present. The base of some flows may contain a concentration of larger clasts (bed load). Sorted sand and silt layers may be common, associated with reworking by meltwater on top of individual flows. Some flow packages may have erosional bases, and small folds may also be present in certain flow types. In general, they are a highly variable and diverse sediment.

8.1.6 Sublimation Till

In cold arid environments, such as those in Antarctica, ablation may occur by sublimation. Sublimation is the direct vaporisation of ice without it passing through a liquid phase. It is common in parts of Antarctica where temperatures rarely exceed freezing. It may occur both in subglacial and supraglacial locations, and the characteristics of sublimation till depend upon whether the process occurs in a subglacial or supraglacial location. The properties given here relate to subglacial sublimation till, and these are summarised in Table 8.1. The clasts within subglacial sublimation till are typical of basal transport, being rounded, spherical, striated and faceted. These characteristics are similar to those of basal debris, but are less pronounced than those of lodgement till since little post-depositional modification of clasts takes place. Again the particle size distribution is typical of basal debris transport, being either bimodal or multimodal. Particle fabric is strong in the direction of ice flow, although it may show a greater range in orientation than a typical lodgement till. Typically sublimation till has a low bulk density and is loose and friable. Clast lithology may have an inverse superposition: the youngest, least-travelled clasts occur at the base of a deposit and the oldest, furthest-travelled clasts occur at the top. Sublimation till is usually stratified and may preserve englacial fold structures.

8.1.7 Distinguishing Tills in the Field

Although each of the six main types of till form in different ways and in theory should have very different properties, distinguishing between them in the field is often difficult. This is particularly true of ancient lithified tills, known as **tillites,** which record periods of ancient glaciation within the geological record. Traditionally the problem is approached by the analysis of the **internal sedimentary properties** of a till unit. Internal sedimentary properties include clast fabric, clast shape, particle size distribution, composition, and sedimentary structures.

However, recent developments have stressed the **external relationships**, that is the context of the till unit and its association with other sediments (**lithofacies**) and landforms (**landsystems**).

Of the internal properties, clast fabric is the most valuable. The **particle fabric** of till is relatively easy to measure, although laborious, in the field (Box 8.3). Fabric is measured by two properties: (1) the compass orientation of elongated particles; and (2) the dip or angle at which those particles are inclined within the sediment. This information is normally portrayed on a rose diagram or on a stereograph (Box 8.3). A variety of different statistical techniques have been used to determine the presence of a preferred particle orientation (fabric), and to assess the distribution of particles about this preferred orientation. Lodgement tills normally have a strong fabric in the direction of ice flow, and the deviation or scatter of particles about that mean fabric is small. Similarly subglacial melt-out and sublimation tills have a strong particle fabric parallel to the ice flow direction but usually show a greater scatter of particles about the mean orientation. This reflects the fact that individual particles are disturbed as they melt-out from the basal ice in which the fabric parallels the ice flow direction. Deformation till may possess well-developed particle fabrics orientated in the direction of tectonic transport, usually similar to the ice flow direction. The fabric strength or variation around the mean direction depends on the thickness of the deforming layer and the level of deformation: a thin deforming layer with a low level of deformation or shear has a strong fabric, while a thick deforming layer which experiences a high level of deformation will have a weaker fabric. Supraglacial melt-out and flow tills do not possess consistent particle fabrics. Particle orientation varies throughout the deposit. Within flow till deposits made up of a number of flow packages, the fabric will vary from one package to the next. If the flows were highly mobile the fabric may be pronounced in the direction of sediment flow. Similarly in supraglacial melt-out tills strong particle orientations may be recorded, but these only reflect the orientation of former ice slopes down which the debris accumulated. Consequently, when these deposits are sampled at several points, a random or scatter fabric is usually recorded. These differences have only recently been quantified with data taken from modern till-forming environments, despite the fact that till fabric analysis is traditionally one of the most commonly undertaken forms of sedimentological analysis on glacial sediments.

Recent work using modern analogue data and a sophisticated statistical method of analysing particle fabric data based on eigenvectors and eigenvalues has developed a methodology by which particle fabric can be used as a quantitative tool in determining till genesis. Eigenvector analysis is a computer-based statistical tool by which the mean orientation and the scatter of clasts about the mean can be determined. Three eigenvectors (V1, V2 and V3) describe the orientation of the clasts within the sample. Eigenvector V1 refers to the direction of maximum clustering (preferred orientation), and V3 to that of minimum clustering. Eigenvector V2 defines a vector that is perpendicular to both V1 and V3. Eigenvalues (S1, S2 and S3) summarise fabric strength or the degree of clustering around each vector. A sample with no preferred clast orientation would have three equal eigenvalues. In contrast, a strong fabric would give a high value in the direction of maximum clus-

BOX 8.3: THE MEASUREMENT AND ANALYSIS OF TILL FABRIC

Till fabric analysis involves recording the compass orientation and dip of elongated clasts within a till. Generally only clasts with a pronounced long axis relative to the short and intermediate ones are analysed. Suitable clasts are carefully excavated from a cleared face of undisturbed till and the dip or inclination of each particle, along its long axis, is then measured using a compass clinometer. The orientation of each particle, in the direction in which the long axis dips, is also recorded with a compass. A sample of at least 25 clasts is required before a reliable indication of the till fabric is obtained.

The data can then be plotted in a number of different ways. The simplest method is to plot a rose diagram of particle orientation. This is simply a frequency bar chart of particle orientation in which the horizontal axis has been curved through 360°. However, the best and most sophisticated method of plotting the data is to use a stereographic plot, like that used in the analysis of geological structures. This is based on the principle that if a sphere was put around

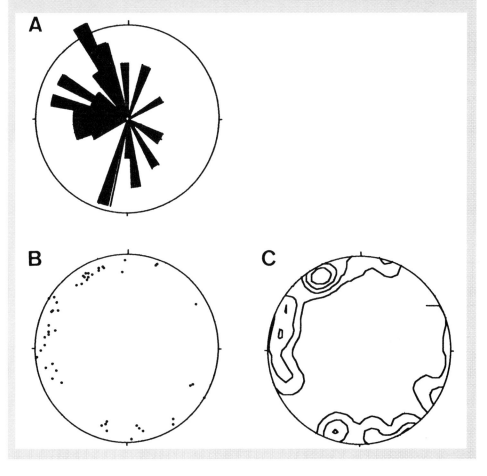

the original undisturbed clast, a point could be placed on the outer shell of the sphere in line with the long axis of the clast. This point records both the orientation and dip of the clast. A plot of this form is produced by plotting the orientation and dip of each particle on stereographic or equal-area graph paper. If the points cluster on the plot they have similar orientations and dips. For clarity, the number of dots per unit area on the graph is usually contoured to emphasise the clustering present. The particle orientation within a lodgement till is shown above as: (A) a rose diagram; (B) an equal-area stereographic scatter plot; and (C) an equal-area stereographic contour plot, contoured at 1%, 2% and 5% point density per 1% plot area. The sample contains 50 clasts.

tering (S1) and a low value in the direction of least clustering (S3). Plots of S1 and S3 eigenvalues have been found to give good discrimination between different types of till (Figure 8.8). At present, however, there are still insufficient data from modern till-forming environments to constrain the range of possible values associated with each type of till and with the processes.

Of the other internal criteria there are few absolute rules. As we saw in Chapter 7, debris transported without coming into contact with the glacier bed has very different particle shape and size characteristics from that transported in contact with the bed. Lodgement, subglacial melt-out and subglacial sublimation till all contain only subglacially transported debris and therefore have its characteristics—rounded, striated, faceted and spherical clasts with a bimodal size distribution (Figure 7.7). In contrast, supraglacial melt-out tills and flow tills contain debris

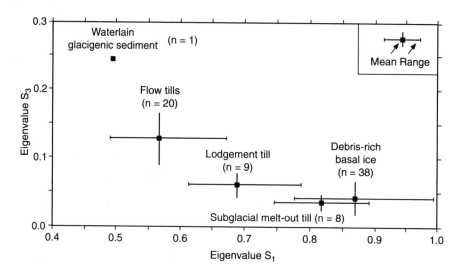

Figure 8.8 *Plot of two eigenvalues (S_1 and S_3) used to describe the particle fabric of modern glacial sediments. The plot shows how different types of till can be separated using fabric analysis and eigenvalues. [Modified from: Dowdeswell et al. (1985) Journal of Sedimentary Petrology **55**, Fig. 7, p. 697]*

transported at high levels within the glacier and possess its characteristics—angular, non-spherical clasts with a coarse unimodal grain-size distribution (Figure 7.7). Unfortunately these tills may also contain subglacial debris if the transfer of basal debris by thrusting occurs at the glacier margin. Consequently, it can be difficult to distinguish tills on this basis alone, although in general an assessment of the proportion of striated, rounded and faceted clasts is useful. If the proportion is high then it may indicate a subglacial origin. Particle-size sorting within flow tills may also be useful in their recognition. In very fluid flows sediment sorting may occur, and dewatering in other flows may give rise to pockets of sorted and occasional layers of silts and sands. In addition, coarse boulder horizons may also be identified beneath some flow packages. These represent the tractional component moved by a slow-moving flow.

Particle composition may also be of some value in distinguishing till genesis. For example, lodgement tills tend to be dominated by local lithothogies, while supraglacial melt-out and flow tills may contain a higher proportion of far-travelled lithologies, which are transported a long way on the surface of the glacier. Deformation tills will contain lithologies which reflect the material that has been assimilated in their formation and may often contain rafts or bodies of original sediment that has not been fully mixed into the till unit (Box 8.1).

Structures within tills also provide important diagnostic information. In particular, lodgement tills are associated with smudges (brecciated clasts), low-angle shear planes and foliations, sole marks caused by erosion of the substrate and the extrusion of till into the underlying rock-head. Deformation till may contain tectonic structures, although these tend only to be preserved at low levels of deformation. The presence of rafts of sediment which have only partially been mixed into the deforming layer is also highly diagnostic. Sublimation tills and subglacial melt-out tills may possess weak stratification inherited from the basal ice.

In practice, therefore, there are very few hard and fast rules with which to interpret till genesis on the basis of internal properties alone. More importantly, the analysis of internal criteria often involves time-consuming field measurements (e.g. fabric analysis) or the use of a laboratory (e.g. grain size).

In many situations the external setting or the context of a till unit is often more useful. In particular, the association with surface landforms can be particularly diagnostic. For example, lodgement till is likely to be found beneath subglacial landforms such as drumlins and flutes (see Section 9.2). In contrast, flow tills and thick deposits of supraglacial melt-out till are likely to be associated with areas of hummocky moraine (see Section 9.1.3). Consequently the geomorphological context (land system) of the upper boundary or contact of a till unit may provide an important insight into its origin.

More recently, the introduction of **facies analysis (lithofacies)** has added a further criterion. This is based on the interpretation of the complete depositional sequence and not just upon the interpretation of individual units. It is based on the premise that most depositional environments can be characterised by distinctive associations or combinations of sediments or **facies. Sedimentary facies** are bodies of sediment that are the product of a particular depositional environment or process. For example, the process of subglacial lodgement gives rise to a lodgement facies. The relationship of one facies to another gives us the ability to assemble a picture of the depositional environment in which sedimentation occurred. At a modern glacier one deposit does not continue infinitely in any one direction, but will grade into other deposits. For example, a lodgement till surface may be dissected by a meltwater stream in which glaciofluvial sediments are being deposited. By studying the relationships between one deposit and the next we can reconstruct the depositional environment.

To understand the relationship of one facies to another one must understand **Walther's principle**. Walther studied recent sedimentary facies and their relationship to the environment in which they were deposited. From these observations he deduced that environments are not static through time and that as environments shift position, the respective sedimentary facies of adjacent environments or processes succeed each other in a vertical profile (Figure 8.9). Therefore, in sequences where there is no apparent break in the sedimentary record, the vertical profile of sedimentary facies is equivalent to the lateral variation of facies at any one time. To put this crudely, if one was to turn a vertical profile on its side then it would give a picture of the lateral variation in the depositional environments present during the period of time represented by the vertical profile. In this way a vertical profile or section of glacial sediments can be translated into a picture of the particular glacial environment in which they were deposited. The formation of individual sediment units can then be explained. This is the principle by which **facies analysis** works. Vertical logs through sediment sections are constructed using a standard coding system to facilitate international comparison (Figure 8.10 and Table 8.2). In this way the whole sequence is documented and then compared as a whole to modern depositional environments. In a crude sense one simply turns the sedimentary log on its side and looks for a glacial environment that would give the same pattern of sediments. Emphasis is therefore placed on interpreting the whole sequence of sediments and the environment in which they formed and not simply on the interpretation of specific units or components. Figures 8.11 and 8.13–8.15 show typical vertical facies logs for common types of glacial environments. Each is characterised by a different pattern of sediment units and structures and is discussed in Section 8.1.8.

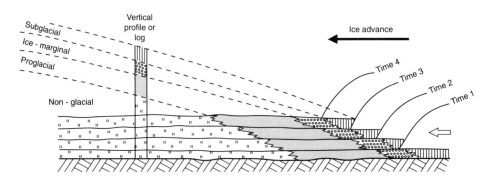

Figure 8.9   *Illustration of Walther's principle in the context of glacial environments*

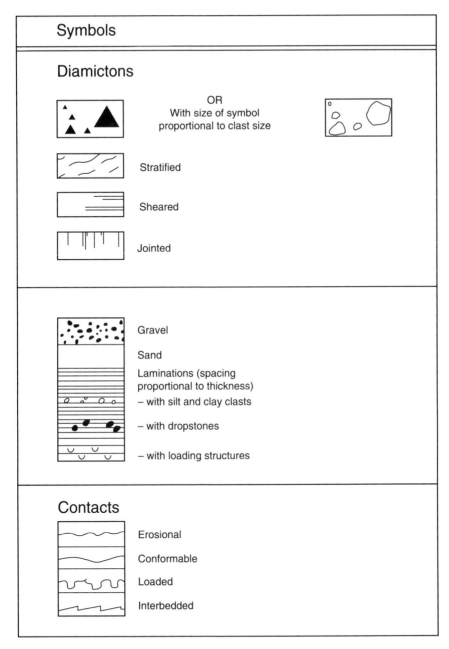

Figure 8.10 *Lithofacies coding for glacial diamictons. [Modified from: Eyles et al. (1983) Sedimentology, 30, Table 2, p. 396]*

Table 8.2   *Diagnostic criteria for recognition of common diamicton lithofacies. Key to common terms: matrix-supported = the matrix dominates the sediment and the clasts are set within it; clast-supported = clasts dominate and matrix infills spaces between the clasts; massive = structureless; grading = variation in particle size; flow noses = small folds or curved beds formed by slumps or flows; stringer = thin discontinuous layer of sand, silt or clay. [Modified from: Eyles et al. (1983) Sedimentology **30**, Table 3, p. 397]*

Code	Facies	Description
Dmm	Matrix-supported, massive	Structureless mix of mud, sand and pebbles
Dmm(r)	Dmm with evidence of resedimentation	Initially appears structureless but careful cleaning reveals subtle textural variability and fine structure (e.g. stringers of silt or clay with small flow noses). Stratification less than 10% of the unit thickness
Dmm(c)	Dmm with evidence of current reworking	Initially appears structureless but careful cleaning reveals subtle textural variability and fine structure produced by water flow (e.g. isolated ripples). Stratification less than 10% of the unit thickness
Dmm(s)	Matrix-supported, massive, sheared	Initially appears structureless but careful cleaning reveals shear planes, foliation and orientated clasts. Breccitated clasts may be present
Dms	Matrix-supported, stratified diamicton	Obvious textural differentiation or structure within the diamicton. Stratification more than 10% of the unit
Dms(r)	Dms with evidence of resedimentation	Flow noses frequently present; diamicton may contain rafts of deformed silt/clay laminae and abundant silt/clay stringers. May show slight grading. Often contain high clast contents, which often form clusters. Clast fabric random or parallel to bedding. Erosion along the base of the unit may be present
Dms(c)	Dms with evidence of current reworking	Diamicton often coarse, due to removal of fines. May be interbedded with sand, silt and gravel beds showing evidence of flowing water (e.g. ripples and cross-bedding). Abundant stringers within the diamicton. Units may have channelised base
Dmg	Matrix-supported, graded	Diamicton exhibits variable vertical grading in either matrix or clast content
Dmg(r)	Dmg with evidence of resedimentation	Clast imbrication common

A

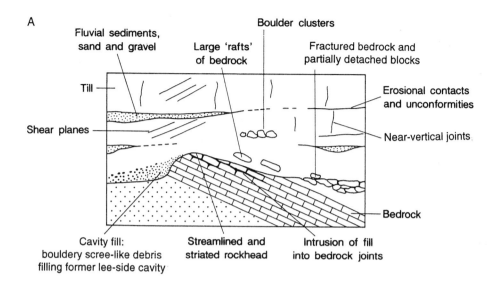

Fluvial sediments,
sand and gravel

Large 'rafts'
of bedrock

Boulder clusters

Fractured bedrock and
partially detached blocks

Till

Shear planes

Erosional contacts
and unconformities

Near-vertical joints

Bedrock

Cavity fill:
bouldery scree-like debris
filling former lee-side cavity

Streamlined and
striated rockhead

Intrusion of fill
into bedrock joints

B

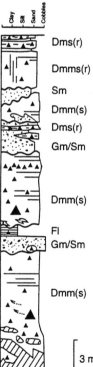

Interpretation
Upper till layers may show evidence of weathering and sediment reworking by solifluction.
Discontinous units of glaciofluvial sediment, may contain lenses of till deposited by the collapse of channel walls.
Concentrations of boulder / clasts may occur locally forming boulder pavements.
Discontinous units of glaciofluvial sediment deposited in subglacial channels. Upper contact may be eroded by over riding ice.
Shear laminations caused by the shearing out of soft clasts. Brecciated clasts may also be present. Foliations and shear planes occur throughout.
Deformation of the rockhead. Inclusion of blocks and rafts of bedrock. Injection of till into bedrock joints.

Clay Silt Sand Cobbles

Dms(r)

Dmms(r)

Sm

Dmm(s)

Dms(r)

Gm/Sm

Dmm(s)

Fl
Gm/Sm

Dmm(s)

3 m

Figure 8.11   *Facies model for a warm-based glacier moving over a rigid bed.* **A:** *Typical facies associated with lodgement tills.* **B:** *Typical vertical log of a lodgement till facies.* [**B** *modified from: Eyles (1983)* Glacial Geology, Pergamon Press, Fig. 1.8, p. 16]

The use of facies analysis provides a powerful field-based tool with which to interpret glacial sediments and the origin of till sequences. It does not, however, replace the need to examine the internal evidence present within each sediment unit. Interpreting glacial sediments and distinguishing between different tills and depositional environments is difficult and success is based largely on experience.

8.1.8 Till Facies and Basal Thermal Regime

Glaciers with different thermal regimes deposit sediment in different ways. They give rise to different depositional environments and therefore different facies patterns or architecture. We can recognise three main types of thermal regime, each of which is associated with a different facies pattern: (1) warm- or wet-based glaciers; (2) cold- or dry-based glaciers; and (3) glaciers with mixed thermal regimes (polythermal). The type of glacial deposition and facies patterns typical of each is discussed below.

1. **Deposition within warm-based glaciers.** Where warm-based glaciers move over rigid beds most of the debris is transported in the basal layers. Lodgement of basal debris occurs over most of the glacier bed. The lodgement facies will vary from areas associated with direct particle lodgement to those deposits laid down in cavities, where the sediment will be locally more diverse and may not possess a strong particle fabric. The process of lodgement may be interrupted locally by subglacial rivers which rework the sediment to give units of sand and gravel. These subglacial rivers are usually ephemeral and flow may switch on and off suddenly. The location of these rivers also varies through time. Changes in ice flow will also affect the continuity of the depositional processes and may cause erosional breaks. For example, different units of lodgement till, perhaps with different lithological clast contents, may be superimposed on top of one another. This occurs where different ice streams or ice lobes, from different areas compete with one another in a lowland area (Box 8.4). As each lobe of ice waxes or wanes in strength, a different unit of lodgement till may be deposited. In summary, deposition beneath a warm-based ice sheet will produce a sequence of lodgement till units with thin fluvial interbeds. The upper surface of the till sequence will be characterised by subglacial landforms. The facies characteristics of lodgement tills and warm-based glaciers are shown in Figure 8.11 along with a typical vertical log.

   Where a warm-based glacier moves over soft deformable substrates a different sedimentary assemblage may result. Subglacial deformation of a soft substrate may generate a deformation till. The thickness of the deforming layer will depend on the properties of the deforming sediment and on the shear stress applied to it by the glacier. As shown in Figure 8.5, low levels of deformation are associated with simple overturning and folding, high levels of deformation involve intense folding and the development of tectonic laminations and boudins, while very high levels of deformation produce a homogeneous diamicton. Sediment is transported within the deforming layer from areas of

BOX 8.4: LODGEMENT TILL FACIES: COMPETING ICE LOBES

Eyles *et al.* (1982) describe sections of lodgement till exposed in cliffs along the coast of north-east England. These glacial sequences consist of units of lodgement till separated by discontinuous layers of silt and sand. Traditionally these sequences have been interpreted in terms of multiple ice advances, the sands being deposited during ice-free episodes. This interpretation was supported by differences in the lithological composition of clasts within the till units. Eyles *et al.* (1982) proposed an alternative explanation. They argued that the whole sequence was deposited by a single episode of glaciation. The layers of sand were interpreted as the product of subglacial meltwater and represent the location of subglacial streams. This interpretation is consistent with the discontinuous form of these sand layers. Eyles *et al.* (1982) explained the changes in clast composition of the lodgement till units in terms of competing ice lobes, each derived from a different source and therefore with a different suite of erratics or clasts. The lateral interplay of these lobes of ice would give rise to vertical changes in clast composition, as illustrated in the diagram below. A similar explanation was developed by Broster and Dreimanis (1981) to explain variations in the composition of lodgement till in British Columbia.

This work illustrates the importance of careful field observation and the

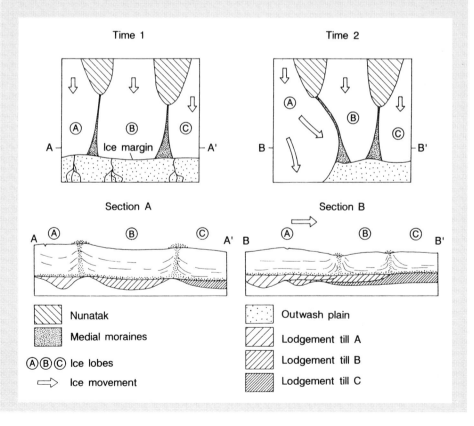

application of modern analogues in the interpretation of glacial sediments. In this case the interpretation has profound importance for the glacial stratigraphy of north-east England.

Sources: Broster, B. E. & Dreimanis, A. 1981. Deposition of multiple lodgement tills by competing glacial flows in a common ice sheet: Cranbrook, British Columbia. *Arctic and Alpine Research* **13**, 197–204. Eyles, N., Sladen, J. A., & Gilroy, S. 1982. A depositional model for stratigraphic complexes and facies superimposition in lodgement tills. *Boreas* **11**, 317–333. [Diagram modified from: Eyles *et al.* (1982) *Boreas* **11**, Fig. 10, p. 325]

extending flow to areas of compressional flow (Figure 8.6). Beneath areas of extending ice flow deformation is excavational; new material is added to the deforming layer by down-cutting of the base of deformation. Increases in basal shear stress cause the base of deformation to descend down through the sediment pile, incorporating new material into the deforming layer. Each successive deformation stage is superimposed as the level of deformation increases. The contact between the deforming layer and the undeformed sediment is usually sharply defined by a slip horizon, known as a **décollement** surface. In contrast, in areas of compressive flow the deforming layer thickens by the accumulation of till from up-ice areas. As a consequence the deforming pile thickens by the addition of sediment at the top of the sequence and each successive tectonic state is preserved above each other in the sequence. At the base of the section there will be undeformed sediment, above which there will be lightly folded and deformed sequences, followed by highly deformed sediment with tectonic laminations, boudins and other evidence of intense shear, and ultimately a homogenised diamicton may occur at the top of the sequence if the level of deformation is sufficient. Figure 8.12 shows a model of the different deformation facies that might exist beneath an ice sheet. In practice, however, the sedimentary facies produced by subglacial deformation may be extremely complex, depending on the deformation history and the character of the sediment that is being assimilated into the deforming layer.

2. **Deposition within cold-based glaciers.** Cold-based glaciers do not usually possess well-developed basal debris layers due to the absence of widespread glacial erosion and basal debris entrainment. Basal debris is derived in one of two ways: (1) by overriding of the frontal apron of fallen ice blocks and debris in front of the glacier snout; and (2) by freezing-on of water and debris draining from warm-based areas of the ice sheet. The debris within these glaciers is usually frozen-on in layers and highly folded and attenuated (Figure 8.13A). During glacier recession the stratified basal debris layers decay *in situ* and englacial debris is lowered onto the basal substrate. Interstitial ice is lost in cold arid areas by sublimation. Large areas of sublimation or subglacial melt-out till will result, often retaining the crude stratification and folded structure of the englacial debris from which it is derived. A typical vertical profile of the facies associated with cold-based glaciers is given in Figure 8.13B.

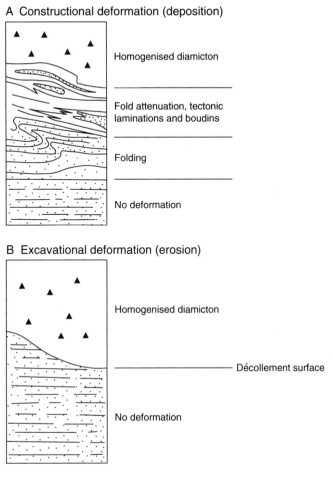

Figure 8.12   *Variation in subglacial glaciotectonic (deformation) facies beneath an ice sheet. Constructional deformation occurs in areas of compression, while excavational deformation is common under areas of extending flow (see Figure 8.6)*

3. **Deposition by mixed regime glaciers.** Many glaciers exhibit very thick basal and englacial debris zones in response to repeated freezing-on of subglacial meltwater draining from warm- to cold-based areas of the glacier and to thrusting at the warm–cold interface. A mixed thermal regime is, however, not the only way in which this type of sediment assemblage may develop, since intense folding of a thin horizon of basal debris may generate thick debris layers within both warm- and cold-based ice. If this is also associated with glacial thrusting, thick englacial sequences may result. This not only occurs when warm-based glaciers have a thin frozen ice margin, but also where ice flows against steep bedrock highs or escarpments. Where the ice is cold-based, deposition at the base of the glacier occurs through basal melt-out. If this sediment is saturated it may become prone to remobilisation and subglacial flow. Thickness

A

Advancing

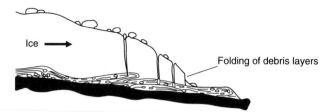

Ice ➡

Debris
apron

Thick debris layers due to
sustained freezing-on

Retreating

Ice ➡

Folding of debris layers

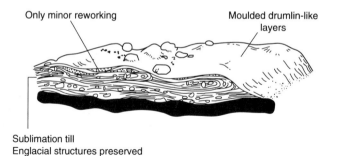

Only minor reworking

Moulded drumlin-like
layers

Sublimation till
Englacial structures preserved

B

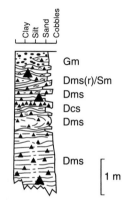

Interpretation
May contain some evidence of till flow, sliding/ slumping or free fall. Occasional signs of sediment reworking by meltwater
Fold structures may be picked out by the stratified till. Clast orientation reflects structure
Stratified till, may be picked out by alternating patterns of clast size or density

Clay
Silt
Sand
Cobbles

Gm

Dms(r)/Sm
Dms
Dcs
Dms

Dms

1 m

variations in the melt-out till reflect the fold and thrust structures within the englacial ice. Where the ice is warm-based, lodgement till may be deposited. Surface melt-out of the englacial debris produces a large thickness of supraglacial melt-out till, which is frequently resedimented as flow till. A complex surface ice topography results through differential debris insulation and surface melting (Figure 8.14A). Fluvial action on this topography is commonplace and much of the supraglacial debris may be reworked and deposited over the ice surface. A complex irregular topography of depositional landforms (hummocky moraine) results when all the buried ice melts. One of the main characteristics of this type of depositional environment is the diverse nature of tills present and their intimate association with fluvial deposits. Multiple till sequences, often separated by layers of sand and gravel and produced by a single glacial episode (Box 8.5), are common. A typical vertical profile of the facies associated with glaciers with mixed thermal regimes is given in Figure 8.14B.

This picture is complicated for valley glaciers, which typically possess thick supraglacial debris layers due to rockfall from adjacent valley sides (Figure 8.15). This supraglacial debris gives rise to a thick irregular drape of coarse angular supraglacial melt-out or moraine till, which is often reworked as flow tills and by fluvial action (Box 8.2). This drape of debris transported at high level within the glacier will mantle the subglacial sediment facies.

8.2 FLUVIAL SEDIMENTATION

Glacial meltwater entrains and transports sediment which is subsequently deposited on, within, beneath or beyond the glacier. Sedimentation on or within the glacier may occur in surface channels and in either englacial or subglacial tunnels. Sedimentation in surface (supraglacial) channels is inhibited by the steep channel gradients and the smooth ice walls which provide little frictional drag on the flow and its sediment load (Figure 4.1A). However, deposition does occur due to changes in the discharge of such channels, since sediment may lodge temporally in active channels as discharges fall. Sediment may also be deposited in channels where the gradient falls and the sediment load is high. Supraglacial channels are also often short-lived and frequently abandoned, dramatically stranding sediment within them.

Frictional drag is greater in subglacial tunnels that are in contact with the glacier bed. Sediments deposited within tunnels consist of sheet-like units of stratified

Figure 8.13 *Facies model for a cold-based glacier. **A:** Debris structure within a cold-based glacier that has experienced extensive freezing-on. Compression during glacier retreat causes the deformation of debris-rich ice stratification. **B:** Typical vertical log of the associated sedimentary facies with this environment. [Modified from: **A:** Shaw (1977) Canadian Journal of Earth Sciences **14** Fig. 1, p.1241; **B:** Eyles (1983) Glacial Geology, Pergamon Press, Fig. 1.8, p. 16]*

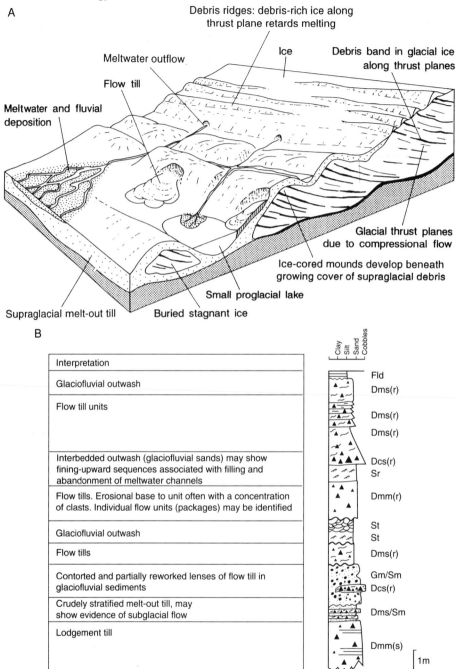

Figure 8.14   *Facies model for a glacier with a mixed thermal regime or high englacial debris content. **A:** Topography of ice-cored moraines or debris ridges, the structure of which reflects the structure of thrusts and debris bands within the glacier. A complex depositional environment involving fluvial deposition, flow tills and supraglacial melt-out till results. **B:** Typical vertical log of the sedimentary facies associated with this environment. [Modified from: **B:** Eyles (1983) Glacial Geology, Pergamon Press, Fig. 1.8, p. 16]*

BOX 8.5: THE INTERPRETATION OF MULTIPLE TILL SEQUENCES

Where thick sequences of supraglacial debris cover the glacier surface, multiple layers of till (lodgement tills, melt-out tills and flow tills) may be superimposed. In certain situations these may be separated by units of sand and gravel. A typical section consists of a till layer at the base, above which there is a sequence of sand and gravel which is in turn capped by a second till layer. This type of section is often referred to as a **tripartite till sequence**. Traditionally such sequences were interpreted as the product of multiple glaciations. The ice advanced to deposit the first till then retreated, depositing the sand and gravel before readvancing to deposit the upper till. Many of these sequences have now been reinterpreted in terms of a single episode of glaciation. Ice advances over the area to deposit a basal lodgement till. As the ice retreats a topography of ice-cored ridges develops, between which glaciofluvial sands and gravels are deposited. As the ice retreats further, these outwash rivers become abandoned and the glaciofluvial deposits are covered by flow till derived from the adjacent ice-cored ridges. Melt-out of the buried ice inverts the topography to give a tripartite till sequence. The correct interpretation of each till layer and of the facies present is therefore very important.

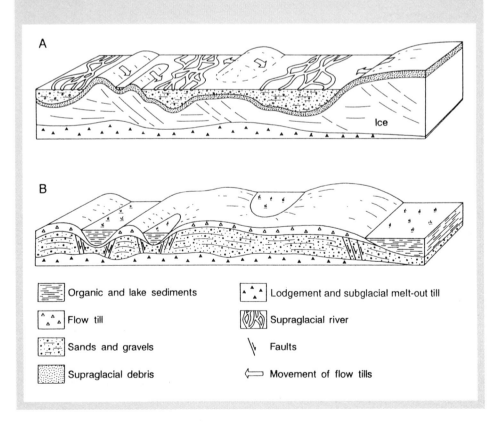

Organic and lake sediments	Lodgement and subglacial melt-out till
Flow till	Supraglacial river
Sands and gravels	Faults
Supraglacial debris	Movement of flow tills

One of the first sequences to be reinterpreted in this way was that at Glanllynnau in North Wales (Boulton 1977). This represents an important step forward in the interpretation of till sequences and glacial stratigraphy.

Source: Boulton, G. S. 1977. A multiple till sequence formed by a late Devensian Welsh ice cap: Glanllynnau, Gwynedd. *Cambria* **4**, 10–31. [Diagram modified from: Addison *et al.* (1990) *North Wales: Field Guide*. Quaternary Research Association, Cambridge, Fig. 17, p. 41]

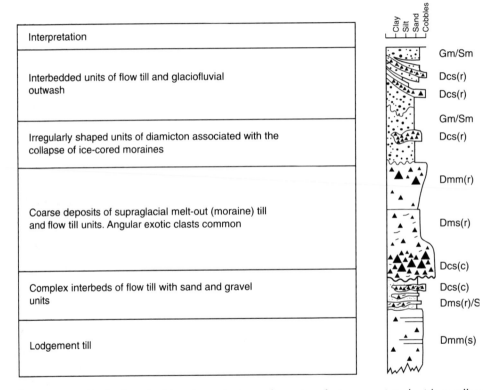

Figure 8.15 *Typical vertical log through the sedimentary facies associated with a valley glacier. [Modified from: Eyles (1983) Glacial Geology, Pergamon Press, Fig. 1.8, p. 16]*

sand and gravel in which secondary bedforms such as ripples, dunes and graded beds may occur. Deposition within such tunnels can be understood to some extent by the application of theory developed for the deposition of solids within pipes. Four flow regimes have been identified within pipes: (1) at low flow velocities a stationary bed with little or no transport occurs; (2) as flow velocities increase, material begins to slide over the bed as a single unit; (3) at higher velocities suspension of all particles occurs, although the coarser fraction is still transported close to the tunnel floor; and (4) at very high velocities all particles move in suspension and no size sorting is present. If the flow velocity was to fall rapidly during this final flow regime a massive, heterogeneous non-sorted sediment would result. Despite this work very little is known about the processes of sedimentation within subglacial tunnels.

Subglacial meltwater may also deposit thin coatings of precipitated solutes in subglacial rock cavities (Figure 8.16). As we saw in Section 5.3.2, subglacial meltwater may dissolve soluble components and transport them. These may in turn be precipitated to form thin coatings. These coatings may either infill shallow depressions or be concentrated into small linear ridges. These precipitates are best developed on carbonate rocks such as limestone or chalk, where they are composed of calcium carbonate. In Norway, areas of bedrock that were once covered by basal cavities frequently contain a brown staining which results from the precipitation of iron oxides from subglacial meltwater in the cavities. The precipitation of silica has also been noted. In general these precipitates are confined to former subglacial cavities and are believed to be the product of regelation. As ice flows against a bedrock obstacle, pressure melting occurs on the upstream side and refreezing or regelation occurs on the downstream side (see Section 3.3.2). Refreezing concentrates the solutes within the meltwater, leading to their eventual precipitation in the lee of the bedrock obstacles. The linear form of many precipitates is due to smearing out of solute-rich meltwater by the flowing ice. These coatings are quickly dissolved and removed by weathering when the rock surfaces are exposed on deglaciation. Their long-term preservation potential is therefore small.

Figure 8.16 *A subglacial calcium carbonate precipitate in front of the Glacier de Tsanfleuron, Switzerland. These precipitates are strung out parallel to the former ice flow, which was from right to left. [Photograph: M. Sharp]*

The processes of sedimentation beyond the glacier margin are much better understood. Sedimentation beyond the glacier occurs in the same way as conventional fluvial deposition, except for the following differences:

A. The water is generally colder, denser and therefore more viscous. The viscosity of water increases with a fall in temperature. Increased viscosity reduces the settling rate for particles in suspension and allows a greater volume of suspended sediment to be transported.

B. The water and sediment discharge is highly seasonal. Water discharge beneath the Glacier d' Argentière for example varies between 0.1 m^3 s^{-1} in the winter to 11 m^3 s^{-1} in the summer. On Nisqually Glacier, in North America, the sediment transported during just five minutes in the month of June is equal to the whole sediment yield for the month of January. Similarly 60% of the annual sediment load from the Decade Glacier on Baffin Island was discharged during just 24 hours in 1965. Sediment discharge is therefore highly seasonal. It also varies diurnally. Most discharge occurs during the period of nival floods early in the melt season (see Section 4.3).

Sediment is transported both in suspension and as bedload (traction and saltation). A number of studies have attempted to record the sediment load of meltwater streams and its variation through time. Suspended sediment can be relatively easily measured by water sampling; the water sample is filtered or evaporated to determine the sediment content. Results show that during the winter, when discharge is negligible, meltwater contains only a few milligrams of sediment per litre, but during the summer this rises to several grams per litre. Suspended sediment content reaches a peak early in the summer as the fluvial system within the glacier is flushed clean (see Section 4.4). Suspended sediment content also varies diurnally and peaks prior to the maximum daily discharge on many glaciers (Figure 4.8). In contrast, accurate estimates of the sediment moving as bedload are much more difficult to obtain and seasonal variations are less well understood at present (Box 8.6). Estimates of the relative importance of these two components, suspended sediment versus bedloads, vary from as little as 40% suspended load to over 90% of the total sediment discharge, depending on the particular characteristics of the glacier. Sedimentation in front of a glacier can be divided into three zones, although the boundaries between each are somewhat unclear. The three zones are: (1) the proximal zone; (2) the medial zone; and (3) the distal zone.

1. **The proximal zone.** In this zone sedimentation is dominated by: (1) the changing position and geometry of the ice margin and/or any ice-cored ridges present (Figure 8.14A); (2) the rate of supply or availability of supraglacial melt-out till; and (3) the seasonal flood regime. Braided stream flow is only one part of the total hydraulic system. Resedimentation of supraglacial melt-out till as mud, debris and subaqueous flows is common and may dominate the depositional process. The availability of large quantities of readily transported sediment and the rapid build up and decay to and from flood discharges has a strong effect on the character of the fluvial sediments deposited. Melt streams

BOX 8.6: SEDIMENT TRANSPORT IN GLACIAL MELTWATER STREAMS

Reliable estimates of bedload transport within meltwater streams are difficult to obtain, due to the high flow magnitudes and large sediment volume involved. One of the first studies to provide reliable estimates of bedload transport was by Østrem (1975). Two techniques were used to measure the bedload of a proglacial stream in front of the glacier Nigardsbreen in Norway.

The first method involved the construction of a 50 m steel fence across the main meltwater stream during the summer of 1969. The mesh size was such that it trapped all sediment larger than 20 mm in diameter. The accumulation of coarse bedload trapped by the fence was measured by probing the depth twice a day at 176 points along the fence. The fence survived for three weeks before being destroyed in a flood. Between 24 May and 19 June, 400 tonnes of material were trapped. Samples of meltwater were also taken during this period to determine the suspended sediment content. During this period, approximately 1200 tonnes of suspended sediment were discharged. This suggests that during the study period bedload transport accounted for 25% of all the material transported, although it must be noted that sediment in the size range of 1–20 mm in diameter was not trapped, so this value must be regarded as a low estimate.

The second method involved the annual survey of a delta formed as the meltwater from Nigardsbreen enters a lake about 1 km in front of the glacier. Most if not all of the bedload in transport is deposited on this delta. The average annual accumulation on this delta is 11 200 tonnes, giving a crude estimate of the total bedload moved each year by the meltwater streams.

In recent years the sophistication of bedload traps has improved and accurate estimates of bedload transport are more commonplace. However, recent work simply confirms the high levels of bedload transport within meltwater streams as first quantified in detail by Østrem.

Source: Østrem, G. 1975. Sediment transport in glacial meltwater streams. In: Jopling, A. V. & McDonald, B. C. (Eds) *Glaciofluvial and Glaciolacustrine Sedimentation*, The Society of Economic Paleontologists and Mineralogists, Special Publication 23, 101–122

are often incompetent to transport the available till load and may simply redistribute it as structureless, matrix-supported outwash beds with particle size distribution little different from the parent till. During flood phases all available particle sizes may be transported and deposited simultaneously. Poor stratification at the top of these massive beds of outwash may develop during the waning flood stage. It usually consists of individual lamenae or sediment layers several clasts thick formed by the removal of the fines (**winnowing**) to give an armoured, often scoured, surface. Proximal outwash is frequently found interbedded with units of flow till and other mass flow deposits, particularly where melt streams are bordered by ice-cored debris ridges. Consequently, the massive unstratified proximal outwash typical of this zone is difficult to distinguish from supraglacial melt-out till.

Deposition also frequently occurs on buried ice, the melt-out of which will cause subsidence structures to form. These usually consist of normal or extensional faults and synclinal fold or sag structures (Figure 8.17). If melt-out and subsidence occur while deposition is still taking place these subsidence structures are referred to as **syn-sedimentary**. Evidence for subsidence may not be present on the surface, since meltwater deposition is concentrated in the subsiding areas and infills them (Figure 8.18). If melt-out occurs after deposition has finished then the landsurface is deformed and parallels the bedding (Figure 8.18). In this case the subsidence structures are **post-sedimentary** (see Section 9.3).

2. **The medial zone.** Away from the ice margin a **braided river** pattern develops, in which ephemeral **bars (sandwaves)** and channels dominate (Figure 8.19). Individual channels and bars vary in size from a few metres to hundreds of metres in width. The depth of channels is typically only a few metres. The braided pattern develops in response to the large sediment load, high discharge and the steep gradient of outwash surfaces. The particle size of sediment deposited falls rapidly with distance from the glacier margin, which reflects the decline in stream power and the processes of clast attrition in highly turbulent channels (Figure 8.20). Three types of bar or sandwave can be identified, although they are not unique to glacial outwash systems (Figure 8.21):

Figure 8.17 *Faulted sands within outwash sediments in front of Skeidararjökull, Iceland. These normal faults formed during subsidence associated with the melt-out of buried ice. [Photograph: M. R. Bennett]*

A Syn-sedimentary subsidence

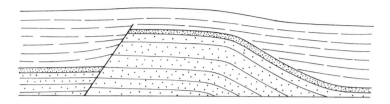

B Post-sedimentary subsidence

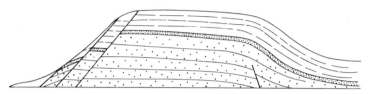

Figure 8.18 *Syn-sedimentary and post-sedimentary subsidence structures associated with the melt out of buried ice*

Figure 8.19 *Braided outwash plain, Skaftafellsjökull, Iceland. [Photograph: M. R. Bennett]*

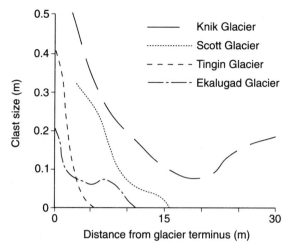

Figure 8.20   *Relationship between clast size and distance from the glacier margin for a variety of glacial outwash systems. [Modified from: Drewry (1986) Glacial Geologic Processes, Arnold, Fig. 10.10, p. 159]*

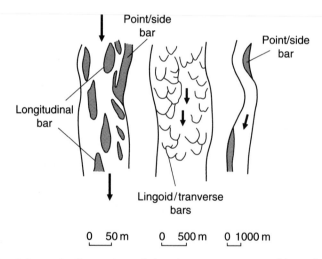

Figure 8.21   *Schematic illustration of the three main types of bars found in braided outwash channels. [Modified from: Drewry (1986) Glacial Geologic Processes, Arnold, Fig. 10.11, p. 160]*

A. The **longitudinal bar**. This forms in a mid-channel position when the coarsest part of the stream load is deposited as steam flow falls and loses competency. These bars are orientated roughly parallel to the current flow. They are small when first formed, but continue to grow in length and height as fine sediment is deposited downstream in the lee of the bar. Grain size tends to decrease downstream.

B. The **linguoid** or **transverse bar**. This type of bar is orientated transverse to the direction of stream flow and may possess a diamond or rhombic shape with a steep downsteam face. They develop under high flow and extend by particles avalanching down their front face.

C. The **lateral, side** or **point bar.** These bars are typically very large and develop on the sides of stream channels in quieter water. They are attached to the sides of the channel.

With distance from the glacier margin the proportion of linguoid or transverse bars increases, as does the sand component within the fluvial system.

The sediments which result from this braided channel system range from boulders to sands. The sedimentary structures within these sediments reflect five processes which operate in the braided channel system: (1) the formation of bars (sandwaves); (2) the formation of bedforms such as ripples and dunes in finer material; (3) the erosion (scour) and fill of channels; (4) the deposition of finer sediments during low flows, particularly in backwaters; and (5) overbank sedimentation during the falling stage of flood flows. Table 8.3 shows the types of sedimentary facies associated with these processes.

Table 8.3 *Diagnostic criteria for recognition of common glaciofluvial stream deposits. For a detailed description of the types of cross-bedding referred to in the table the reader is referred to the volume by Collinson, J. D. & Thompson, D. B. 1989. Sedimentary Structures. Unwin Hyman, London. [Modified from: Eyles et al. (1983) Sedimentology **30**, Table 1, p.395]*

Code	Facies	Interpretation
Gms	Massive, matrix-supported gravel with no sedimentary structure	Debris flow
Gm	Massive or crudely bedded gravel with horizontal bedding and clast imbrication	Longitudinal bars and channel lag deposits
Gt	Stratified gravel with trough cross-beds	Minor channel fills
Gp	Stratified gravel, planar cross-beds	Transverse or linguoid bars
St	Medium to very coarse pebbly sand with solitary or grouped cross-beds	Dunes
Sp	Fine to very coarse pebbly sand with solitary or grouped planar cross-beds	Transverse or linguoid bars
Sr	Very fine to coarse sand, with ripple marks	Ripples, low flow regime
Sh	Fine to very coarse often pebbly sand, with horizontal laminations	Planar bed flow, high flow regime
Ss	Fine to coarse sand, may be pebbly, with broad shallow scour structures	Minor channels or scour hollows
Fl	Sand, silt and mud, with ripple marks	Waning flood deposits and overbank deposits
Fm	Mud and silt with desiccation cracks	Drape deposits formed in pools of standing water

3. **The distal zone.** With distance away from the glacier margin the proportion of fine-grained sediment involved increases dramatically. During normal flows the main flow is concentrated in a single channel, although at peak flow a braided pattern may still be adopted. The glacial influence decreases and the flow regime is less dominated by the seasonal patterns of ice melt. There is a gradual transition into a more conventional fluvial system.

The pattern of sedimentation in the proximal, medial and distal zones will be disrupted if the glacier is subject to catastrophic floods or jökulhlaups. The nature of the sedimentation will be partly controlled by the shape of the flood hydrograph. Floods triggered by volcanic activity beneath a glacier are typically of high magnitude but of short duration, in contrast to floods caused by the drainage of ice-dammed lakes which tend to be of longer duration. Research to date suggests that jökulhlaup flows result in sediment sequences which consist of massive, poorly sorted, non-graded or inversely graded sediments, the latter being characterised by large surface boulders, which may be channelled. If present, these channels tend to contain boulder lags, fields of megaripples and streamlined boulder hummocks. The main unit is interpreted as the product of hyperconcentrated fluid–sediment mixtures which are sufficiently dense to transport large boulders on their surface and prevent any size sorting or grading from developing. These hyperconcentrated flows are associated with the initial flood surge. As the flow stage declines the sediment solidifies and dewaters. This water may then form a series of more fluid flows on the surface of the main deposit. If these fluid flows are sufficiently large they may scour channels and deposit the lag horizons. The large massive units of boulder-rich gravel typical of jökulhlaups provide a sharp contrast with the deposits produced by normal glacial discharges.

8.3 SUMMARY

On land, glaciers deposit sediment either directly or via meltwater. The product of direct glacial sedimentation is a glacial till. A till is defined as a sediment whose components are brought into contact by the direct agency of glacier ice, and which has not been disaggregated, although it may suffer glacially induced flow either in the subglacial or supraglacial environment. There are six types of glacial till: lodgement tills, subglacial melt-out tills, deformation tills, supraglacial melt-out or moraine tills, flow tills, and sublimation tilis. The properties of each are distinct and are dependent upon the transport pathway followed by the debris within the glacier. The glacial facies produced by a glacier is primarily controlled by its basal thermal regime. Deposition by meltwater occurs either within the glacier or beyond its margins. The type of sediment and sedimentary facies produced by glacial meltwater is primarily controlled by the distance from the ice margin. Close to the ice front the sediments are coarse, poorly sorted and often chaotic. With distance from the ice margin the sediments rapidly become finer and are deposited first in a braided channel sequence and then increasingly within single channel flows. Subglacial meltwater may also deposit coatings or precipitates in subglacial cavities.

8.4 SUGGESTED READING

The literature on glacial deposition is considerable. There are a number of useful anthologies of papers that contain important contributions: Goldthwait (1971), Legget (1976), Schlücter (1979), Evenson *et al.* (1983), Goldthwait & Matsch (1988), Van der Meer (1987), and Kujansuu & Saarnisto (1990). In the early 1970s, Boulton wrote a series of highly influential papers on till formation based on observations made in Svalbard (Boulton 1968, 1970a,b, 1971, 1972a,b, 1975, 1977, 1982; Boulton & Paul 1976).

Eyles *et al.* (1982) and Broster & Dreimanis (1981) discuss the characteristics and facies patterns associated with lodgement tills. The morphology of clasts and other characteristics of lodgement tills are discussed by Sharp (1982), Krüger (1984) and Clark & Hansel (1989). Paul & Eyles (1990) discuss the formation and preservation of melt-out tills, while Shaw (1977, 1988) deals with the formation of sublimation till. The processes of subglacial deformation and the formation of deformation tills are covered in Boulton & Jones (1979), Alley *et al.* (1986), Boulton & Hindmarsh (1987), Hart *et al.* (1990), Hart & Boulton (1991), Hart & Roberts (1994), and Eyles *et al.* (1994). Flow tills are discussed by Lawson (1981, 1982), while the sedimentation of supraglacial melt-out till is covered by Eyles (1979) and by Boulton & Eyles (1979).

The recognition of different types of till is covered in several papers: Marcussen (1973, 1975), Krüger & Marcussen (1976), McGown & Derbyshire (1977), Mills (1977), Krüger (1979), and Dreimanis (1990). The measurement and analysis of till fabric is covered in papers by Andrews (1971), Mark (1973) and Woodcock (1977), while the application of fabric in determining till genesis has been used or discussed by Rose (1974), Lawson (1979), Rappol (1985), Dowdeswell & Sharp (1986), and Hart (1994). The application of fabric analysis in the interpretation of ancient tills (tillites) is covered by Dowdeswell *et al.* (1985). The use of facies analysis (lithofacies) in the interpretation of glacial sediments is introduced in the paper by Eyles *et al.* (1983), which should be read in conjunction with the work of Martin (1980) and Shaw (1987). The reaction and comment to the paper by Eyles *et al.* (1983) also makes interesting reading (Karrow *et al.* 1984; Shaw *et al.* 1986). The paper and book by Brodzikowski & Van Loon (1987, 1991) also provide a useful overview of glacial sedimentary facies.

The volume edited by Jopling & McDonald (1975) deals with most aspects of glaciofluvial sedimentation; in particular, papers by Church & Gilbert (1975) and Boothroyd & Ashley (1975) are of note. The subglacial precipitation of solutes is discussed by Hallet (1975, 1976) and Hallet *et al.* (1978). Discussion of the characteristics of proximal outwash can be found in Boulton & Eyles (1979), while comprehensive information about the sedimentology and structure of braided outwash rivers can be found in the papers by Miall (1977, 1978). Sedimentation by jökulh-laups is covered in the paper by Maizels (1989).

Alley, R. B., Blankenship, D. D., Bentley, C. R. & Rooney, S. T. 1986. Deformation of till beneath ice steam B, West Antarctica. *Nature* **322**, 57–59.

Andrews, J. T. 1971. Methods in the analysis of till fabric. In: Goldthwait, R. P. (Ed.) *Till: A Symposium.* Ohio State University Press, Ohio, 321–327.

Boothroyd, J. C. & Ashley, G. M. 1975. Processes, bar morphology, and sedimentary structures on braided outwash fans, northeastern Gulf of Alaska. In: Jopling, A. V. & McDonald, B. C. (Eds) *Glaciofluvial and Glaciolacustrine Sedimentation.* The Society of Economic Paleontologists and Mineralogists, Special Publication **23**, 193–222.

Boulton, G. S. 1968. Flow tills and related deposits on some Vestspitsbergen glaciers. *Journal of Glaciology* **7**, 391–412.

Boulton, G. S. 1970a. On the origin and transport of englacial debris in Svalbard glaciers. *Journal of Glaciology* **9**, 213–229.

Boulton, G. S. 1970b. On the deposition of subglacial and melt-out tills at the margins of certain Svalbard glaciers. *Journal of Glaciology* **9**, 231–245.

Boulton, G. S. 1971. Till genesis and fabric in Svalbard, Spitsbergen. In: Goldthwait, R. P. (Ed.) *Till: A Symposium.* Ohio State University Press, Ohio, 41–72.

Boulton, G. S. 1972a. Modern Arctic glaciers as depositional models for former ice sheets. *Journal of the Geological Society* **128**, 361–393.

Boulton, G. S. 1972b. The role of thermal regime in glacial sedimentation. In: Price, R. J. & Sugden, D. E. (Eds) *Polar Geomorphology.* Institute of British Geographers, Special Publication 4, 1–19.

Boulton, G. S. 1975. Processes and patterns of subglacial sedimentation: a theoretical approach. In: Wright, A. E. & Moseley, F. (Eds) *Ice Ages: Ancient and Modern.* Seel House Press, Liverpool, 7–42.

Boulton, G. S. 1977. A multiple till sequence formed by a late Devensian Welsh ice cap: Glanllynnau, Gwynedd. *Cambria* **4**, 10–31.

Boulton, G. S. 1982. Subglacial processes and the development of glacial bedforms. In: Davidson-Arnot, R., Nickling, W. & Fahey, B. D. (Eds) *Research in Glacial, Glacio-fluvial and Glacio-lacustrine Systems.* Proceedings of the 6th Guelph Symposium on Geomorphology, Geo Books, Norwich, 1–31.

Boulton, G. S. & Eyles, A. S. 1979. Sedimentation by valley glaciers: a model and genetic classification. In: Schlüchter, Ch. (Ed.) *Moraines and Varves.* Balkema, Rotterdam, 11–23.

Boulton, G. S. & Hindmarsh, R. C. A. 1987. Sediment deformation beneath glaciers: rheology and geological consequences. *Journal of Geophysical Research* **92(B9)**, 9059–9082.

Boulton, G. S. & Jones, A. S. 1979. Stability of temperate ice caps and ice sheets resting on beds of deformable sediment. *Journal of Glaciology* **24**, 29–44.

Boulton, G. S. & Paul, M. A. 1976. The influence of genetic processes on some geotechnical properties of glacial tills. *Quarterly Journal of Engineering Geology* **9**, 159–194.

Brodzikowski, K. & Van Loon, A. J. 1987. Glacigenic sediments. *Earth Science Reviews* **24**, 297–381.

Brodzikowski, K. & Van Loon, A. J. 1991. *Glacigenic Sediments.* Elsevier, Amsterdam.

Broster, B. E. & Dreimanis, A. 1981. Deposition of multiple lodgement tills by competing glacial flows in a common ice sheet: Cranbrook, British Columbia. *Arctic and Alpine Research* **13**, 197–204.

Church, M. & Gilbert, R. 1975. Proglacial fluvial and lacustrine environments. In: Jopling, A. V. & McDonald, B. C. (Eds) *Glaciofluvial and Glaciolacustrine Sedimentation.* The Society of Economic Paleontologists and Mineralogists, Special Publication 23, 22–100.

Clark, P. U. & Hansel, A. K. 1989. Clast ploughing, lodgement and glacier sliding over a soft glacier bed. *Boreas* **18**, 201–207.

Dowdeswell, J. A. & Sharp, M. J. 1986. Characterisation of pebble fabrics in modern terrestrial glacigenic sediments. *Sedimentology* **33**, 699–710.

Dowdeswell, J. A., Hambrey, M. J. & Wu, R. 1985. A comparison of clast fabric and shape in Late Precambrian and modern glacigenic sediments. *Journal of Sedimentary Petrology* **55**, 691–704.

Dreimanis, A. 1990. Formation, deposition, and identification of subglacial and supraglacial tills. In: Kujansuu, R. & Saarnisto, M. (Eds) *Glacial Indicator Tracing*. Balkema, Rotterdam, 35–59.

Evenson, E. B., Schlüchter, Ch. & Rabassa, J. 1983. *Tills and Related Deposits*. Balkema, Rotterdam.

Eyles, N. 1979. Facies of supraglacial sedimentation on Icelandic and Alpine temperate glaciers. *Canadian Journal of Earth Science* **16**, 1341–1361.

Eyles, N., Sladen, J. A. & Gilroy, S. 1982. A depositional model for stratigraphic complexes and facies superimposition in lodgement tills. *Boreas* **11**, 317–333.

Eyles, N., Eyles, C. H. & Miall, A. D. 1983. Lithofacies types and vertical profile models: an alternative approach to the description and environmental interpretation of glacial diamict and diamictite sequences. *Sedimentology* **30**, 393–410.

Eyles, N., McCabe, A. M. & Bowen, D. Q. 1994. The stratigraphic and sedimentological significance of Late Devensian ice sheet surging in Holderness, Yorkshire, UK *Quaternary Science Reviews* **13**, 727–759.

Goldthwait, R. P. 1971. *Till: A Symposium*. Ohio State University Press, Ohio.

Goldthwait, R. P. & Matsch, C. L. 1988. *Genetic Classification of Glacigenic Deposits*. Balkema, Rotterdam.

Hallet, B. 1975. Subglacial silica deposits. *Nature* **254**, 682–683.

Hallet, B. 1976. Deposits formed by subglacial precipitation of $CaCO_3$. *Geological Society of America Bulletin* **87**, 1003–1015.

Hallet, B., Lorrain, R. & Souchez, R. 1978. The composition of basal ice from a glacier sliding over limestones. *Geological Society of America Bulletin* **89**, 314–320.

Hart, J. K. 1994. Till fabric associated with deformable beds. *Earth Surface Processes and Landforms* **19**, 15–32.

Hart, J. K. & Boulton, G. S. 1991. The interrelation of glaciotectonic and glaciodepositional processes within the glacial environment. *Quaternary Science Reviews* **10**, 335–350.

Hart, J. K. & Roberts, D. H. 1994. Criteria to distinguish between subglacial glaciotectonic and glaciomarine sedimentation, I. Deformation styles and sedimentology. *Sedimentary Geology* **91**, 191–213.

Hart, J. K., Hindmarsh, R. C. A. & Boulton, G. S. 1990. Different styles of subglacial glaciotectonic deformation in the context of the Anglian ice sheet. *Earth Surface Processes and Landforms* **15**, 227–241.

Jopling, A. V. & McDonald, B. C. 1975. *Glaciofluvial and Glaciolacustrine Sedimentation*. The Society of Economic Paleontologists and Mineralogists, Special Publication 23.

Karrow, P. F., Dreimanis, A., Kemmis, T. J., Hallberg, G. R., Eyles, N., Miall, A. D. & Eyles, C. H. 1984. Discussion: lithofacies types and vertical profile models: an alternative approach to the description and environmental interpretation of glacial diamict and diamictite sequences. *Sedimentology* **31**, 883–898.

Krüger, J. 1979. Structures and textures in till indicating subglacial deposition. *Boreas* **8**, 323–340.

Krüger, J. 1984. Clasts with stoss-lee forms in lodgement tills: a discussion. *Journal of Glaciology* **30**, 241–243.

Krüger, J. & Marcussen, I. B. 1976. Lodgement till and flow till: a discussion. *Boreas* **5**, 61–64.

Kujansuu, R. & Saarnisto, M. 1990. *Glacial Indicator Tracing*. Balkema, Rotterdam.

Lawson, D. E. 1979. A comparison of the pebble orientations in ice and deposits of the Matanuska glacier, Alaska. *Journal of Geology* **87**, 629–645.

Lawson, D. E. 1981. Distinguishing characteristics of diamictons at the margin of Matanuska Glacier, Alaska. *Annals of Glaciology* **2**, 78–84.

Lawson, D. E. 1982. Mobilisation, movement and deposition of active subaerial sediment flows, Matanuska glacier, Alaska. *Journal of Geology* **90**, 279–300.

Leggett, R. F. 1976. *Glacial Till*. Royal Society of Canada. Special Publication.

Miall, A. D. 1977. A review of the braided-river depositional environment. *Earth Science Reviews* **13**, 1–62.

Miall, A. D. 1978. Lithofacies types and vertical profile models in braided river deposits: a summary. In: Miall, A. D. (Ed.) *Fluvial Sedimentology.* Canadian Society of Petroleum Geology, Memoir 5, 597–604.

Maizels, J. 1989. Sedimentology, paleoflow dynamics and flood history of jökulhlaup deposits: paleohydrology of Holocene sediment sequences in southern Iceland sandur deposits. *Journal of Sedimentary Petrology* **59**, 204–223.

Marcussen, I. B. 1973. Studies on flow till in Denmark. *Boreas* **2**, 213–231.

Marcussen, I. B. 1975. Distinguishing between lodgement and flow till in Weichselian deposits. *Boreas* **4**, 113–123.

Mark, D. M. 1973. Analysis of axial orientation data, including till fabrics. *Geological Society of America Bulletin* **84**, 1369–1374.

Martin, J. H. 1980. The classification of till: a sedimentologist's view point. *Quaternary Newsletter* **32**, 1–13.

McGown, A. & Derbyshire, E. 1977. Genetic influences on the properties of tills. *Quarterly Journal of Engineering Geology* **10**, 389–410.

Mills, H. H. 1977. Differentiation of glacier environments by sediment characteristics: Athabasca glacier, Alberta, Canada. *Journal of Sedimentary Petrology* **47**, 728–737.

Paul, M. A. & Eyles, N. 1990. Constraints on the preservation of diamict facies (melt-out tills) at the margins of stagnant glaciers. *Quaternary Science Review* **9**, 51–68.

Rappol, M. 1985. Clast fabric strength in tills and debris flows compared for different environments. *Geologie en Mijnbouw* **64**, 327–332.

Rose, J. 1974. Small-scale spatial variability of some sedimentary properties of lodgement tills and slumped till. *Proceedings of the Geologists' Association* **85**, 223–237.

Schlücter, Ch. 1979. *Moraines and Varves.* Balkema, Rotterdam.

Sharp, M. 1982. Modification of clasts in lodgement tills by glacial erosion. *Journal of Glaciology* **28**, 475–481.

Shaw, J. 1977. Tills deposited in arid polar environments. *Canadian Journal of Earth Science* **14** 1239–1245.

Shaw, J. 1987. Glacial sedimentary processes and environmental reconstruction based on lithofacies. *Sedimentology* **34**, 103–116.

Shaw, J. 1988. Sublimation till. In: Goldthwait, R. P. & Matsch, C. L. (Eds) *Genetic Classification of Glacigenic Deposits.* Balkema, Rotterdam, 141–142.

Shaw, J., Dreimanis, A., Eyles, N., Eyles, C. H. & Miall, A. D. 1986. Discussion: lithofacies types and vertical profile models: an alternative approach to the description and environmental interpretation of glacial diamict and diamictite sequences. *Sedimentology* **33**, 151–155.

Van der Meer, J. J. M. 1987. *Tills and Glaciotectonics.* Balkema, Rotterdam.

Woodcock, N. H. 1977. Specification of fabric shapes using an eigenvalue method. *Geological Society of America Bulletin* **88**, 1231–1236.

9
Landforms of Glacial Deposition on Land

This chapter describes the morphology, formation and significance of landforms produced by both glacier ice and glacial meltwater acting on land. One can recognise landforms produced by: (1) the direct action of glacier ice; and (2) the action of meltwater (**glaciofluvial landforms**). Table 9.1 provides a simple classification of these landforms based on the distinction between those which form along an ice margin and those which form beneath the glacier. This distinction is of particular importance in the interpretation of glacial landscapes. Ice-marginal landforms allow us to reconstruct the changing position of a glacier through time, for example during deglaciation. In contrast, subglacial landforms provide information about the boundary conditions beneath a glacier, such as the direction of ice flow, the basal thermal regime, and palaeohydrology. Four main groups of landform can be recognised: (1) ice-marginal moraines formed by glacier ice; (2) subglacial landforms formed by ice or subglacial sediment flow; (3) ice-marginal landforms formed by glaciofluvial processes; and (4) subglacial landforms formed by glaciofluvial processes. Each of these groups of landforms is discussed in turn below.

9.1 ICE-MARGINAL MORAINES

Ice-marginal landforms produced directly by the action of a glacier are known as **ice-marginal moraines**. They may form by the action of six processes: (1) ice pushing, (2) englacial and proglacial thrusting, (3) rockfall or debris flow onto or against the ice margin, (4) ice dumping, (5) ice surface/marginal melt-out, and (6) subglacial melt-out and lodgement. In practice, moraines may form by a combination of all five processes, but it is possible to identify three broad categories of moraine: (1) glaciotectonic moraines, (2) dump moraines, and (3) ablation moraines.

Table 9.1   *Classification of terrestrial glacial landforms*

	Ice-marginal	Subglacial
Glacial	Glaciotectonic moraines Dump moraines Ablation moraines	Flutes Megaflutes Drumlins Rogens Mega-scale glacial lineations Geometrical ridge networks (crevasse-squeeze ridges)
Glaciofluvial	Outwash fans Outwash plains Kame terraces Kames Kame and kettle topography	Eskers Braided eskers

9.1.1 Glaciotectonic Moraines

Glaciotectonic moraines encompass a broad range of different types of moraine formed by the tectonic deformation of ice, sediment and rock. The simplest form of tectonic moraine is the **push moraine** formed when a glacier advances into proglacial sediment and bulldozes it up to form a ridge (Figure 9.1). This may involve simply sweeping up the supraglacial debris which has been dumped from the ice margin, or it may involve deep-seated deformation of proglacial sediment. There are two types of push moraine; those which form during short seasonal readvances, known as **seasonal** or **annual push moraines**, and those which form during more sustained readvances, termed **composite push moraines**.

Seasonal readvances may occur at glaciers which are either stationary or experiencing net retreat (negative mass balance). Seasonal fluctuations of the ice margin occur where winter ice flow exceeds winter ablation. In these circumstances the ice margin must advance during the winter. In the summer months ablation exceeds glacier flow and the ice margin will continue to retreat. The presence or absence of seasonal fluctuations depends therefore on the balance between winter ablation and winter ice flow. If winter ablation exceeds ice flow there will be no advance. As a consequence, seasonal push moraines tend only to form in maritime areas where glaciers have relatively high ablation gradients and levels of glacier activity. In more continental climates with lower mass balance gradients and levels of glacier activity, seasonal push moraines are usually absent. In this case, even though winter ablation is small, snout velocities are also very small due to the low mass balance gradients. It is important to note that the moraine may be pushed up some distance in front of the ice margin if a snow patch or frozen lake surface separates the ice margin from the glacier fore-field. In this case the advancing glacier transmits the forward stress to the sediment via the snow patch or frozen lake surface.

At modern glacier margins seasonal push moraines are typically 1–5 m high and tend to be asymmetric in cross-section with a shallow proximal and a steep distal

Figure 9.1 *Small seasonal push moraines at the margin of Skeidararjökull, Iceland. The ice is flowing from right to left in **A** and from left to right in **B**. [Photograph: M. R. Bennett]*

flank (Figures 9.1 and 9.2). They are frequently lobate in plan, with the intervening re-entrants marked by a linear concentration of boulders which collect in longitudinal crevasses on the ice margin (Figure 9.2). The continuity of the moraine and its detailed plan form is determined by the degree to which the ice margin is crevassed: the greater the crevasse density, the more fragmentary the moraine plan form becomes (Figures 9.2 and 9.3). Moraines may merge or bifurcate along the ice margin, as different sections advance by different amounts each season (Figure 9.2). The spacing between push moraines is a function of the amount by which the ice margin retreats each summer and is therefore sensitive to climate. The greater the spacing between moraines, the greater the summer ablation.

Push moraines are typically composed of subglacial till, although outwash sediments and other proglacial debris may be incorporated. Sedimentary structures, such as folds and thrusts, are not usually well preserved in these moraines due to the coarse nature of the sediment involved, although where finer material is incorporated, folds, thrusts and faults may be evident. In locations where meltwater flows or seeps through the moraine, fines may be washed out and deposited in a small fan in front of the moraine (Figure 9.2). The detailed formation of these ridges is both complex and highly variable. Field observations have recorded the following: (1) extrusion of water soaked till from beneath the glacier as it advances; (2) simple pushing, sweeping together surface sediment; (3) stacking of imbricate slabs of subglacial and proglacial sediment (Box 9.1).

Larger, more complex push moraines may develop where a more sustained advance takes place, as a result of a positive mass balance or a glacier surge (see Sections 3.5 and 4.3). In this case a series of ridges may be pushed together to form a composite push moraine. A good example of this type of ridge is found in front of Holmstrømbreen in Svalbard (Figure 9.4). This ridge was probably formed during a glacier surge and is composed of highly folded till and outwash sediments. In push moraines such as that at Holmstrømbreen, numerous ridges are produced by a single glacier advance and the glacier margin was not located along each ridge crest (Figure 9.4).

The size of the push moraine produced does not necessarily equate with the size of the advance, but instead equates with the character of the sediment in front of the glacier and more importantly with the effectiveness with which the stress imposed by the glacier advance is transferred to the sediments in front of it. For example, if a glacier is advancing up a reverse slope it will transfer the forward stress effectively into the slope, pushing up a larger moraine. A glacier on a horizontal surface will not have the same impact because much of the forward stress is not transferred into the proglacial sediment (Figure 9.2). In such situations large moraines will only result where large amounts of debris are stacked or dumped against the ice margin. Alternatively, a glacier margin that is frozen to its bed seasonally may also help transfer forward glacier stress to the bed. A large push moraine at Mydralsjökull in south Iceland is forming as a consequence of seasonal freezing of the ice margin to its bed. The ice margin is stationary, although subject to seasonal readvances. Each winter the thin ice at the glacier margin becomes cold-based and freezes to the underlying lodgement till. As the glacier advances at the end of each winter it rips-up a slab of lodgement till and places it on the

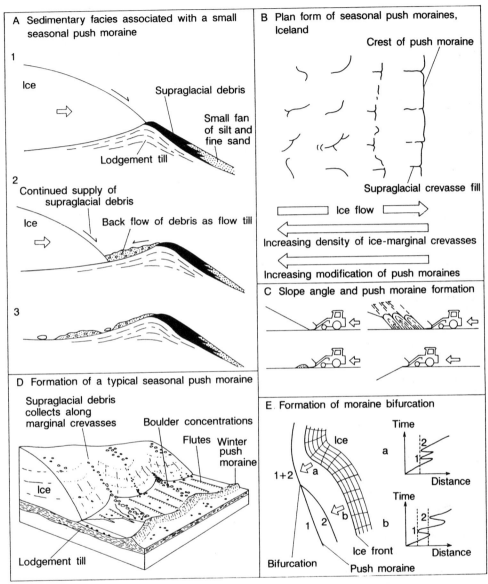

Figure 9.2 *Seasonal push moraines.* **A:** *The sedimentary facies associated with a typical seasonal push moraines.* **B:** *Plan form of seasonal push moraines, showing increasing modification with the intensity of ice-marginal crevasses.* **C:** *The size of push moraines in relation to proglacial slope angle. A glacier which bulldozes sediment over a flat surface or a slope which dips away from the glacier will produce only a small moraine. However, a glacier pushing into a slope which dips towards the glacier will produce a large push moraine since the forward stress of the glacier is effectively transferred into the slope.* **D:** *The morphology of a typical seasonal push moraine.* **E:** *The formation of a moraine bifurcation, due to differential ice retreat along an ice margin. Different sections of ice front commonly retreat at different rates*

Figure 9.3 *Push moraine in front of Brieðamerkurjökull in Iceland. Note the complex crenulated plan form caused by the crevassed and uneven ice margin. [Photograph: G. S. Boulton]*

BOX 9.1: IMBRICATE PUSH MORAINES IN ICELAND

Humlum (1985) and Krüger (1985) describe a push moraine at the margin of Höfdabredkkujökull in southern Iceland. This glacier consists of a piedmont lobe extending from the south-eastern part of the Mydalsjökull ice cap. The margin has been advancing since 1979 and has pushed up a 5–10 m high push moraine. This push moraine consists of folded blocks of till which have been stacked in an imbricated fashion on top of one another, as shown in Diagram A which shows a section cut through part of the moraine. The formation of the moraine is illustrated in Diagram B. Glacier pushing bulldozes up a slab of lodgement till or outwash sediment [Time 1: Slab 1]. This slab in turn folds and thrusts up further blocks [Time 2: Slabs 2–4]. The continued advance of the glacier causes shearing and the upper part of the glacier may advance over part of the moraine [Time 3]. Throughout this period the deposition of supraglacial debris from the ice front occurs, modifying the moraine's morphology. These blocks of sediment have retained a remarkable degree of coherence despite being unfrozen. This moraine also illustrates how moraines form by more than one process, since the dumping or flow of supraglacial sediment from the ice margin forms an important part of the moraine.

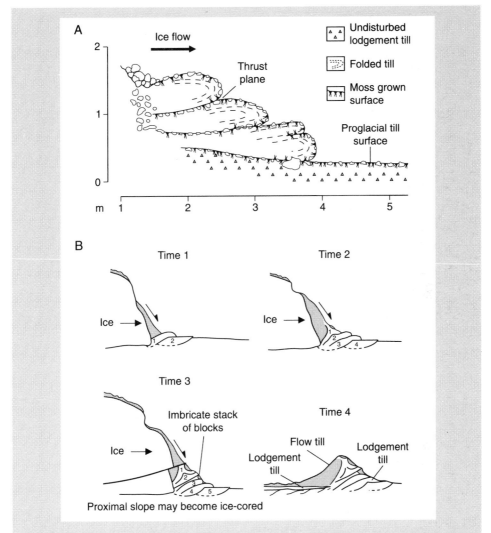

Sources: Krüger, J. 1985. Formation of a push moraine at the margin of Höfdabrekkujökull, south Iceland. *Geografiska Annaler* **67A**, 199–212. Humlum, O. 1985. Genesis of an imbricate push moraine Höfdabrekkujökull, Iceland. *Journal of Geology* **93**, 185–195. [Diagram modified from: Krüger (1985) *Geografiska Annaler* **67A**, Figs 13a and 13b, pp. 208 and 210]

moraine in front of the glacier, where it becomes detached during the summer as the ice margin becomes warm-based. In this way the moraine is formed by the seasonal addition of slabs of lodgement till. A push moraine will not simply continue to grow in front of a glacier as it advances since in most cases it will eventually be overridden by the glacier.

It has been suggested that the location of push moraines may be controlled by the presence of nuclei against which a glacier may push. On a flat glacier fore-field

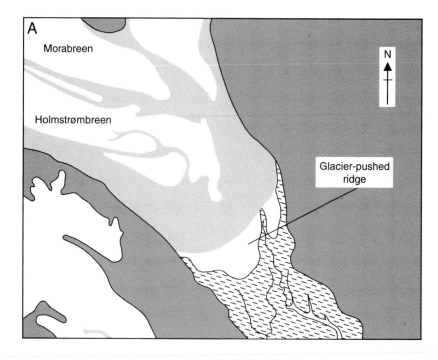

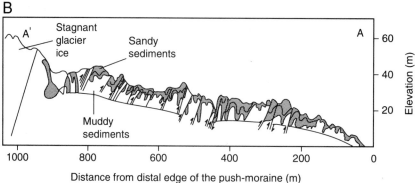

there are very few such nuclei. However, if the ice margin was stable, outwash fans might grow, providing the nuclei for push moraine formation if the glacier was to subsequently advance. This would explain why at many modern glaciers push moraines are frequently found on the proximal face of outwash fans and why many fans are completely or partially deformed into push moraines (see Section 9.3).

If we first consider a glacier experiencing a period of long-term advance, due to a positive mass balance, continuous frontal advance would plough up only a small push moraine given a horizontal glacier fore-field. Its size will remain limited by the loss of material beneath the advancing glacier and it will only grow where there is an upstanding or resistant sediment mass against which the glacier may push effec-

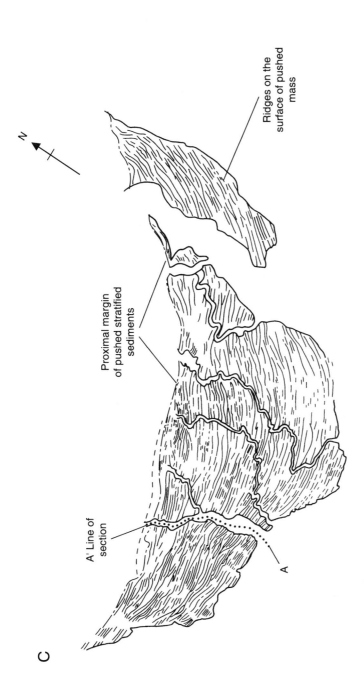

Figure 9.4 *Large composite push moraine in front of Holmstrømbreen in Svalbard. This moraine was probably formed by a single glacier surge.* **A:** *Location of the moraine in front of Holmstrømbreen.* **B:** *Cross-section through the moraine along line A–A' showing the glaciotectonics within.* **C:** *Trace of ridge crestlines within the moraine. The ice did not sit adjacent to each of these ridges, but pushed the whole mass up in the same way as a carpet is 'rucked-up' when it slides over a polished floor. [Modified from: Boulton & Meer (1989) Preliminary Report on an Expedition to Spitsbergen in 1984 to Study Glaciotectonic Phenomena (Glacitecs' 84), Universiteit van Amsterdam, Figs 2, 19 and 22, pp. 4, 36 and 43]*

tively. If, however, the glacier's ice-marginal position was to stabilise due to a short amelioration in climate, ice contact fans could develop. Further glacier advance would deform these fans into large push moraines, because the glacier now has something against which to push. The push moraines would be ultimately overridden given continued glacier advance (Figure 9.5). In contrast, if we now consider a glacier that is retreating, due to a negative mass balance, temporary stabilisation of the ice margin would be required for ice contact fans to develop. A readvance, due to a short deterioration in climate, would cause these fans, or at least their proximal faces, to be deformed into push moraines (Figure 9.5). It follows, therefore, that widespread push moraine formation during a phase of glacier advance would require a temporary amelioration of climate, whilst a deterioration in climate is required to produce push moraines during a period of prolonged glacier retreat.

So far we have only considered relatively small composite push moraines pushed up in front of advancing glaciers. However, very large push moraines may develop, involving not just glacial sediment but deformation of bedrock in front of an advancing ice sheet, given suitable circumstances. These ridges may be up to 200 m high, over 5 km wide and may extend for over 50 km. They involve major deep-seated deformation of bedrock beneath an advancing ice sheet. One of the most famous examples is Møns Klint in Denmark, where large thrust-blocks of Cretaceous chalk and glacial sediment form a series of ridges, the structure of which is revealed in spectacular coastal cliff sections. This large pushed complex was formed by the expansion of the Scandinavian Ice Sheet into Denmark during the last glacial cycle. Similar, although older, chalk rafts are exposed in coastal cliffs at Sidestrand, Norfolk, England. Other examples include the Dirt Hills in Saskatchewan, Canada. These glaciotectonic hills have been pushed up over 300 m and have been partly overridden by an ice sheet advance during the last glacial cycle. They consist of literally hundreds of parallel ridges, transverse to the former ice flow, between which there are small linear lakes. The ridges are formed from blocks of sandstone bedrock which have been thrust forward over a décollment (slip) surface located in softer clay-rich beds beneath. It has been suggested that rapid glacial loading of the competent sandstone bedrock and the saturated incompetent mudstone and clay-rich strata beneath caused the sandstone to fracture and be thrust up in a complex series of blocks along the margins of the advancing ice sheet. The magnitude of these bedrock blocks and the size of the glaciotectonic hills produced give some indication of the huge power of glacier ice.

Glaciotectonic moraines may also form by englacial thrusting (Figures 7.11 and 7.12). Thrusting is an important process in the deformation of sediment within push moraines, but can also occur within an ice margin to generate landforms. Significant rafts of sediment may be incorporated englacially by thrusting within an ice margin. On ablation this debris may form moraines. The processes by which these rafts are incorporated along the thrust is unclear, but probably involve the freezing-on of debris to the glacier bed prior to thrusting (see Section 7.3).

Englacial thrusts are particularly common at glaciers with complex basal thermal regimes at their margins, in particular those with a cold-based ice margin but warm-based interior. Here basal sliding occurs in the warm-based zone but not at the cold-based ice margin, which results in large compressional forces at the ice

225

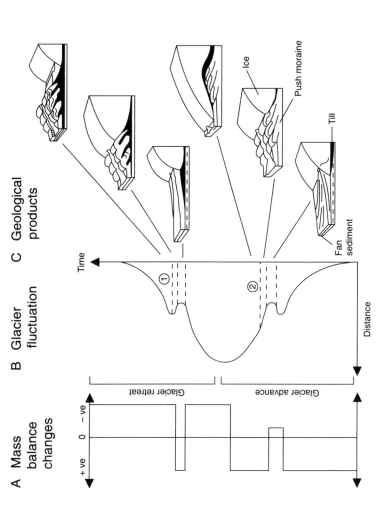

Figure 9.5 *Push-moraine-forming episodes within a glacial cycle. 1, Glacier decay is halted by a period of positive mass balance and an outwash fan forms which is subsequently pushed up by a glacier advance due to the continued positive mass balance. As a negative mass balance re-establishes itself and ice retreat continues, the glacier decays away from the moraine and meltwater is trapped between it and the moraine to form kame terraces. 2, A period of glacier advance is halted by a short spell of negative mass balance. An outwash fan first forms at the stationary ice margin before being pushed up into a push moraine as a positive mass balance regime re-establishes itself and the glacier advances. In summary, during a glacier advance, push-moraine formation is initiated by a climatic amelioration, while during decay it is stimulated by a deterioration in climate. [Modified from: Boulton (1986) Sedimentology **33**, Fig. 17, p. 695]*

margin which cause thrusts to develop. This type of polythermal ice margin is particularly common in Svalbard. Compression against a subglacial bedrock scarp or reverse slope may also lead to englacial thrusting and thrusts are also common in surge-type glaciers (see Section 7.3).

Figure 9.6 illustrates a model of moraine formation by thrusting at a polythermal glacier margin. Thrusts first develop within the ice margin, due to flow compression, and they then form progressively in front of each other as the ice margin advances. Each thrust may or may not be linked to a basal décollment surface or **sole thrust**. Thrusting of proglacial sediment and the melt-out of debris within englacial thrusts produces a multi-crested ridge complex in front of the glacier, not dissimilar in morphology to the large composite push moraines formed by glacier surges like that at Holmstrømbreen (Figure 9.4). This has implications for the recognition of surge-type glaciers, since composite push moraines may be formed by thrusting at non-surging ice margins.

In other cases thrusting is confined to the ice margin and does not propagate into the sediment in front of the glacier. Here rafts of subglacial sediment along individual thrusts may vary in thickness from a few millimetres to several tens of metres (Figure 7.13). Figure 9.7 shows a moraine ridge forming through the melt-out of debris along an englacial thrust at the margin of an outlet glacier of the Greenland Ice Sheet. Debris along the thrust plane melts out where it outcrops on the glacier surface and avalanches forward to insulate the ice beneath the thrust plane. In this way a wedge-shaped moraine composed of both buried ice and debris is formed as the glacier ablates (Figure 9.7). In this example the raft of debris within the thrust plane is relatively thin, but in other cases it may be considerably thicker. Figure 9.8 shows a series of thrust moraines in front of the Kongsvegen in Svalbard. Here rafts of gravel and diamicton, several tens of metres thick, were thrust up into the body of the glacier as it surged in 1948. Ablation and glacier retreat has revealed these rafts of debris as a series of linear ridges over 400 m long and 40 m high in places. These large slabs of debris are still underlain by buried ice (Figure 9.8). As the buried ice melts, sediment flow transfers debris from the ridge

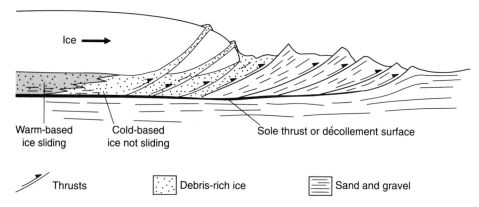

Figure 9.6   *Model of moraine formation by glacial thrusting at a polythermal ice margin.*
[Modified from: Hambrey (1994) Glacial Environments, UCL Press, Fig. 2.36, p. 80]

Figure 9.7 *Debris melting out along a thrust plane to form a moraine in front of Nord Glacier, an outlet glacier of the Greenland Ice Sheet, southern Greenland. Note that the moraine does not reflect the geometry of the ice margin. [Photograph: N. F. Glasser]*

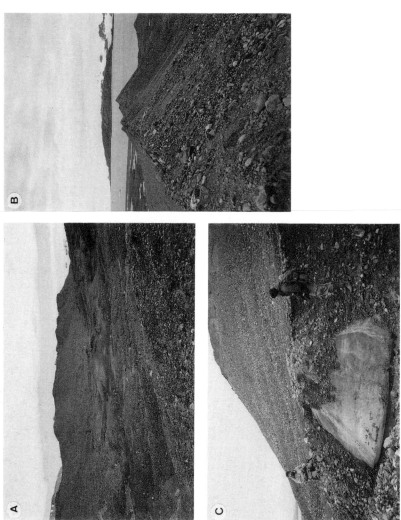

Figure 9.8 *Large thrust moraines in front of Kongsvegen, Svalbard. Linear ridge formed originally by englacial thrusting (**A** and **B**). The ice around this large slab of sand and gravel has subsequently melted although ice is still present within the ridge core (**C**). [Photographs: M. R. Bennett]*

to the intervening hollows and a complex topography of mounds and ridges is formed, which retains little of the original morphology of the englacial thrust system. Moraines formed by thrusting do not provide an accurate picture of the geometry of former ice margins. Instead they contain information about the location and geometry of thrust structures within the ice, although these tend to be broadly transverse to the ice flow direction.

9.1.2 Dump Moraines

Dump moraines form by the delivery of debris to an ice margin where it accumulates along the side or in front of the glacier to form a ridge (Figure 9.9). Their formation requires a stationary ice front. The size of moraine produced is a function of: (1) ice velocity – the faster the ice flow, the higher the rate of debris delivery to the ice margin; (2) debris content within the ice – the greater the debris content, the larger the moraine formed; and (3) the rate of ice-marginal retreat – if the ice margin retreats quickly then any debris will be distributed widely, but if it is slow or punctuated by long still stands ice-marginal debris will become concentrated into moraines. Large dump moraines will therefore result when: (1) ice velocity is high; (2) the rate of retreat is slow; and (3) the debris content in the glacier is high. A steep ice margin is also important to ensure the effective transfer of the debris away from the glacier so that it does not simply accumulate on the ice surface to form an ablation-type moraine (see Section 9.1.3). At a glacier terminus these circumstance may not always be met, particularly since stationary ice margins often

Figure 9.9 *A dump moraine forming at the margin of Skeidararjökull, Iceland.* *[Photograph: M. R. Bennett]*

have low gradients and supraglacial debris consequently tends to accumulate on the glacier surface. However, the lateral margins of a glacier are usually steep and relatively stationary, since ice flow is parallel to the margin, not perpendicular to it. Here **lateral moraines** develop primarily by the dumping of glacial debris, although some debris may also be derived from the valley sides. The formation of these moraines is illustrated in Figure 9.10. The dumping process tends to develop a strong fabric and crude bedding of platy boulders within the moraine, similar to that found in talus slopes (Figure 9.10). Typically this fabric and bedding dips towards the valley wall at between 10° and 40°. The dumping process may be seasonal, material being stored on the glacier during summer and deposited in a single pulse during a winter readvance (Figure 9.10). Debris in lateral moraines is derived from both the supraglacial and subglacial transport pathways and is consequently highly variable. The detailed morphology of the moraine and the sedimentary facies present within it depends upon the relative importance of debris

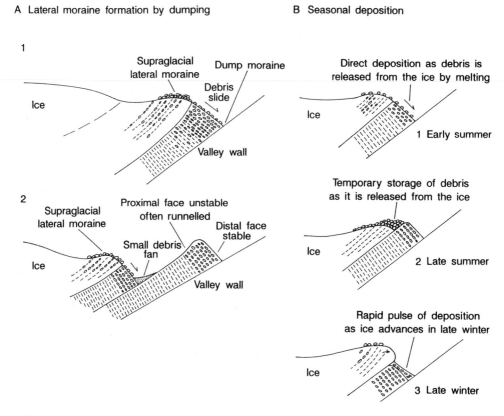

Figure 9.10 *Formation of lateral moraines by dumping of supraglacial debris.* **A:** *Lateral moraine formation. Note the scree-like bedding.* **B:** *Seasonal cycles in the growth of lateral moraines. [Modified from: Small (1987) In: Gurnell, A. M. & Clark, M. J. Glacio-fluvial Sediment Transfer, John Wiley, Figs 8.2 and 8.5, pp. 173 and 180]*

derived from the glacier relative to that from the valley side (Figure 9.11). If the valley-side component is high then the moraine will tend to have a more bench-like form or may be absent. This is particularly true where a debris or talus cone delivers material directly to the lateral moraine. In these situations the debris cone is often drawn out or extended in the direction of glacier flow. In contrast, if the direct supply of valley side debris is less important then a distinct ridge bordering the glacier may develop (Figure 9.11). The morphology of the moraine may also be modified by ice-marginal meltwater. This may deposit material between the lateral moraine and the valley wall, or alternatively small kame terraces (see Section 9.3) may be superimposed upon the proximal slopes of lateral moraines as the ice retreats (Figure 9.11). Lateral moraines often display cross-glacier asymmetry, the lateral moraine on one side of a valley being larger than that on the other. This asymmetry reflects the distribution of debris within the glacier, which is normally a function of the distribution of rock faces in the upper reaches of the glacier.

9.1.3 Ablation Moraines

Material on the surface of a glacier will become concentrated at the ice margin. This results from the supply of supraglacially transported debris and the transfer of subglacial and englacial material to the ice surface by upward-flowing ice and along thrust planes (see Section 7.3). As supraglacial debris accumulates it initially accelerates ice melt, since darker surfaces absorb solar radiation more effectively than light surfaces, which are more reflective. However, as the thickness of debris accumulates, surface melting is retarded, as it insulates the ice from surface heating. If the debris cover is sufficiently high, the glacier margin may become detached from the main body of the glacier and become stagnant (Figures 9.12 and 9.13). This will result in an **ice-cored moraine** or **ablation moraine**. The morphology of ablation moraines is controlled by the distribution of debris on and within the glacier, which may be either a product of the glacier structure (i.e. thrusts and folds within it), or alternatively the result of the surface distribution of supraglacial debris (see Section 7.1). The supraglacial debris distribution is controlled by the location of medial moraines on the glacier and the presence of patches of rockfall debris.

The type of landform produced depends on two main variables: (1) the debris concentration, and (2) the nature of the debris supply, continuous or discontinuous. If the debris concentration on the glacier surface is low and it is concentrated along the ice margin, discrete moraines will tend to form by dumping and ablation of a narrow belt of buried ice along the ice margin (Figure 9.13). However, if the debris concentration is high and it is spread over a larger part of the glacier, an area of **hummocky moraine** will result (Figure 9.14). Hummocky moraine—an irregular collection of mounds and enclosed hollows—is formed by the melt-out of ice-cored debris (Box 8.2). It may form a continuous sheet of irregular hummocks if the debris on the ice surface is continuous; alternatively, discrete belts of hummocky moraine may result if the debris on the glacier surface is not continuous or

232

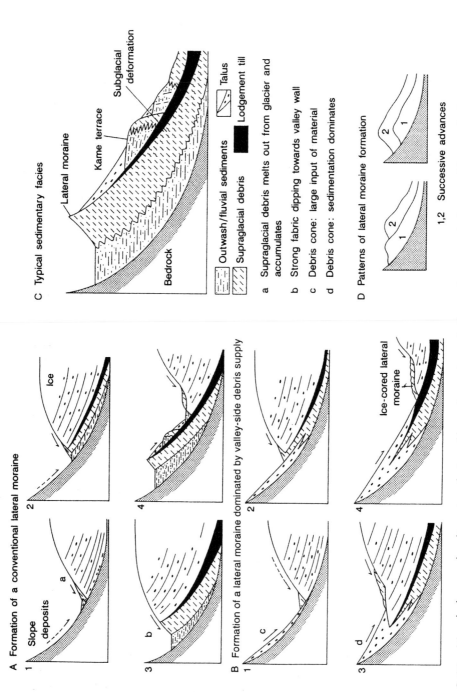

Figure 9.11 *Morphology and sedimentary facies of lateral moraines. **A:** Formation, morphology and facies of a lateral moraine where the supply of supraglacial debris is greater than the supply of valley-side debris. **B:** Formation, morphology and facies of a lateral moraine where the supply of supraglacial debris is less than the supply of valley-side debris. **C:** Typical sedimentary facies of a lateral moraine. **D:** Possible modes of growth for a lateral moraine formed by successive ice advances*

Figure 9.12 *Ablation moraines along the margin of Fjallsjökull, Iceland. These are formed by the accumulation of rockfall debris on the ice surface. The debris cover is thin and much of the moraine morphology is formed by buried ice. [Photograph: M. R. Bennett]*

of a constant thickness (Figure 9.13). This may result if the supply of debris is either pulsed, due for example to discontinuous episodes of rockfall activity, or controlled by the spatial pattern of thrusts within the glacier (Figure 9.13). These broad ablation moraines may have a strong morphology while they retain their ice core, a morphology that may reflect the englacial structure of the glacier ice (Figure 9.13). This morphology is often lost, however, when the ice core ablates and the size of the moraine is radically reduced.

9.1.4 Distinguishing Types of Ice-marginal Moraine

Ice-marginal moraines provide information about the geometry and pattern of retreat of former ice margins. It is important, however, to note that each moraine ridge need not necessarily reflect the position of a former ice margin. For example, the individual ridges of a composite push moraine do not reflect individual ice-marginal positions. Similarly, ridges formed by englacial thrusting reflect the geometry of the thrusts within the ice margin and not its extent. In most cases, however, a broad indication of the pattern of ice retreat can be obtained from ice-marginal moraines irrespective of their mode of formation. This is important since determining the genesis of individual moraines formed in deglaciated regions is

234

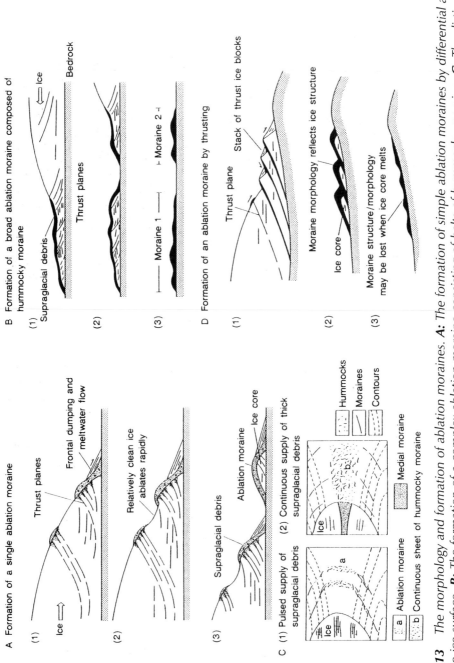

Figure 9.13 *The morphology and formation of ablation moraines.* **A:** *The formation of a single ablation moraine.* **B:** *The formation of a complex ablation moraine consisting of belts of hummocky moraine. The distinction between ablation moraines, belts of hummocky moraines, and a continuous spread of hummocky moraine is controlled by the rate of supply of supraglacial debris.* **D:** *The formation of an ablation moraine by thrusting*

Figure 9.14 *Hummocky moraine in front of austra Brøggerbreen, Svalbard. [Photograph: M. R. Bennett]*

often difficult—the different types of moraine are not always morphologically distinct, particularly after a prolonged period of post-glacial modification. Figure 9.15 shows a sequence of moraines formed by the retreat of small valley glaciers which existed in Cumbria and the Cairngorm Mountains in Britain 10 000 years ago, during the Younger Dryas. These moraines appear to reflect the geometry of a retreating ice margin, but do not contain clear sedimentological or morphological clues as to their genesis. They illustrate the practical difficulty frequently involved in distinguishing between tectonic, dump and ablation style moraines in the field.

9.2 SUBGLACIAL LANDFORMS FORMED BY ICE OR SEDIMENT FLOW

We can divide this category of glacial landforms into those which have been ice-moulded and those which have not. Ice-moulded landforms are significant as they provide information about the direction of glacier flow.

9.2.1 Ice-moulded Subglacial Landforms

Three broad families of ice-moulded subglacial landforms (bedforms) have been identified on the basis of size (Figure 9.16). Each family may be genetically distinct.

Figure 9.15 *Ice-marginal moraines in Britain formed by the last glaciers to exist in Britain, which did so during a period known as the Younger Dryas, 10 000 years ago.* **A:** *Ice-marginal moraines at Hause Gill, Cumbria [Photograph: M. R. Bennett].* **B:** *Ice-marginal moraines at Seathwaite, Cumbria [Photograph: A. F. Bennett].* **C:** *Moraines*

formed by the retreat of a small valley glacier on High Street, in Cumbria. Note the well-defined lateral moraine on the right [Photograph: A. F. Bennett]. **D:** *Ice-marginal moraines in Glen Geusachan, in the Cairngorm Mountains, Scotland. [Photograph: J. Hunt]*

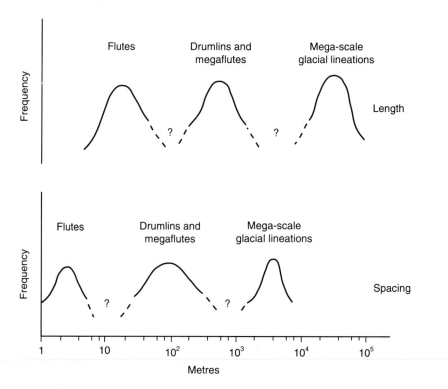

Figure 9.16   *Schematic representation of the principal spatial frequencies and lengths of streamlined subglacial landforms. The data suggest that there are three populations of subglacial bedform: (1) flutes, (2) drumlins and megaflutes, and (3) mega-scale glacial lineations. [Reproduced with permission from: Clark (1993)* Earth Surface Processes and Landforms **18**, Fig. 6, p. 9]

1. **Flutes.** Typically these are low (< 3 m), narrow (< 3 m), regularly spaced ridges which are usually less than 100 m long and are aligned parallel to the direction of ice flow (Figure 9.17). They have a uniform cross-section and usually start from either: (1) a large boulder; (2) a collection of boulders; or (3) a bedrock obstacle. They are typically composed of lodgement till, although they may also contain fluvial sands and gravels. Clusters of boulders may occur within the body of the flute. They are a common landform in front of many glaciers today.

2. **Drumlins, megaflutes and rogens. Drumlins** are typically smooth, oval-shaped or elliptical hills composed of glacial sediment (Figure 9.18). They are between 5 and 50 m high and 10–3000 m long. They have length to width ratios which are less than 50. The steep, blunter end usually points in the up-ice direction (Box 9.2). They are composed of a variety of constituents, including: (1) lodgement till; (2) bedrock; (3) deformed mixtures of till, sand and gravel; and (4) unde-formed beds of sand and gravel. They tend to occur in distinct fields or drumlin 'swarms' and are not uniformly distributed beneath a glacier. The formation of

Figure 9.17   *A glacial flute in Torridon, northern Scotland. [Photograph: M. R. Bennett]*

Figure 9.18   *Oblique air photograph of a drumlin swarm in Langstrathdale, northern England. Ice flow was from left to right. [Photograph: Cambridge Air Photograph Library]*

drumlins by small glacial readvances suggest that drumlins may form rapidly. **Megaflutes** are taller (< 5 m), broader and longer (> 100 m) than flutes and are distinguished from drumlins by having a length to width ratio in excess of 50. Their long axis is parallel to the direction of basal ice flow and they typically have a uniform cross-section. Occasionally they may start from a large bedrock obstacle, but most do not. **Rogens** are drumlinised ridges that are transverse to the direction of ice flow (Figure 9.19). In plan they often have a lunate form, with the concave elements pointing in a down-ice direction.

3. **Mega-scale glacial lineations.** In recent years lineations composed of glacial sediment have been recognised on satellite images (Figure 9.19). Typical lengths range between 8 and 70 km, widths between 200 and 1300 m and the spacing between lineations may vary between 300 and 5000 m. On the ground their morphology is often difficult to detect.

BOX 9.2: DRUMLIN SHAPE

Chorley (1959) examined the shape of drumlins and made some interesting observations. He noted that drumlins have a shape similar to the cross-sectional form of hydrofoils or aircraft wings, both of which have blunt ends which point towards the flow direction. This is the most streamlined or efficient shape possible. Chorley recognised that in designing aircraft the greater the elongation in the cross-sectional shape of a wing, the greater the air speed it could withstand. The higher the air speed therefore the more elongated the wing cross-sectional shape must be. Consequently, he argued that elongated drumlins should form under fast-flowing ice. He made the same observation in a more unusual way with reference to birds' eggs which also have a streamlined form similar to a drumlin. A bird's egg is usually laid with the blunt end first; the pressure at this end is greatest, and therefore the shape of the egg is adjusted to this. Birds that lay large eggs relative to their size (e.g. bitterns, great crested grebes) tend to lay the most elongated eggs, while birds which lay relatively small eggs lay ones that are much more rounded (e.g. golden plover). The less pressure on the eggs as they are laid, the more rounded they should be. By analogy, drumlins formed under ice that is flowing slowly should be more rounded than those formed under fast-flowing ice. This analogy should not be carried too far since other factors may control the shape of a drumlin, such as the availability of subglacial sediment, but it does provide an interesting idea which is supported by drumlin observations.

Source: Chorley, R. J. 1959. The shape of drumlins. *Journal of Glaciology* **3**, 339–344.

Subglacial bedforms may occur superimposed one on top of another: smaller bedforms rest on the backs of larger ones (Box 2.2 and Figure 9.19). For example, megaflutes may be superimposed on the backs of larger drumlins, while in many of the drumlin fields of northern England small drumlins are found on larger drumlins. This results from a decrease in the energy regime of the glacier (Box 9.3). The orientation of the two sets of bedforms can either be coincident or discordant. In the latter case the bedform patterns may provide evidence of changing patterns of ice flow. In a hypothetical ice sheet, ice will flow from the ice divide to the margin and this pattern will be recorded by subglacial bedforms. Within this ice sheet most geomorphological activity will occur where the ice velocity is greatest, close to the equilibrium line (Figure 3.15), and will decrease beneath the ice divide. If the location of the ice divide changes then the pattern of ice flow within the ice sheet will be reorganised and a new set or population of bedforms will begin to form parallel to the new pattern of ice flow. These would alter or be superimposed on the original set of bedforms (Figures 9.19 and 9.20). Beneath the new ice divide little modification would occur, due to low glacier velocity. However, beneath the equilibrium line the old set of bedforms would be quickly eroded and replaced by a new set with an orientation consistent with the new flow pattern. In between

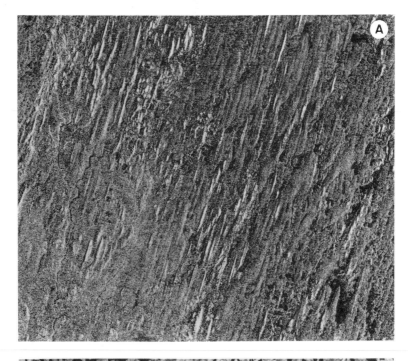

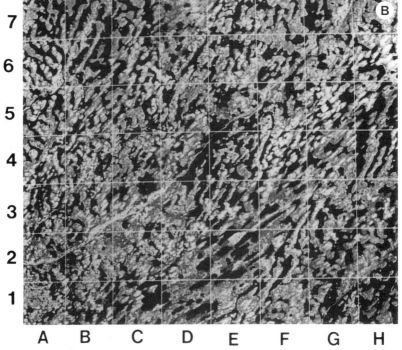

BOX 9.3: SUPERIMPOSED DRUMLINS

Working in drumlin fields near Glasgow and in eastern Cumbria in Great Britain, Rose and Letzer (1977) recorded small drumlins superimposed on the backs of larger drumlins, as illustrated in the diagram below.

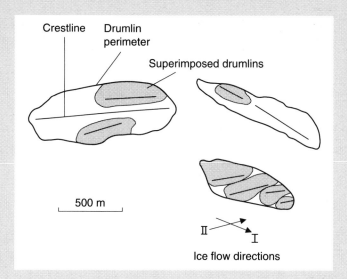

They went on to consider the circumstances in which two populations of bedforms could coexist. They envisaged that drumlins form at the ice–sediment interface due to subglacial deformation. For a drumlin to form, the glacier must have sufficient energy to deform the sediment. They argued that this energy will vary as the ice changes in thickness and velocity. The resistance of the sediment to deformation may also change. It may decrease with the commutation of debris or with changes in subglacial hydrology which may reduce interparticle packing and friction. It may increase due to the incorporation of new debris or due to changes in subglacial hydrology causing the particles to be more effectively packed and locked together. As illustrated below, superimposed drumlins

Figure 9.19   *Continental-scale glacial lineations in Canada. A: Large-scale glacial lineations in northern Quebec displayed on ERS-1 radar image. Ice flow was from NNE to SSW and the image is approximately 50 km wide. North is at the top of the page. [Photograph: European Space Agency, processed by C. D. Clark]. B: Cross-cutting glacial lineations and rogen moraines in Canada shown on a Landsat TM image. The image is approximately 40 km wide. The dominant ice flow direction is recorded by the large-scale glacial lineations 10–20 km in length, trending in a NE–SW direction. North is at the top of the page. Superimposed upon these is evidence of a subsequent ice flow phase; note the cross-cutting pattern in square E2 and the superimposed drumlins in E3. Rogen moraines are also present in squares A5, A6 and A7. A large esker runs from square A2 to G7. [Photograph: processed by C. D. Clark]*

reflect diminishing glacial energy regimes such as those found during deglaciation as ice thins and slows. In contrast, single populations reflect a relatively
constant and persistent glacial energy condition. For a single population to survive deglaciation without having drumlins superimposed, it must occur rapidly
and ice must be unable to overcome the resistance of the bed.

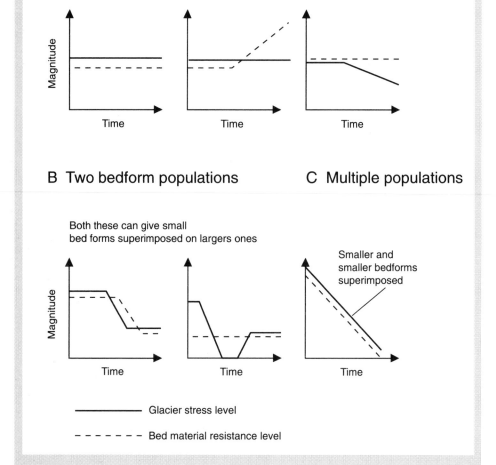

A Single bedform populations

B Two bedform populations   C Multiple populations

Both these can give small
bed forms superimposed on largers ones

Smaller and
smaller bedforms
superimposed

——————— Glacier stress level

– – – – – – – Bed material resistance level

Source: Rose, J. & Letzer, J. M. 1977. Superimposed drumlins. *Journal of
  Glaciology* **18**, 471–480. [Diagrams modified from: Rose & Letzer (1977) *Journal
  of Glaciology* **18**, Figs 3 and 6, pp. 474 and 478]

these two locations two populations of bedforms coexist in a superimposed or cross-cutting fashion (Figure 9.20). Cross-cut bedforms therefore hold important information about changing patterns of ice flow.

The formation of subglacial bedforms has interested and intrigued successive generations of glacial geologists and has been the subject of numerous research papers. Despite this interest there is still no single accepted theory for their formation. There is, however, a general consensus about the formation of glacial flutes. Their formation is explained by the presence of a large boulder or bedrock obstacle at their up-ice end. As the ice flows around the obstacle a cavity or area of low pressure will form in its lee. The pressure on the sediment imposed by the ice will be higher on either side of this area of low pressure. Consequently, subglacial sediment may flow along this pressure gradient—high to low pressure—forming a linear ridge in the lee of the boulder (Figure 9.21). The flute grows in length, down-ice, as the low-pressure area extends in front of the sediment ridge. Recent observations suggest that subglacial meltwater flow within the cavity may accentuate the morphology of the flute by eroding sediment along its flanks. Although this model is widely accepted, not all flutes have boulders or bedrock obstacles at their up-ice ends. There are two possible explanations: (1) the boulder that was there has been subsequently removed by ice flow; or (2) there was never a boulder at the head of the flute and an alternative explanation for the formation of these flutes is required, perhaps in the same way that megaflutes and drumlins form.

In contrast to the formation of flutes, there is little consensus about the formation of drumlins, megaflutes and rogens. The diversity of these landforms is such, particularly in the context of their internal composition, that some researchers have suggested that there may not be a single mechanism responsible for their formation. This concept is known as **equifinality**: different processes give the same morphological products. The acceptance of such an idea should, however, be consequent only upon our failure to find a universal theory. A general model for the formation of drumlins, megaflutes and rogens must be able to explain the following:

A. Variables in the theory must be able to account for the different subspecies of subglacial landform: megaflutes, drumlins and rogens.
B. It must account for the different composition and structure of drumlins, megaflutes and rogens; in particular, for the presence of the three main types of drumlin core commonly found: (1) bedrock, (2) till, and (3) bedded sands and gravels.
C. It must account for the spatial distribution of bedforms: why do they only occur beneath certain parts of an ice sheet?
D. It must account for the rapid rates of formation observed at modern glacier margins.

Most attention has focused on explaining the formation of drumlins. Numerous models, hypotheses and explanations have been proposed. These can be grouped into four broad categories:

1. **Drumlins as the product of subglacial deformation.** This idea is based on the presence of a deforming layer of till which is in transport beneath the glacier

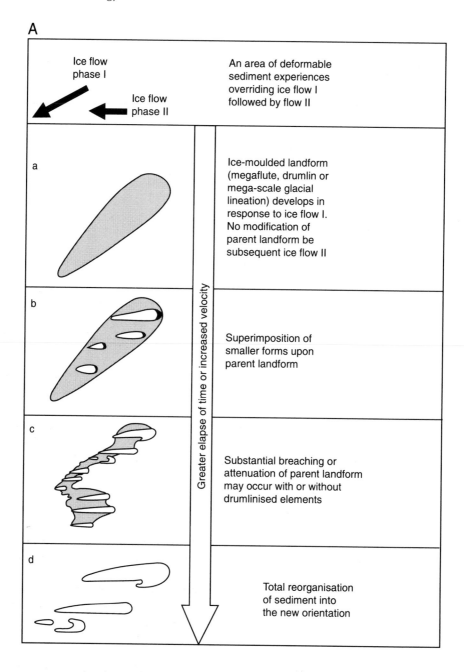

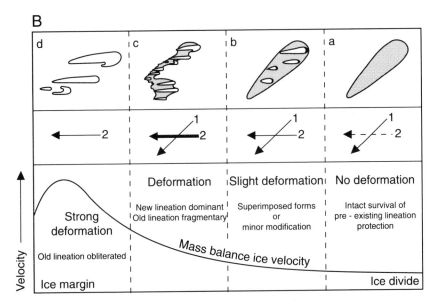

Figure 9.20 *Pattern of cross-cutting relationships within subglacial bedforms formed by two phases of ice flow beneath an ice sheet.* **A:** *Types of observed cross-cutting or superimposed relationships found beneath former ice sheets.* **B:** *Distribution of types of cross-cut bedforms in relation to ice geometry and mass balance velocity (see Figure 3.15). Under the ice divide a single population of bedforms is present which belongs to the first phase of ice flow. At the ice margin a single population of bedforms exists which reflects the second phase of ice flow. Between these two extremes, two populations of bedforms coexist. [Reproduced with permission from: Clark (1993) Earth Surface Processes and Landforms* **18***, Figs 12 and 13, pp. 22 and 23]*

(Box 3.3). This deforming layer moulds itself around subglacial obstacles such as bedrock bumps, boulder clusters, folded sediment and non-deforming sediment to form the drumlin.

2. **Drumlins as the product of subglacial lodgement.** These theories explain the origin of drumlins by the accretion of lodgement till around subglacial obstacles.

3. **Drumlins as the product of melt-out of debris-rich ice.** Beneath cold-based glaciers, basal ice rich in sediment may thicken irregularly due to folding and thrusting. The melt-out of this irregular layer of basal debris may produce an undulating or drumlinised surface (Figure 8.13A). This model has in particular been used to explain the formation of rogens.

4. **Drumlins as fluvial infills or erosional remnants of subglacial floods.** This model stems from the apparent similarity between the morphology of drumlins and the shape of scour marks produced by fluvial erosion, although the scale of the two landforms is very different. It has been suggested that large subglacial floods may scour the base of an ice sheet, creating cavities similar in morphology, but much larger, to scour marks. These are then infilled by fluvial

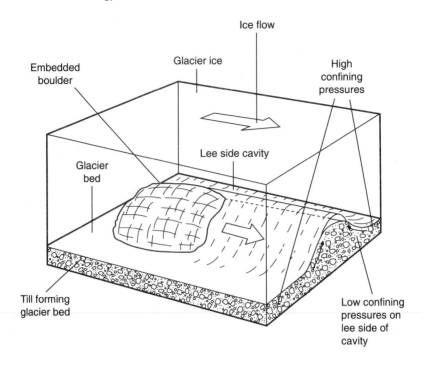

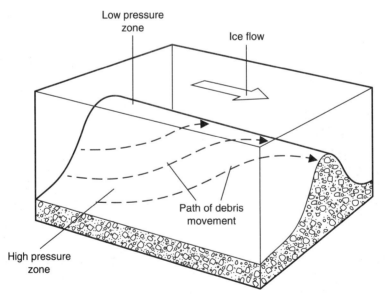

Figure 9.21   *The formation of a glacial flute*

sediment or by subglacial till. Alternatively, subglacial meltwater erosion during a flood may dissect the bed and leave scour remnants which are subsequently streamlined by ice flow into drumlins. In recent years this explanation has attracted many advocates because it explains the presence of fluvial sands and gravels which have not been deformed by flowing ice (Box 9.4). Unfortunately, however, the mechanism for these large floods remains unexplained, nor is there strong independent evidence for them.

BOX 9.4: DRUMLINS FORMED BY SUBGLACIAL MELTWATER

Over the last ten years J. Shaw has championed the idea that drumlins are formed by subglacial floods. Two distinct mechanisms are envisaged. First, flow separation within turbulent subglacial sheet flows erodes/melts cavities in the base of the glacier which are then infilled by sediment transported in hyperconcentrated flows. The second mechanism involves the development of horseshoe vortices within the subglacial flows which erode the substrate to leave remnant

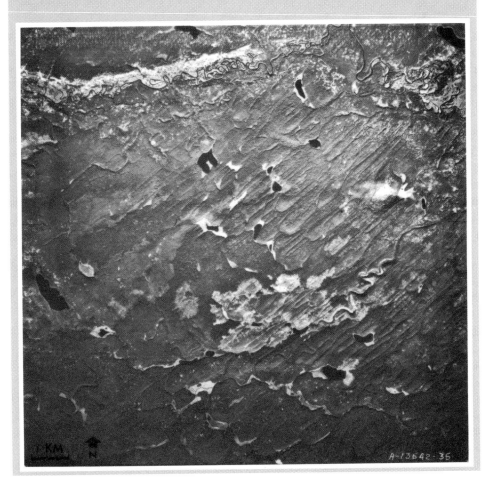

ridges of sediment mass that form drumlins. According to Shaw (1994), the adjacent air photograph above shows drumlins defined by large hairpin erosion marks near Prince George, British Columbia (compare with Figure 6.4D).

This model has the important implication that drumlins may not be orientated in the direction of glacier flow. In the context of the former Laurentide Ice Sheet in North America, the widespread floods necessary to generate the huge fields of drumlins present are explained by the drainage of large proglacial lakes and of subglacial water bodies. The mechanisms by which floods are generated elsewhere are not explained. This whole hypothesis is based primarily on the morphological similarity between erosional scour structures cut in bedrock surfaces and the morphology of many larger drumlins.

Source: Shaw, J., Kvill, D. & Rains, B. 1989. Drumlins and catastrophic subglacial floods. *Sedimentary Geology* **62**, 177–202. Shaw, J. 1994. Hairpin erosional marks, horseshoe vortices and subglacial erosion. *Sedimentary Geology* **91**, 269–283. [Photograph: Air Photograph A-13542–35 from the National Air Photograph Library, Department of Energy, Mines and Resources Canada]

Of these four explanations the most likely is that drumlins form through the process of subglacial deformation. The detailed mechanisms of this are still unclear but G. S. Boulton has proposed a preliminary theory, based on subglacial deformation, which fits the requirements of a general model outlined above. This theory is reviewed below.

When a glacier flows over soft deformable sediment Boulton suggests that three horizons may be identifiable. At the surface is the A horizon, which is rapidly deforming and in which material is in transport. Beneath this there is a slowly deforming horizon (B_1 horizon), below which the sediment is not deforming but stable (B_2 horizon) (Figure 8.12). The boundaries between the three horizons should not be considered to be planar but will vary, as will the thickness of each horizon, due to the changing properties of the sediment. For example, the presence of the slowly deforming horizon (B_1 horizon) is dependent on the **rheology** or stiffness of the sediment. If the sediment is stiff and not easily deformed it may be absent. Sediment rheology is controlled by a range of variables of which **pore water pressure** is of particular importance. Pore water pressure is the pressure of the water in the pores or interstices within the sediment and helps determine intergranular friction. If the pore water pressure is high, individual grains of sediment are pushed further apart and the friction between one grain and the next is reduced. The lower the level of intergranular friction, the more easily the sediment will deform. Fine-grained sediments tend to have a higher pore water content and pressure than coarse sediments and will therefore deform more easily. Pore water pressure may also be reduced by increasing the effective normal pressure imposed on a sediment, since this tends to drive off water, provided it can drain away. This will increase intergranular friction within the sediment. The shape and size of individual grains of sediment also help determine intergranular friction.

The nature of the boundary between the A and B horizons may either be erosional or depositional, depending upon where the glacier is experiencing extending or compressional flow. As we saw in Section 8.1.3, where a glacier is experiencing extending flow the deforming layer may down-cut (erosion) into the sediment pile beneath, assimilating new sediment (B horizon) into the deforming layer (A horizon). In this case the junction between the A and B horizons is erosional. In areas of compressive flow the deforming layer will grow by accumulation of till transported laterally in the deforming layer from up-ice areas. In summary, therefore, the nature of the deforming layer is a function of compressive or extending ice flow (Figure 8.6) and of variation in the rheology of the deforming sediment beneath the glacier.

Boulton developed a semi-quantitative flow model for the deformation of the rapidly deforming A horizon on the basis of field observations (Box 3.3). Using this model he was able to predict how the rapidly deforming A horizon would become moulded around an obstacle to form a drumlin. Figure 9.22A shows the flow lines within a layer of soft deforming sediment. There are two zones of enhanced sediment flow either side of the obstacle and a zone of slower flow in its lee. This pattern of sediment flow produces a sheath of soft sediment around the core, as shown in Figure 9.22B. The sediment within this sheath is not stationary, although the shape of the sheath is, since sediment is added at the up-glacier side and removed down-glacier. If the glacier decays and/or the stress field beneath it changes then the deforming A horizon will become stationary around the core to form a drumlin.

The range of drumlin forms present may be explained by the deformation of this A horizon around either fixed obstacles or mobile obstacles. The obstacles need not, however, necessarily be visible at the surface but simply provide a rigid or stiffer area within the deforming bed. Three types of obstacle were considered in the theory: (1) bedrock obstacles (Figure 9.23); (2) folds within the B_1 horizon (Figure 9.24); and (3) undeformed areas of sand and gravel (Figure 9.25).

According to Boulton's model, deforming sediment will thicken in front of a bedrock scarp due to the high effective pressures present here and form drumlinoid noses. These noses form because the high pressures generated by ice flowing against such scarps (see Section 4.3 and Figure 4.6) expel water from the sediment and therefore reduce its ability to deform. The reduced rate of deformation (i.e. sediment transport) causes the sediment to accumulate. Deforming sediment will also form in the lee of bedrock knobs, due to the decrease in flow of the deforming sediment (Figure 9.23). If the supply of deforming sediment is large then the tail of sediment around the bedrock obstacle will remain like a stationary or standing wave. In this case there is a constant throughput of sediment within the drumlin; it is added upstream as quickly as it is removed downstream. However, if the supply of deforming sediment is small, for example if it is just a patch of deforming sediment, then the drumlin formed around the bedrock knob will move past the obstacle as sediment is removed from the up-glacier face but not replaced. The pattern of drumlins created by bedrock obstacles depends largely on the availability of sediment and the roughness of the bedrock surface (Figure 9.23).

Folding along the boundary between the B_1 horizon and the A horizon may provide a focus for a drumlin. If the properties of the B_1 horizon vary in the

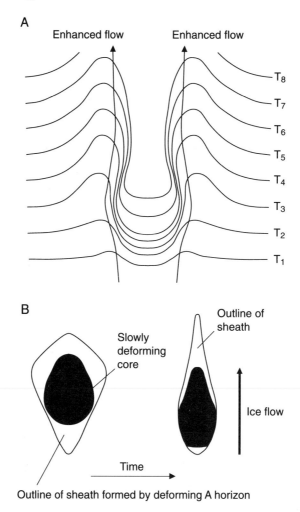

A

Enhanced flow   Enhanced flow

T8
T7
T6
T5
T4
T3
T2
T1

B

Slowly
deforming
core

Outline of
sheath

Ice flow

Time

Outline of sheath formed by deforming A horizon

Figure 9.22   *Drumlins by subglacial deformation. **A:** The pattern of flow within a deforming layer passing around a rigid cylinder. The progressive deformation of an originally straight transverse line is followed from T1 to T8. Note the reduced rate of deformation in the lee of the cylinder and the enhanced flow along its flanks. **B:** Outline of the shape formed by a sheath of soft deforming sediment around a slowly deforming core. [Modified from: **B:** Boulton (1987) In: Menzies, J. & Rose, J. (Eds) Drumlin Symposium, Balkema, Fig. 11, p. 49]*

direction of flow, for example if the sediment becomes stiffer down-ice, then its rate of deformation will change (high to low) which may lead to compression and folding. Deformation of the A horizon around such folds may form a drumlin as shown in Figure 9.24. Repeated folding and refolding of the original fold may cause it to be de-rooted, in the same way that a piece of chewing gum may be stretched and stretched until it finally breaks. Once the fold has been de-rooted it

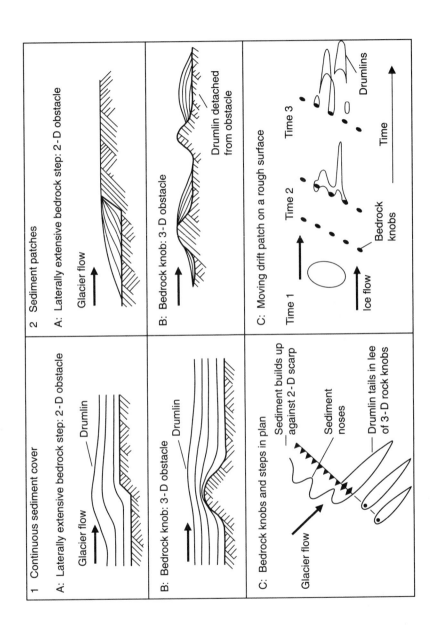

Figure 9.23 *Morphology of a deforming layer moving over an irregular bedrock surface. Accumulations of deforming sediment, drumlins, are static where the sediment supply or cover is continuous, but mobile where the sediment cover is patchy and the supply is therefore discontinuous. [Modified from: Boulton (1987) In: Menzies, J. & Rose, J. (Eds) Drumlin Symposium, Balkema, Fig. 27, p. 72]*

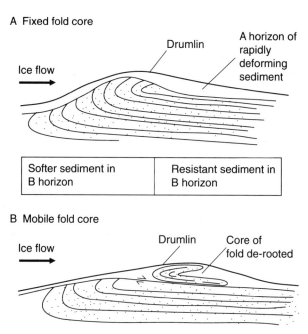

A Fixed fold core

Drumlin

A horizon of rapidly deforming sediment

Ice flow

| Softer sediment in B horizon | Resistant sediment in B horizon |

B Mobile fold core

Ice flow

Drumlin   Core of fold de-rooted

Figure 9.24 *Morphology of a deforming layer around a fold generated at the interface between the A and B horizons. **A:** Fixed fold core and static drumlin. **B:** De-rooted, mobile fold core and therefore mobile drumlin. [Modified from: Boulton (1987) In: Menzies, J. & Rose, J. (Eds) Drumlin Symposium, Balkema, Fig. 27, p. 73]*

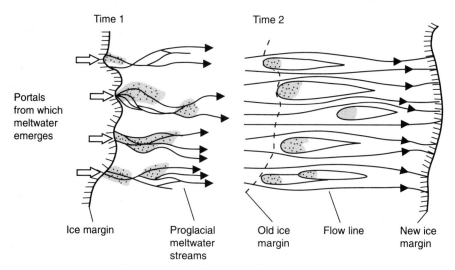

Time 1   Time 2

Portals from which meltwater emerges

Ice margin   Proglacial meltwater streams   Old ice margin   Flow line   New ice margin

Figure 9.25 *Drumlins initiated around cores of stiff undeforming sediment, in this case coarse gravel bars. Time 1: Deposition of coarse gravel close to meltwater portals. Time 2: Ice advance, subglacial deformation around the coarse gravel produces stream-lined drumlins with cores of undeformed gravel. They are analogous to boudins within a highly deformed rock body. [Modified from: Boulton (1987) In: Menzies, J. & Rose, J. (Eds) Drumlin Symposium, Balkema, Fig. 27, p. 73]*

is able to move in the direction of glacier flow. The drumlin will also be able to migrate.

The final type of drumlin core considered by Boulton was one of undeformed sand and gravel. Figure 9.25 shows a glacier fore-field in which coarse gravel is deposited close to the meltwater portal of an ice front. If this ice front was to advance over the outwash sediment the sediment beneath it would deform it. The coarse free-draining gravels would be less likely to deform due to the low pore water pressures within them and would however remain as fixed undeformed cores around which finer-grained sediment, less well-drained and therefore with higher pore water pressures, can deform. The core is shaped and eroded by the deformation of the A horizon over it and by erosion at the base of the deforming layer. In this way drumlins with undeformed cores of bedded fluvial sands and gravels may be generated (Box 9.5). In areas of strongly extending flow, erosion occurs at the boundary between the A and B horizons as deformation cuts down through the sediment pile. The deforming A horizon may be very thin and in this case a drumlin may effectively be formed by erosion along the interface between the A and B horizons.

Within this model megaflutes are simply a subtype of drumlin produced by rapid rates of glacier flow and subglacial deformation, which would tend to produce more elongated forms. Rogen moraines are considered within this model to be formed by the re-moulding of earlier linear bodies of sediment, perhaps formed by earlier ice flow directions (Figure 9.26).

The strength of Boulton's model lies in the fact that it is able to explain all the requirements of a general theory. The model explains the presence of different subspecies of subglacial landform—megaflutes, drumlins and rogens. The model is able to explain the range of different compositions and structures found within drumlins, megaflutes and rogens; in particular, the presence of drumlin cores composed of: (1) bedrock; (2) till; and (3) bedded sands and gravels. The model explains the spatial distribution of bedforms: they only occur where subglacial deformation is possible. Finally the model can explain the rapid rates of drumlin formation observed in some work. It represents a first attempt to develop a unified model of drumlin formation by subglacial deformation and will no doubt be revised with time.

It is not, however, without its opponents. Several researchers have argued that at present there is no direct evidence to suggest that subglacial deformation is a pervasive processes beneath ice sheets. To date, the direct field observation of subglacial deformation is restricted to fast-flowing Antarctic outlet glaciers and some Icelandic glaciers. Opponents of subglacial deformation point to the absence of widespread evidence of subglacial tectonic structures, such as folds and thrusts within glacial sediments, and point to homogeneous till layers and undisturbed sediment sequences. However, as we saw in Section 8.1.3, intense subglacial deformation is not marked by folds and faults but by homogeneous till units. The absence of widespread tectonic structures does not therefore provide evidence against subglacial deformation, since they would only be present in areas subject to low levels of deformation.

BOX 9.5: DRUMLINS AND SUBGLACIAL DEFORMATION

Evidence for the theory of drumlin formation by subglacial deformation is provided by Boyce and Eyles (1991). They examined the Peterborough drumlin field in central Canada, which was formed beneath a lobe of ice at the margin of the former Laurentide Ice Sheet during the last glacial cycle. They examined the morphology and internal composition of the drumlins along a line parallel to the direction of glacier flow (a flow line). Along the flow line the drumlins change from elongate to oval forms as they approach the limit of the ice lobe (A to C). The elongated drumlins are composed of till resting on bedrock, while the oval forms are eroded from outwash sediments. The drumlins with cores of outwash sediments are mantled with till which was derived from the deformation and incorporation of underlying sediment to form a deformation till. A transition therefore occurs along the flow line from depositional to erosional forms close to the ice limit. The down-ice evolution of drumlin form was interpreted by Boyce and Eyles (1991) as a function of the time available for subglacial deformation during the advance of the ice lobe. The duration of deforming bed conditions was greatest up-glacier, where the drumlins are elongated and

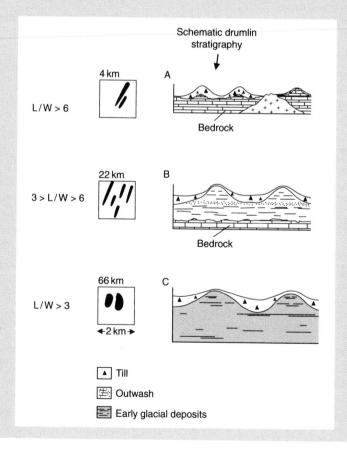

where all the pre-existing sediment has been eroded and incorporated into the deformation till which forms the drumlins. Towards the limit of the ice lobe, where subglacial deformation will only have occurred for a short period, the drumlins are more oval in shape, with cores of undeformed sediment. Insufficient time elapsed here to produce highly streamlined forms or for sub-glacial deformation to cut down and remould the outwash sediment. These observations not only support the subglacial deformation model of drumlin formation but provide rare insight into the internal composition of drumlins, and illustrate how their morphology may vary along a glacier flow line.

Source: Boyce, J. I. & Eyles, N. 1991. Drumlins carved by deforming till streams below the Laurentide ice sheet. *Geology* **19**, 787–790. [Diagram modified from: Boyce & Eyles (1991) *Geology* **19**, Fig. 2, p. 788]

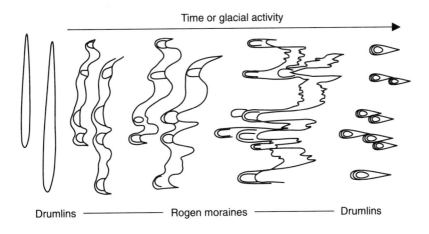

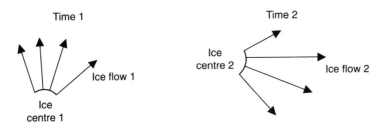

Figure 9.26 *Schematic illustration of the way in which transverse drift ridges may be progressively transformed by deformation into rogen moraine and drumlins. Original drift ridges, mega-scale glacial lineations or large drumlins may reflect an earlier direction of ice flow and are deformed into a rogen moraine by a new ice flow direction associated with a shift in the ice divide or ice centre (see Box 2.2). [Modified from: Boulton (1987) In: Menzies, J. & Rose, J. (Eds) Drumlin Symposium, Balkema, Fig. 28, p. 75]*

Other opponents of subglacial deformation have pointed to the juxtaposition of eskers (see Section 9.3 and Figure 9.19B) with drumlins, and argue that they could not have formed at a deforming bed where subglacial water flow is believed to occur in shallow canals cut within the sediment and not in subglacial tunnels. This is only valid provided that: (1) the eskers formed in subglacial cavities; and (2) they formed at the same time as the drumlins. The eskers may have formed in englacial tunnels and have been subsequently lowered to the glacier bed during deglaciation, alternatively, they may have formed after the glacier bed had ceased to deform, perhaps during deglaciation. Significantly, work on the distribution of eskers at a continental scale has shown that they are restricted to areas that are unlikely to have experienced subglacial deformation (Box 9.6).

Few explanations have yet been proposed for mega-scale glacial lineations but it has been suggested that they may simply represent patches of deforming subglacial sediment that have become fixed in place by deglaciation. The argument relies upon the fact that one would not necessarily expect a deforming layer to be continuous. Where sediment was scarce (for example, if a deforming patch of sediment was moving over a hard substrate), it would tend to be organised into linear sediment patches orientated in the direction of ice flow. Therefore, they may simply be part of the continuum of subglacial bedforms formed by subglacial deformation.

To conclude, there is a vast literature on the formation of subglacial bedforms and a proliferation of theories. In this section we have concentrated on the idea of subglacial deformation, which at present seems to offer the most plausible explanation. Only time will show whether this is correct.

9.2.2 Non-ice-moulded Subglacial Landforms

Subglacial landforms may form in some situations where subsequent ice flow is minimal, and consequently no ice-moulding occurs. One of the most important of these landforms are **geometrical ridge networks** or **crevasse-fill ridges**. The non-genetic term geometrical ridge network is used here in preference to crevasse-fill ridges. They are low (1–3 m) ridges with a symmetrical cross-section. The ridges form networks, when viewed in plan, that have a distinct geometrical pattern. This geometrical pattern is often similar to the pattern of crevasses on the adjacent ice margin. The ridges are normally composed of basal till. Traditionally they are believed to form by the squeezing of basal till into subglacial crevasses. If crevasses penetrate to the base of a glacier then basal debris may be squeezed into them to reach an englacial position. Squeezing may also occur into subglacial tunnels. Survival of these ridges is only possible if the ice is stagnant or becomes cold-based immediately after ridge formation. Recent work has stressed the importance of englacial thrusts in the formation of these ridge networks. Small ridges of basal sediment may be pushed up at the base of englacial thrusts (Figure 9.27). In some cases these small thrust ridges form geometrical networks. It is probable therefore that geometrical ridge networks form by a variety of different mechanisms, including thrusting and the flow of sediment into basal crevasses.

BOX 9.6: ESKERS AND SUBGLACIAL DEFORMATION

It has frequently been argued that the presence of eskers, assumed by many to form only in subglacial tunnels, is inconsistent with subglacial deformation. Subglacial deformation would not only destroy eskers as they form but subglacial drainage probably occurs in deforming beds via shallow channels, known as canals, cut into the deforming sediment and not into R-channels cut into the ice. Consequently, the presence of eskers has frequently been used to argue against the existence of a pervasive deforming bed beneath the former mid-latitude ice sheets in the northern hemisphere.

Clark and Walder (1994) examined this idea. They started from the assumption that eskers should be rare where subglacial bed deformation occurred and that they would be more common where the bed was rigid. On a rigid bed subglacial drainage is likely to be via a series of R-channels cut into the ice which can become blocked by sediment to form eskers. They went on to argue that bed deformation would be most likely where the subglacial till was relatively continuous, fine-grained, and of low permeability; a situation most common in regions where it is derived primarily from underlying sedimentary rock. In contrast, bed deformation would be unlikely where discontinuous, coarse-grained, high-permeability till derived from underlying crystalline bedrock occurs. To test these ideas they constructed maps of the distribution of eskers formed during the last glacial by both the North American and Scandinavian ice sheets using air photographs and satellite images. They found a close correlation between the outcrop of crystalline bedrock with a discontinuous cover of coarse-grained high-permeability till and the distribution of eskers. Eskers were not commonly recorded where the sediment was fine-grained and therefore likely to have experienced subglacial deformation.

This work carries the interesting implication that the distribution of eskers within an ice sheet is primarily controlled by the underlying geology.

Source: Clark, P. U. & Walder, J. S. 1994. Subglacial drainage, eskers, and deforming beds beneath the Laurentide and Eurasian ice sheets. *Geological Society of America Bulletin* **106**, 304–314.

At modern glaciers, geometrical ridge networks are commonly associated with glacier surges. The rapid ice flow during the surge opens up basal crevasses and causes steep thrusts to form, which become filled with basal sediment as the ice lobe formed by the surge becomes stagnant. Exceptionally good examples of crevasse-squeezed ridges are to be found on the island of Corahomen in the Ekmanfjorden fjord, Svalbard. As shown in Figure 9.28, the Sefstrømbreen glacier surged forward into the Ekmanfjorden fjord and part of the ice margin became grounded on the island of Corahomen. As the heavily crevassed ice margin settled into the marine mud it had transported on to the limestone island, from the floor of the fjord, sediment was squeezed out in front of the ice margin to form a moraine, while sediment was also squeezed into basal crevasses. On deglaciation,

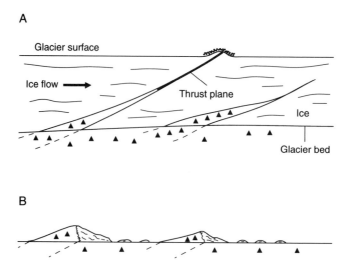

Figure 9.27   *Formation of a small transverse ridge by englacial thrusting. Basal sediment is frozen to the glacier bed prior to thrusting and then transported along the thrust plane from where it melts out to form a ridge. Ridges similar to this can be seen in situ at the base of the thrusts shown in Figure 7.11*

a series of crevasse-squeezed ridges was uncovered behind a distinct moraine ridge (Figure 9.28).

Crevasse-squeezed ridges are widely recognised from glacier margins that have surged and are regarded by many as a landform that is diagnostic of a surging glacier. However, the preservation potential for these landforms is poor due to their low morphological form, and few examples have been recorded with confidence from former glaciers (Box 9.7).

9.3 GLACIOFLUVIAL ICE-MARGINAL LANDFORMS

The principal glaciofluvial ice-marginal landforms are **outwash fans**, **kames** and **kame terraces**. As glacial meltwater emerges from the glacier, the pressure gradient which drives it from beneath the glacier drops, and its velocity falls. As a consequence, it deposits sediment rapidly to form a variety of **outwash landforms**. The morphology of these landforms depends on: (1) their location with respect to the ice margin; (2) the presence or absence of buried ice; and (3) the total amount of sediment in transport within the meltwater. Deposition by meltwater does not only occur beyond the ice margin but frequently starts on or builds back over the glacier, incorporating large amounts of buried ice.

The morphology of ice-marginal glaciofluvial landforms depends on the rate at which any buried ice melts relative to the rate at which the drainage system evolves. As a glacier retreats the drainage pattern will change, and on the outwash surface river channels and bars are abandoned. The morphology of these aban-

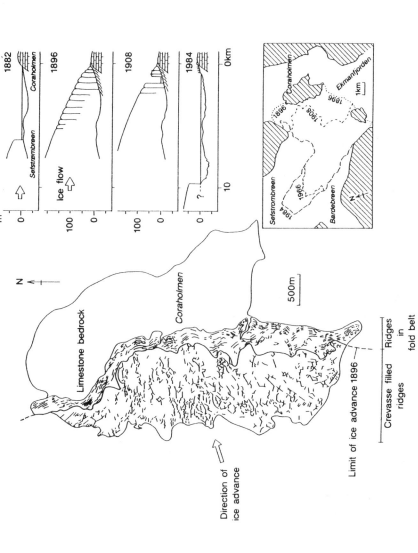

Figure 9.28 *Crevasse-squeezed ridges on the island of Coraholmen in Ekmanfjordan fjord, Svalbard. Sefstrømbreen surged or advanced rapidly onto Coraholmen in 1896. The limit of the advance is marked by a moraine or fold belt formed not by frontal push-ing but by the squeezing out of subglacial sediment. Behind this moraine, crevasse-squeezed ridges occur with a pronounced rectilin-ear pattern in plan form. [Modified from: Boulton & Meer (1989) Preliminary Report on an Expedition to Spitsbergen in 1984 to Study Glaciotectonic Phenomena (Glacitecs' 84), Universiteit van Amsterdam, Figures 62, 68 and 73, pp. 98, 108 and 119]*

BOX 9.7: LINEAR TILL RIDGES BENEATH COLD BASE ICE?

Kleman (1988) describes a series of unusual linear till ridges in the southern Norwegian–Swedish mountains. These ridges are located high on gentle mountain slopes and usually have an ordered appearance in plan form. Rectilinear zigzag patterns are particularly common. In some areas the pattern of ridges resembles patterns typical of subglacial drainage networks. The ridges are, however, all composed of homogeneous basal till. The ridges have no clear relationship to the landforms of deglaciation within the area. Kleman (1988) suggests that these ridges are a form of crevasse-fill produced by subglacial till being squeezed into basal crevasses and meltwater tunnels. For this to occur the ice must have been warm-based. The preservation of these landforms is only possible therefore if: (1) they formed close to the ice margin and melted out almost immediately after formation, or (2) immediately after formation the ice became cold-based and remained so until deglaciation was complete. Kleman (1988) argues that the latter is more likely since the features bear no relationship to landforms of deglaciation within the area. The mechanism and cause of such a rapid change in basal thermal regime is unclear however. The paper is of considerable importance as it illustrates how landforms may be preserved during deglaciation if the ice is cold-based.

Source: Kleman, J. 1988. Linear till ridges in the southern Norwegian–Swedish mountains—evidence for a subglacial origin. *Geografiska Annaler* **70A**, 35–45. [Photographs: J. Kleman]

doned outwash surfaces depends upon the amount of buried ice within them and whether it has melted out before they were abandoned.

The conceptual model in Figure 9.29 shows two glacier margins at which an outwash surface is being deposited. As the two glaciers retreat, the drainage system evolves and this outwash surface is abandoned. At the glacier margin with no buried ice, the surface morphology reflects the fluvial depositional processes which deposited it: a morphology of bars, channels and river terraces. In contrast, the ice margin with buried ice may follow two different evolutionary paths, depending upon whether the buried ice has melted out before or after the outwash surface is abandoned. If the ice melts out after the fan is abandoned then the outwash surface will be deformed by subsidence. This may be confined to the occasional **kettle hole** (an enclosed hollow formed by the melt-out of buried ice) if the proportion of buried ice is small, but if it is large then the whole surface will be deformed into an area of **kame and kettle topography**. In contrast, if the melt-out of the buried ice is complete prior to the outwash surface being abandoned then there will be little evidence of subsidence on the surface, although it will be evident in the sedimentary structure of the landform (Figure 8.18). As kettle holes form, the meltwater streams will tend to be diverted into them. Deposition will proceed rapidly within these kettle holes due to the standing water within them and the velocity reduction thereby caused in streamflow. In this way areas of sub-

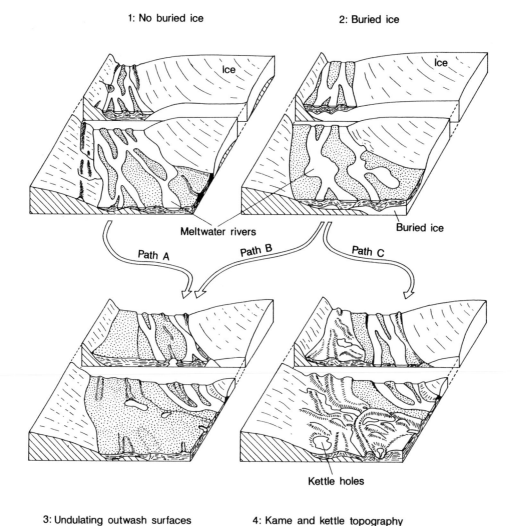

1: No buried ice

2: Buried ice

Meltwater rivers

Buried ice

Path A   Path B   Path C

Kettle holes

3: Undulating outwash surfaces   **4: Kame and kettle topography**

Figure 9.29 *Conceptual model showing the evolution of two outwash surfaces, one of which is underlain by buried ice, the other is not. The key control on the outwash morphology that results is the timing between the melt-out of the buried ice and the abandonment of the outwash surface. If the surface is abandoned before all the ice has melted out then a kame and kettle topography results. Alternatively, if the buried ice melts out before the surface is abandoned its presence may not be visible in the outwash surface since meltwater streams tend to infill the kettle holes as they form*

sidence are infilled as they form. The two cases illustrated in Figure 9.29 simply illustrate end members of a continuum; the character of the landsurface that results will depend on: (1) the rate at which the buried ice melts; (2) the rate of glacier retreat and the rate at which the outwash surfaces are abandoned; and (3) the rate of fluvial deposition and its distribution across the outwash surface. Two groups of outwash landforms can therefore be recognised: (1) outwash plains and fans, and (2) kames and kame terraces.

Outwash fans build up in front of a stationary ice margin (Figure 9.30). The apex of each fan is centred at the point at which the meltwater emerges. Fans develop because deposition of coarse material occurs relatively close to the meltwater portal, while the finer fraction is transported further. Outwash fans may merge away from the glacier and grade into large braided river sequences, forming an **outwash plain (sandur)**. The ice-contact face of outwash fans is usually underlain by buried ice which melts when the fan is abandoned to give a kame and kettle topography typical of that caused by subsidence due to the melt-out of buried ice (Figure 9.30). The distal flanks of the fan may be pitted by occasional kettle holes. The surface of the fan contains a shallow relief formed by abandoned channels and bars. Outwash fans develop whenever the ice margin is stationary for a period of time

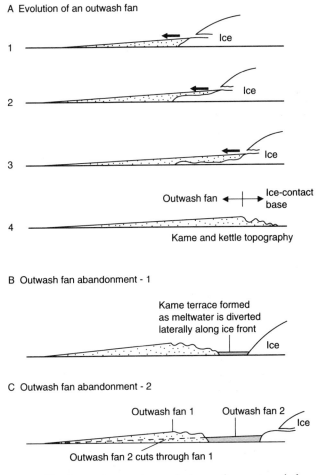

Figure 9.30   *The formation and morphology of an outwash fan*

and sediment is able to accumulate over the ice margin (Figure 9.30). As ice retreats away from the ice-contact face of a large fan, proglacial drainage may be diverted laterally along the ice front between the ice margin and the ice-contact face of the fan. In these situations a series of kame terraces may form against the ice-contact face of the fan (Figure 9.30 and see below). Where an ice margin is advancing or retreating at a steady rate there is usually insufficient time for fans to accumulate, and a broad and relatively flat **outwash plain** is formed instead. As an ice margin retreats, its meltwater will dissect earlier outwash surfaces and flights of terraces may form (Figure 9.30). These terraces are erosional features and are commonly paired on either side of the dissecting channel or valley.

The morphology of outwash fans may be modified if they are subject to periodic episodes of catastrophic or jökulhlaup flow. These high-magnitude, low-frequency flows may deposit large amounts of sediment on the fan very rapidly, causing aggradation and fan modification. They may also dissect the fan by cutting large channels that are often lined with large imbricate boulders. These periods of high-magnitude flow are normally associated with major events in the evolution of outwash fans, such as fan abandonment or changes in the direction of proglacial drainage. Jökulhlaups may also result in the formation of distinct pock-marks or rimmed kettle holes on outwash surfaces, formed by the transport and melt-out of large blocks of ice in the flood (Box 9.8). Obstacle or scour marks may also form around large boulders.

Where a glacier rests against a reverse slope, such as a valley side, meltwater is diverted laterally along the ice margin and glaciofluvial sediment will be deposited between the slope and the ice-front. When the ice retreats this sediment accumulation is left as a **kame terrace** (Figure 9.31). The morphology of a kame terrace depends largely upon: (1) the size of the meltwater stream; (2) the steepness of the ice margin; and (3) the angle of the slope or hillside against which the glacier rests. If the meltwater stream is large, the ice margin steep and the valley-side slope gentle then a large broad kame terrace will result which is only underlain by buried ice close to the ice margin (Figure 9.31). In this case, when the ice retreats, a broad terrace surface will be produced which has an outer edge that consists of a narrow belt of kame and kettle topography formed by subsidence of the buried ice at the terrace edge. The sedimentary architecture of these kame terraces is dominated on the ice-contact side of the landform by subsidence structures consisting of extensional faults and folds (Figure 8.17). Alternatively, if fluvial deposition occurs over a large area of the ice margin then a narrow terrace bordered by a very broad belt of kame and kettle topography will result (Figure 9.31). Where kame terraces are formed against steep valley sides by meltwater streams, small irregular fragments of terrace often form. These are often associated with linear **kames** formed in ice-walled channels which parallel the ice margin (Figure 9.31). Internally these kame terraces and kames may not only contain sediment typical of ice-proximal meltwater but also large amounts of supraglacial debris and flow till deposits (see Section 8.1.8).

One of the finest examples of a large assemblage of kame terraces is that along the shores of Loch Etive in Scotland. During the Younger Dryas a valley glacier terminated at the mouth of Loch Etive. Large kame terraces mark the lateral margins

BOX 9.8: BOULDER RING STRUCTURES PRODUCED DURING JÖKULHLAUP FLOWS

Maizels (1992) investigated the origin of a series of ring structures formed on an outwash plain in southern Iceland following a volcanically triggered jökulhlaup in 1918. The ring structures consist of kettle holes with distinct ramparts around their edge. These ramparts are typically 4 m high and the kettle holes are usually about 40 m in diameter. The ramparts are composed of diamicton which dips steeply into the central hollow, which is infilled by laminated sediment. These ring structures are believed to have formed by the *in situ* melting of debris-rich ice blocks transported on the surface of a hyperconcentrated mix of sediment and water during the jökulhlaup.

Maizels (1992) combined field observation with laboratory simulations, which suggest that the morphology of the ring structure is controlled by concentration of debris within the ice and the degree to which the block is submerged within the surrounding sediment. On the basis of the laboratory observations, a sequence of four types of rimmed kettle hole or boulder ring structure can be distinguished. They form a continuum along a line of increasing concentration of debris within the ice blocks, as illustrated below. This

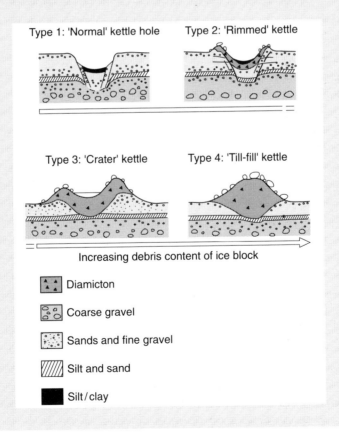

Type 1: 'Normal' kettle hole   Type 2: 'Rimmed' kettle

Type 3: 'Crater' kettle   Type 4: 'Till-fill' kettle

Increasing debris content of ice block

Diamicton

Coarse gravel

Sands and fine gravel

Silt and sand

Silt/clay

research clearly shows how field observations can be combined with simple laboratory simulations.

Source: Maizels, J. 1992. Boulder ring structures produced during Jökulhlaup flows. *Geografiska Annaler* **74A**, 21–33. [Diagram modified from: Maizels (1992) *Geografiska Annaler* **74A**, Fig. 9, p. 31]

A

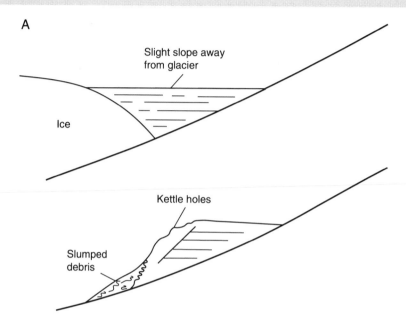

Slight slope away from glacier

Ice

Kettle holes

Slumped debris

B

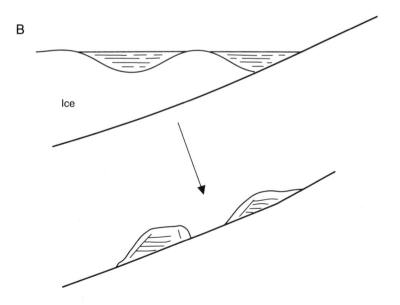

Ice

C

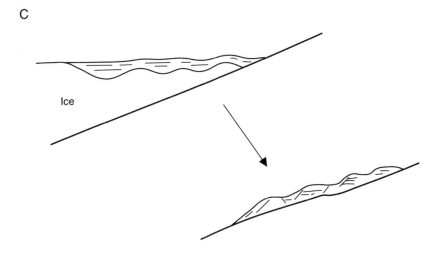

D

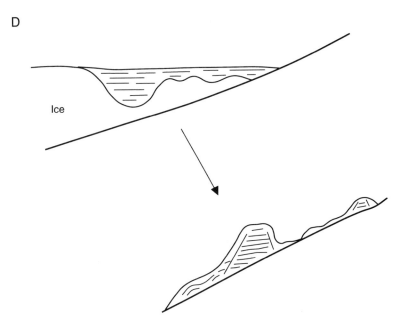

Figure 9.31   *The formation and morphological diversity of kame terraces. The type of terrace that results depends upon the degree to which the outwash is deposited over the edge of the ice margin. If the ice margin is steep then a well-defined terrace may form (**A** and **B**). However, if the ice margin possesses a more gentle profile deposition is more likely to occur on the ice margin, resulting in a more irregular terrace morphology (**C** and **D**)*

of this former glacier. The outside edges of these terraces are heavily kettled and are marked by narrow belts of hummocky topography. This hummocky topography also contains ridges which have a sinuous plan form. These formed when englacial tunnels or supraglacial channels supplying meltwater to the terrace became blocked with sediment. As the glacier retreated and the buried ice melted, these sediment-filled tunnels or channels were lowered to the ground to form ridges of sediment that follow the former course of the channel or tunnel. The terraces at Loch Etive merge down-valley into a large ice-contact outwash fan which partially blocks the entrance to the loch.

Kame terraces need not only form at lateral ice margins but may also form at a glacier snout where it rests against a reverse slope such as the ice-contact face of an outwash fan. Kame terraces can be distinguished from outwash terraces formed by the dissection of outwash fans and plains because: (1) kame terraces are formed by deposition not erosion; (2) kame terraces usually contain kettle holes along their front margins and are higher here than at their valley-side margin, while outwash terraces may be kettled but not preferentially at their front edges and also tend to be highest at their valley-side margins; (3) outwash terraces are frequently altitudinally matched across the valley, whereas kame terraces are not; and (4) kame terraces contain subsidence structures at their margins when viewed in section, while outwash terraces do not.

A kame terrace has only one ice-contact margin whereas a **kame** has two or more. Kames form whenever glaciofluvial deposition occurs within an ice-walled channel or depression. They may form at a lateral ice margin (Figure 9.31), but may also form large areas of kame and kettle topography in front of a glacier. Where an ice margin is experiencing large amounts of ice frontal compression and thrusting, a topography of ice-cored ridges may develop (Figure 8.14A). Meltwater rivers will weave their way in between these ridges, depositing sediment between them. As the ice-cored ridges melt, the sediment-filled depression or channels may either be inverted to form kames, or be spread over the surface by slumping and flow to form an undulating hummocky surface. A continuum exists between these two extremes.

Where well-defined kames form they may reveal the position of former river channels. The strongest meltwater flows will have the least buried ice beneath them due to the rapid ablation caused by flowing water. Consequently they will often form the highest kames. It is therefore sometimes possible to reconstruct the patterns of drainage from the elevation of kames.

The pattern of kames, when viewed from above, may either be controlled or uncontrolled. If the pattern of ice-cored ridges, usually determined by the pattern of debris bands and thrusting within the glacier, is regular then the pattern of kames that will result will also have a regular pattern and will therefore be **controlled**. If, on the other hand, the pattern of ice-cored ridges is irregular or random, perhaps because the distribution of debris on the surface of the glacier is complex, then the resulting topography of kames will also be irregular or **uncontrolled**. The margin of Elisbreen in Oscar II Land, Svalbard, contains a series of ice-cored ridges which parallel the ice margin. Here the kame topography that is currently forming is controlled. The pattern of meltwater rivers and lakes at this ice margin is determined by the pattern of ice-cored ridges. Glaciofluvial sediment is accumulating between these ridges, and with distance from the current ice margin the ice-cored moraines are progressively melting out. As this occurs, the topography is inverted and the former fluvial channels are being left as upstanding mounds or kames which trace the former pattern of drainage and are orientated parallel to the former ice margin and its ice-cored moraines. The sedimentary architecture of kames reflects their history of subsidence and of the removal of lateral ice support. Typical subsidence structures found within kames are illustrated in Figure 9.32.

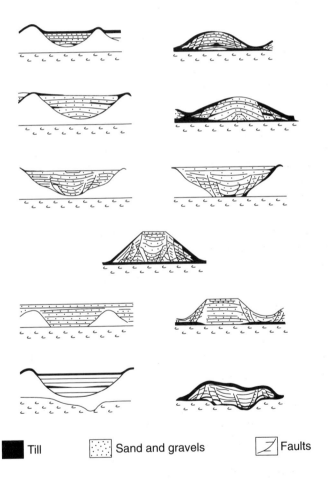

Till   **Sand and gravels**   **Faults**

Figure 9.32 *Typical sedimentary structures found within kames. [Modified from: Boulton (1972) Journal of the Geological Society of London* **128***, Fig. 4, p. 370]*

9.4 GLACIOFLUVIAL SUBGLACIAL LANDFORMS

The principal landform formed by meltwater flow beneath a glacier is the esker. **Eskers** are channel deposits of former subglacial, englacial or supraglacial channels. They are usually slightly sinuous ridges of glaciofluvial sediment which undulate in height along their length (Figure 9.33). Their orientation is controlled by glacier slope and the pattern of water pressure potential within the glacier; they may therefore show little respect for basal topography and need not run downslope.

One can recognise two broad types of esker: (1) single ridge eskers, and (2) braided (anastomosing) eskers. Braided eskers consist of a network of ridges which merge and bifurcate (Figure 9.33). Braided eskers are typically quite short in

Figure 9.33 *Oblique air photograph of an esker system in the Scottish Highlands (Kildrummie kames). The esker consists of a single-crested ridge in the mid-distance and becomes braided or multi-crested in the foreground. [Photograph: Cambridge Air Photograph Library]*

length, the braided reach being usually less than 1 km. They are often associated with areas of kame and kettle topography. Single ridge eskers vary from less than 1 km to hundreds of kilometres in length. Long eskers are typically 400–700 m wide and 40–50 m high, while smaller eskers (< 300 m long) are usually 40–50 m wide and only 10–20 m high.

In general, eskers may vary in cross-profile along their length and may occasionally have a beaded form. This beaded form may simply consist of wider sections at regular intervals along the length of the esker, or consist of a chain of short lengths of esker ridge between which the ridge is barely visible. Eskers occur in a variety of different topographical settings and have no clear relationship with the topography.

In a few cases eskers have been recorded on top of, and therefore infilling, subglacial N-channels. For example, quarry excavations within the Blakeny Esker in Norfolk, England, have revealed a series of small meltwater channels cut into the till surface on which the esker rests. This indicates that the esker formed subglacially and that deposition followed a period of subglacial meltwater erosion. Esker-like bodies of sediment have also been recorded from within units of lodgement till (see Section 8.1.8 and Figure 8.11).

Eskers are typically composed of a core of poorly sorted sand and gravels. Above this core, sorted sands and gravels may occur, possibly with arched bedding dipping out from the centre of the ridge. These sands and gravels are usually well rounded and have palaeocurrent orientations that are parallel to the trend of the esker. The esker may be capped by a thin veneer of till. In general, however, the sedimentology of eskers is highly variable and generalisations are difficult. This perhaps reflects the variety of depositional environments in which they form and the high flow regimes or energy levels present.

Single ridge eskers form when a supraglacial, englacial or subglacial channel or tunnel becomes blocked. Sedimentation in supraglacial channels is easily understood, but sedimentation within englacial or subglacial tunnels is more problematic. By applying theory developed to explain the flow of solids within pipes, it has been suggested that within englacial or subglacial tunnels sediment of all sizes may be in transport as a single mass (see Section 8.2). The high concentration of debris has a buoyant effect, allowing larger particles to be transported more easily. Deposition of these particles is impossible because any constriction of the tunnel caused by deposition would simply increase the water velocity and thereby re-entrain the sediment. Deposition of all the material in transit takes place, however, if the flow in the pipe is suddenly blocked. Obstruction of the tunnel by an ice fall or similar blockage will cause the deposition of large sheets of poorly sorted sand and gravel. Finer-grained arched gravel and sands are predicted by pipe flow theory when discharge falls—a situation that would occur at the end of the melt season. It has been suggested that eskers may form very rapidly due to the very high sediment loads within meltwater streams. A blockage of a tunnel, perhaps by a fall of ice at the ice margin, would cause the deposition of large masses of sand and gravels within the tunnel, while falling discharge may deposit sands and gravels with arched bedding on top. In this way an esker may form rapidly by the deposition of gravel masses associated with permanent or temporary blockage of a tunnel and by the deposition of arched gravel units during declining discharge at the end of the melt season. Subsequently the esker is uncovered by ice retreat or lowered to the ground surface by ablation if the tunnel is located in an englacial position. Modification of the esker during lowering is likely, although observations in Iceland suggest that englacial tunnel fills are lowered quickly, with little re-mobilisation of the esker sediment as the ice retreats.

Similarly, supraglacial channel fills can be infilled with sediment and lowered during ablation. This process has been observed on the margin of Holmstrømbreen in Svalbard (Figure 9.34). Active stream capture on the surface of this glacier causes supraglacial channels to be regularly abandoned. The sediment fill within these channels tends to inhibit ablation of the underlying ice, in contrast to the debris-free ice either side which melts relatively rapidly. The end result is an esker, a ridge of stream sediment standing on an ice core above the flanking ice. These ridges are often flanked by active streams or in some cases by a series of other esker ridges which represent successive generations of the same supraglacial stream.

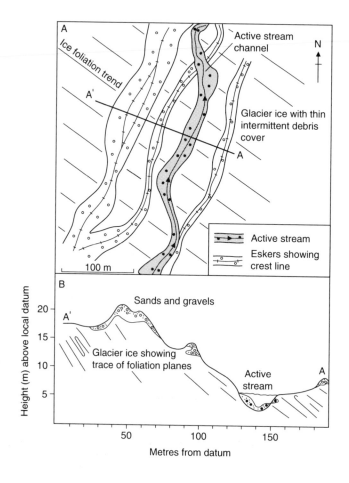

C

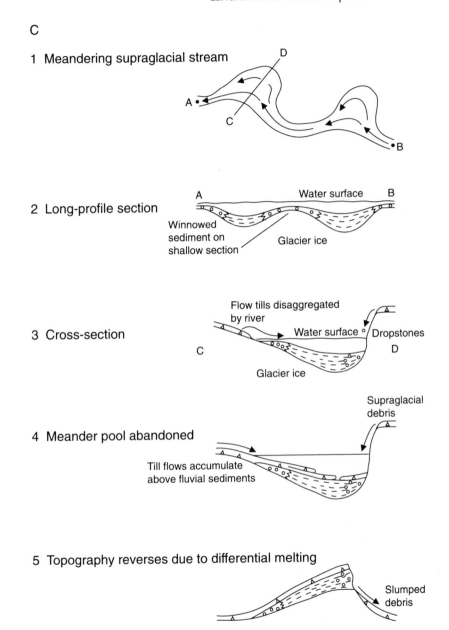

1 Meandering supraglacial stream

2 Long-profile section

3 Cross-section

4 Meander pool abandoned

5 Topography reverses due to differential melting

Figure 9.34   *A series of supraglacial eskers on the surface of Holmstrømbreen in Svalbard.* **A:** *Map of a series of eskers formed by the successive abandonment of channels by a supraglacial stream.* **B:** *Profile across the eskers showing their ice-cored character and their proximity to the current stream channel.* **C:** *The eskers sediments have a wedge-shaped cross-section which reflects the cross-sectional shape of the stream channel.* [Diagram modified from: Boulton & Meer (1989) Preliminary Report on an Expedition to Spitsbergen in 1984 to Study Glaciotectonic Phenomena (Glacitecs' 84). University of Amsterdam, Amsterdam, Figures 10 and 15, pp. 22 and 27]

The formation of braided eskers is a more complex problem since it is difficult to envisage the formation of a braided network of subglacial or englacial tunnels. Although contemporary eskers from Iceland, formed by the infill of englacial and supraglacial tunnels do show simple bifurcation. It has also been suggested that subglacial braided channels and associated eskers may form as a result of catastrophic subglacial floods. This model suggests that when a single channel cannot accommodate the high discharges of water and sediment during a flood event, new channels are produced, causing a multi-channelled subglacial system to develop. The exact mechanisms of this process remain unclear. It is interesting to note that recent observations have described the formation of a linear boulder mass, an esker, within a single subglacial conduit during a jökulhlaup. No evidence of multiple channel formation was observed. An alternative hypothesis is that braided eskers develop supraglacially either by the development of a cross-cutting pattern of time-successive channels (Figure 9.34) or by the lowering (through the melt-out of buried ice) of a supraglacial outwash fan that has deeply incised sediment-filled channels which are inverted during melt-out to form a braided esker.

Preliminary observations from within braided eskers in the Southern Uplands of Scotland support this last idea. The Carstairs kames are a prominent system of braided esker ridges in southern Scotland (Figure 9.35). Within this area there are two large braided esker systems, one centred at Carstairs and the other at Newbiggin. The Carstairs system is over 7.5 km long and several hundred metres wide. The major ridges are generally 10–15 m high, but may locally increase in height to over 30 m. The ridges are sharp-crested and steep-sided. The principal ridges lie within a broad belt of kame and kettle topography, in which low sinuous ridges can occasionally be identified. The pattern of ridges present is quite complex, as illustrated in Figure 9.35. Internally the main ridges appear to be composed of a core of boulder-rich gravel in which 40–50% of the material is over 250 mm in size. Boulders up to 2 m in diameter are also common. The surrounding kames contain much finer-grained sands and gravels more typical of that found on braided outwash surfaces. Faulting within these mounds suggests that the sediment has undergone subsidence associated with the melt-out of buried ice. It has been suggested that the whole sequence formed as a supraglacial outwash fan subject to catastrophic or high-magnitude low-frequency flood events. During periods of normal flow a system of channels and bars developed on the ice-cored fan surface in which fine gravel and sands were deposited. This was followed by a period of catastrophic flow which cut a series of deep channels into the buried ice beneath the fan. These channels were then filled with a coarse boulder gravel as the flow magnitude fell. With the return of normal flow conditions these channels were infilled further by low-magnitude deposits. When the outwash surface was abandoned, perhaps as a consequence of the catastrophic flows, melt-out of the buried ice inverted the topography. The large boulder-filled flood channels were inverted to form the main esker ridges, while the small sediment filled channels produced the kame and kettle topography around them. This model awaits further testing but provides one potential explanation for the formation of some braided eskers. It is interesting to note that this, like other ideas which attempt to explain the formation of braided eskers, involves catastrophic flood events.

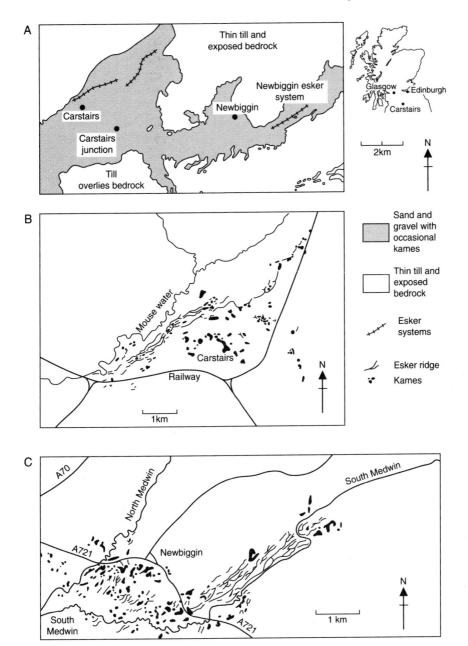

Figure 9.35 The braided eskers of Carstairs and Newbiggin, Southern Uplands, Scotland. *A:* The distribution of sand and gravel in the Carstairs area shows the broad outline of the outwash complex of which the braided eskers are a part. *B:* Morphology of the Carstairs esker system. *C:* Morphology of the Newbiggin esker system. [Based on unpublished work by K. Jenkins]

In summary, eskers may form in a variety of different settings. These can be summarised as: (1) deposition in subglacial tunnels; (2) deposition in englacial tunnels and subsequent lowering; (3) deposition in supraglacial channels and subsequent lowering; and (4) deposition in ice-walled re-entrants at the ice margin.

9.5 SUMMARY

A wide range of landforms are produced by direct glacial deposition on land. These landforms can be divided into two main groups: those which form by the direct action of glacier ice and those which form by deposition from glacial meltwater. Both groups of landforms can be subdivided into those which form along the ice margin and those which form subglacially. Ice-marginal landforms (moraines) form by a combination of processes which include the pushing, dumping and melt-out of glacial debris. The interaction of these processes determines the morphology of the moraines produced, whether it be a push moraine, dump moraine or ablation moraine. Subglacial landforms are produced beneath a glacier and generally have a streamlined form. Subglacial bedforms include flutes, megaflutes, drumlins, rogens and mega-scale lineations, and probably form through the process of subglacial deformation. Non-streamlined landforms, such as crevasse-squeezed ridges, form beneath stagnant or inactive ice.

When meltwater emerges from a glacier, rapid sedimentation occurs in front of the glacier or along its margin. The morphology of the landforms produced depends upon: (1) whether deposition occurs over the ice margin and thereby incorporates buried ice; (2) the rate at which this buried ice melts out in relation to when the outwash surface is abandoned due to glacier retreat; and (3) the rate and distribution of glaciofluvial deposition over the surface. A range of glaciofluvial landforms form at the ice margin, including outwash plains, outwash fans, kame terraces and kames. Glaciofluvial deposition may also occur subglacially to produce eskers. Eskers form either through infill of subglacial and englacial tunnels or by sedimentation in supraglacial channels.

Tables 9.2 and 9.3 summarise the inferences that can be made about the extent, dynamics and glacial conditions under which each depositional landform is normally produced. They provide a key to the clues left about the former glacier that deposited the depositional landscape visible in areas once covered by Cenozoic glaciers.

Table 9.2 *Summary table of the principal landforms of glacial deposition and their significance in the reconstruction of former glaciers*

Seasonal push moraines	Morphology: usually low sediment ridges transverse to the direction of ice flow. Indicative of: the position of the ice margin and of warm-based ice in a maritime climate. They occur where winter ablation is less than winter ice velocity at the snout. Moraine spacing is usually a function of summer ablation and therefore air temperature. The number of moraines along a flow line may provide an estimate of the rate of retreat, assuming annual moraine formation.
Composite push moraines	Morphology: large multi-crested ridges transverse to the ice flow; the ice was not located along the crest of each ridge. Indicative of: position of the ice margin; may also suggest surging behaviour or strong ice compression at the ice margin either due to thermal variation at the snout or due to the presence of a focus for frontal tectonics.
Thrust moraines	Morphology: when ice-cored they consist of single or multi-crested ridges transverse to the ice flow. Indicative of: the tectonic structure, thrust and shear zones, within the ice. They do not provide direct evidence of ice-marginal positions.
Dump moraines	Morphology: usually steep-sided ridges with well-developed scree-like bedding within them. Their morphology may be affected significantly by the withdrawal of lateral ice support. Indicative of: the position of the ice margin. Common as lateral moraines around the margins of warm-based glaciers, although they may occur as frontal moraines, particularly where the ice is cold-based. Cross-valley asymmetry in moraine size may indicate the patterns of debris supply within the glacier basin. Some moraines contain a distinct stratification which may be seasonal in nature.
Ablation moraines	Morphology: variable, ranging from well-defined ridges to belts of mounds, ridges and enclosed hollows. Morphological form may be very strong and organised while buried ice persists and may reflect the structure of thrust and shear planes within the ice. Indicative of: the position of the ice margin. They result from: high supraglacial debris content; high englacial debris content due to a mixed basal thermal regime and freezing-on of abundant debris; or are due to strong compressive thrusting at the ice margin transferring basal debris to the ice surface.
Hummocky moraine	Morphology: mounds, ridges and enclosed hollows with an irregular plan form distribution composed in part of supraglacial till. Indicative of: ice-marginal areas in which the surface cover of debris has prevented ablation. This may result from: high supraglacial debris content; high englacial debris content due to a mixed basal thermal regime and freezing on of

abundant debris; or due to strong compressive thrusting at the ice margin transferring basal debris to the ice surface. It may form as a single area of hummocks or in increments at the ice margin. It is not indicative of widespread glacier stagnation. Any uniform pattern or organisation within the mound reflects the debris structure on or within the ice margin.

Flutes

Morphology: low linear sediment ridges formed in the lee of boulders or bedrock obstacles (L/W>50).
Indicative of: local ice flow directions; thin ice, and the presence of warm-based ice.

Megaflutes

Morphology: linear sediment ridges which may or may not be formed in the lee of bedrock obstacles (L/W>50).
Indicative of: local ice flow directions; thin ice, and the presence of warm-based ice.

Drumlins

Morphology: smooth oval-shaped or elliptical hills composed of glacial sediment (L/W<50). Drumlins may possess other bedforms superimposed upon them, such as small drumlins, megaflutes and flutes.
Indicative of: local ice flow directions; subglacial deformation, and warm-based ice. Superimposed drumlins may record changes in subglacial conditions and ice flow directions.

Rogens

Morphology: streamlined ridges of glacial sediment orientated transverse to the direction of ice flow. The ridge may have a lunate form and be drumlinised.
Indicative of: subglacial deformation and warm-based ice. May provide a record of changing ice flow patterns.

Mega-scale glacial lineations

Morphology: broad, low ridges of glacial sediment which can only be recognised clearly on satellite images (L/W>50). May possess smaller bedforms superimposed upon them.
Indicative of: regional ice flow patterns, subglacial deformation and probably the presence of warm-based ice. Superimposed bedforms record changes in subglacial conditions and ice flow directions.

Crevasse-squeezed ridges
(Geometrical ridge network)

Morphology: low, often straight ridges, with a rectilinear pattern in plan form.
Indicative of: stagnant ice often associated with surging glacier lobes, but may be preserved under cold ice if a rapid change in basal thermal regime occurs post-formation. They may also be used to reconstruct crevasse patterns.

Table 9.3 *Summary table of the principal glaciofluvial landforms and their significance in the reconstruction of former glaciers*

Outwash fans	Morphology: low-angled fan-shaped accumulations of sand and gravel, with braided surface and a fan apex located at a meltwater portal. They usually have a steep ice-contact face which may contain kame and kettle topography. Indicative of: a stationary ice margin with a relatively high meltwater/sediment discharge usually, although not exclusively, associated with a warm-based ice margin.
Outwash plain	Morphology: flat surface of sand and gravel formed by braided river systems. Indicative of: retreating ice margin with a relatively high meltwater/sediment discharge.
Kames	Morphology: irregular collection of mounds and ridges, often with enclosed kettle holes or depressions. Indicative of: areas of outwash deposition in which melt-out of buried ice occurred after the surface had been abandoned by the melt streams. Linear kames, often described as eskers, indicate the location of former channels and therefore stream patterns.
Kame terraces	Morphology: valley-side terraces with outer edges which possess a concentration of kettle holes or belts of kame and kettle topography. Indicative of: the position of the ice margin.
Eskers	Morphology: steep-crested sinuous ridges of variable extent and size. Indicative of: the location of discharge routes within the glacier. If they formed in englacial or subglacial tunnels they should follow the equipotential surface, in which case the eskers may be used to predict the surface slope of the glacier in which they formed.
Braided eskers	Morphology: a multiple series of steep-crested sinuous ridges which form a bifurcating or anastomosing pattern. Indicative of: glaciofluvial sedimentation on the surface of a glacier; may also be indicative of high-magnitude flow events.

9.6 SUGGESTED READING

The morphology and formation of seasonal push moraines is described in the following papers: Hewitt (1967), Price (1969, 1970), Worsley (1974), Birnie (1977), Sharp (1984), Boulton (1986), Humlum (1985) and Krüger (1985, 1993). More complex push moraines are described in papers by Matthews *et al.* (1979), Eybergen (1986), Croot (1987) and Aber *et al.* (1989). The formation of moraines by thrusting at polythermal ice margins is covered by Hambrey & Huddart (1995).

The formation of lateral moraines by dumping is covered in the following papers: Humlum (1978), Small (1983) and Small *et al.* (1984). Cross-valley contrasts in moraine size are discussed in papers by Matthews & Petch (1982), Benn (1989) and Shakesby (1989).

Rains & Shaw (1981) describe the formation of moraines by ablation at cold-based glaciers, and ablation moraines are also discussed by Goldthwait (1951), Weertman (1961), Souchez (1967), Hooke (1970) and Fitzsimons (1990).

The formation of hummocky moraine is discussed by Eyles (1979) and by Hoppe (1952), Gravenor & Kupsch (1959) and Boulton & Eyles (1979).

The identification and interpretation of moraines in formerly glaciated areas is discussed in a wide variety of different papers, a few examples of which are Anderson & Sollid (1971), Pierce (1979), Benn (1992), Bennett (1990, 1994), Bennett & Glasser (1991) and Bennett & Boulton (1993).

The morphology and formation of flutes is discussed in detail in the paper by Boulton (1976) and in the following papers: Hoppe & Schyatt (1953), Rose (1989), Gordon *et al.* (1992), Benn (1994) and Bennett (1995). The literature on drumlins is vast, but much of the early literature has been reviewed and compiled by Menzies (1979, 1984). Recent work on drumlins and rogens is contained within the volume edited by Menzies and Rose (1987) and within two special issues of the periodical *Sedimentary Geology* (volumes 62 and 91). The significance of superimposed drumlins and cross-cut patterns is explored in Rose & Letzer (1977), Boulton & Clark (1990a,b), and in Clark (1993). Some pertinent field observations are contained within the papers by Krüger & Thompson (1985), Krüger (1987) and Boyce & Eyles (1991). The subglacial deformation theory of drumlin formation is covered in Boulton (1987). The idea that drumlins may form by subglacial floods is explored in Shaw (1983, 1989), Shaw & Kvill (1984), Shaw & Sharpe (1987) and in Shaw *et al.* (1989). The significance and formation of mega-scale glacial lineations is dealt with by Clark (1993, 1994) and the formation of crevasse-squeezed ridges is covered by Sharp (1985) and Bennett *et al.* (1996).

The literature on the morphology of glaciofluvial landforms is, in contrast to other depositional landforms, relatively small. The characteristics of outwash fans are discussed by Boulton (1986) and by Price (1969, 1971). Sedimentation on an ice-cored outwash fan is discussed by Thomas *et al.* (1985) and by Hambrey (1984). In his seminal paper, Boulton (1972) describes the development of kames and their characteristic sedimentary architecture. This work is complemented by Paul (1983). The impact of jökulhlaups on outwash surfaces is covered in papers by Desloges & Church (1992), Maizels (1992) and Russell (1994). The kame terraces of Loch Etive, Scotland, are discussed in Gray (1975). General reviews of glaciofluvial landforms are provided by Price (1973) and Gray (1991).

The morphology and formation of eskers is discussed in a variety of different papers; of particular note are Price (1966, 1969), Shreve (1985), Syverson *et al.* (1994) and Warren & Ashley (1994). The characteristics and formation of braided eskers are discussed by Auton (1992) with reference to an example on the Moray Firth in Scotland. The internal composition of eskers is discussed by Banerjee & McDonald (1975) and the presence of N-channels beneath Blakeney Esker in Norfolk, England, is discussed by Gray (1988).

Aber, J. S., Croot, D. G. & Fenton, M. M. 1989. *Glaciotectonic Landforms and Structures*. Kluwer Academic Publishers, Dordrecht.

Andersen, J. L. & Sollid, J. L. 1971. Glacial chronology and glacial geomorphology in the marginal zones of the glaciers Midtdalsbreen and Nigardsbreen, south Norway. *Norsk Geografiska Tidsskrift* **25**, 1–38.

Auton, C. A. 1992. Scottish landform examples—6: the Flemington eskers. *Scottish Geographical Magazine* **108**, 190–196.

Banerjee, I. & McDonald, B. C. 1975. Nature of esker sedimentation. In: Jopling, A. V. & McDonald, B. C. (Eds) *Glaciofluvial and Glaciolacustrine Sedimentation*. The Society of Economic Paleontologists and Mineralogists, Special Publication 23, Tulsa, 132–154.

Benn, D. I. 1989. Debris transport by Loch Lomond Readvance glaciers in Northern Scotland: basin form and the within-valley asymmetry of lateral moraines. *Journal of Quaternary Science* **4**, 243–254.

Benn, D. I. 1992. The genesis and significance of 'hummocky moraine': evidence from the Isle of Skye, Scotland. *Quaternary Science Review* **11**, 781–799.

Benn, D. I. 1994. Fluted moraine formation and till genesis below a temperate valley glacier: Slettmarkbreen, Jotunheimen, southern Norway. *Sedimentology* **41**, 279–292.

Bennett, M. R. 1990. The deglaciation of Glen Croulin, Knoydart. *Scottish Journal of Geology* **26**, 41–46.

Bennett, M. R. 1994. Morphological evidence as a guide to deglaciation following the Loch Lomond Readvance: a review of research approaches and models. *Scottish Geographical Magazine* **110**, 24–32.

Bennett, M. R. 1995. The morphology of glacially fluted terrain: examples from the Northwest Highlands of Scotland. *Proceedings of the Geologist's Association* **106**, 27–38.

Bennett, M. R. & Boulton, G. S. 1993. A reinterpretation of Scottish 'hummocky moraine' and its significance for the deglaciation of the Scottish Highlands during the Younger Dryas or Loch Lomond Stadial. *Geological Magazine* **130**, 301–318.

Bennett, M. R. & Glasser, N. F. 1991. The glacial landforms of Glen Geusachan, Cairngorms: a reinterpretation. *Scottish Geographical Magazine* **107**, 116–123.

Bennett, M. R., Hambrey, M. J., Huddart, D. & Ghienne, J. F. 1996. The formation of geometrical ridge networks (Crevasse-fill ridges) Kongsvegen, Svalbard. *Journal of Glaciology*, in press.

Birnie, R. V. 1977. A snow-bank push mechanism for the formation of some 'annual' moraine ridges. *Journal of Glaciology* **18**, 77–85.

Boulton, G. S. 1972. Modern Arctic glaciers as depositional models for former ice sheets. *Journal of the Geological Society of London* **128**, 361–393.

Boulton, G. S. 1976. The origin of glacially fluted surfaces: observations and theories. *Journal of Glaciology* **17**, 287–309.

Boulton, G. S. 1986. Push-moraines and glacier-contact fans in marine and terrestrial environments. *Sedimentology* **33**, 677–698.

Boulton, G. S. 1987. A theory of drumlin formation by subglacial deformation. In: Menzies, J. & Rose, J. (Eds) *Drumlin Symposium*. Balkema, Rotterdam, 25–80.

Boulton, G. S. & Clarke, C. D. 1990a. A highly mobile Laurentide ice sheet revealed by satellite images of glacial lineations. *Nature* **346**, 813–817.

Boulton, G. S. & Clarke, C. D. 1990b. The Laurentide ice sheet through the last glacial cycle: the topology of drift lineations as a key to the dynamic behaviour of former ice sheets. *Transactions of the Royal Society of Edinburgh* **81**, 327–347.

Boulton, G. S. & Eyles, N. 1979. Sedimentation by valley glaciers: a model and genetic classification. In: Schlüchter, C. (Ed.) *Moraines and Varves*. Balkema, Rotterdam, 11–23.

Boyce, J. I. & Eyles, N. 1991. Drumlins carved by deforming till streams below the Laurentide ice sheet. *Geology* **19**, 787–790.

Clark, C. D. 1993. Mega-scale glacial lineations and cross-cutting ice flow landforms. *Earth Surface Processes and Landforms* **18**, 1–29.

Clark, C. D. 1994. Large-scale ice-moulding: a discussion of genesis and glaciological significance. *Sedimentary Geology* **91**, 253–268.

Croot, D. G. 1987. Glacio-tectontic structures: a mesoscale model of thin-skinned thrust sheets? *Journal of Structural Geology* **9**, 797–808.

Desloges, J. R. & Church, M. 1992. Geomorphic implications of glacier outburst flooding: Noeick River valley, British Columbia. *Canadian Journal of Earth Science* **29**, 551–564.

Eyles, N. 1979. Facies of supraglacial sedimentation on Iceland and alpine temperate glaciers. *Canadian Journal of Earth Science* **16**, 1341–1361.

Eybergen, F. A. 1986. Glacier snout dynamics and contemporary push moraine formation at the Turtmannglacier, Wallis, Switzerland. In: Van der Meer, J. J. M. (Ed.) *Tills and Glaciotectonics*. Balkema, Rotterdam, 217–231.

Fitzsimons, S. J. 1990. Ice-marginal depositional processes in a polar maritime environment, Vestfold Hills, Antarctica. *Journal of Glaciology* **36**, 279–286.

Goldthwait, R. P. 1951. Development of end moraines in east-central Baffin Island. *Journal of Geology* **59**, 567–577.

Gordon, J. E., Whalley, W. B., Gellatly, A. F. & Vere, D. M. 1992. The formation of glacial flutes: assessment of models with evidence from Lyngsdalen, North Norway. *Quaternary Science Review* **11**, 709–731.

Gravenor, C. P. & Kupsch, W. O. 1959. Ice disintegration features in western Canada. *Journal of Geology* **67**, 48–64.

Gray, J. M. 1975. The Loch Lomond Readvance and contemporaneous sea-levels in Loch Etive and neighbouring areas of western Scotland. *Proceedings of the Geologists' Association* **86**, 227–238.

Gray, J. M. 1988. Glaciofluvial channels below the Blakeney Esker, Norfolk. *Quaternary Newsletter* **55**, 8–12.

Gray, J. M. 1991. Glaciofluvial landforms. In: Ehlers, J. Gibbard, P. L. & Rose, J. (Eds) *Glacial Deposits in Great Britain and Ireland*. Balkema, Rotterdam, 443–454.

Hambrey, M. J. 1984. Sedimentary processes and buried ice phenomena in the proglacial areas of Spitsbergen glaciers. *Journal of Glaciology* **30**, 116–119.

Hambrey, M. J. & Huddart, D. 1995. Englacial and proglacial glaciotectonic processes at the snout of a thermally complex glacier in Svalbard. *Journal of Quaternary Science* **10**, 313–326.

Hewitt, K. 1967. Ice-front deposition and the seasonal effect: a Himalayan example. *Transactions of the Institute of British Geographers* **42**, 93–106.

Hooke, R. Le B. 1970. Morphology of the ice-sheet margin near Thule, Greenland. *Journal of Glaciology* **2**, 140–144.

Hoppe, G. 1952. Hummocky moraine regions, with special reference to the interior of Norrbotten. *Geografiska Annaler* **34**, 1–72.

Hoppe, G. & Schyatt, V. 1953. Some observations on fluted moraine surfaces. *Geografiska Annaler* **35**, 105–115.

Humlum, O. 1978. Genesis of layered lateral moraines: implications for palaeoclimatology and lichenometry. Norsk *Geografisk Tidsskrift* **77**, 65–72.

Humlum, O. 1985. Genesis of an imbricate push moraine, Höfdabrekkujökull, Iceland. *Journal of Geology* **93**, 185–195.

Krüger, J. 1985. Formation of a push moraine at the margin of Höfdabrekkujökull, South Iceland. *Geografiska Annaler* **67A**, 199–212.

Krüger, J. 1987. Relationship of drumlin shape and distribution to drumlin stratigraphy and glacial history, Myrdalsjökull, Iceland. In: Menzies, J. & Rose, J. (Eds) *Drumlin Symposium*. Balkema, Rotterdam, 257–266.

Krüger, J. 1993. Moraine-ridge formation along a stationary ice front in Iceland. *Boreas* **22**, 101–109.

Krüger, J. & Thompson, H. H. 1984. Morphology, stratigraphy, and genesis of small drumlins in front of the glacier Myrdalsjökull, South Iceland. *Journal of Glaciology* **30**, 94–105.

Maizels, J. 1992. Boulder ring structures produced during jökulhlaup flow. *Geografiska Annaler* **74A**, 21–33.

Matthews, J. A. & Petch, J. R. 1982. Within-valley asymmetry and related problems of Neoglacial lateral moraine development at certain Jotunheimen glaciers, southern Norway. *Boreas* **11**, 225–224.

Matthews, J. A., Cornish, R. & Shakesby, R. A. 1979. 'Saw-tooth' moraines in front of Bødalsbreen, southern Norway. *Journal of Glaciology* **22**, 535–546.

Menzies, J. 1979. A review of the literature on the formation and location of drumlins. *Earth Science Review* **14**, 315–359.

Menzies, J. 1984. *Drumlins: A Bibliography*. Geo Books, Norwich.

Menzies, J. & Rose, J. 1987. *Drumlin Symposium*. Balkema, Rotterdam.

Paul, M. A. 1983. Supraglacial landsystem. In: Eyles, N. (Ed.) *Glacial Geology*. Pergamon Press, Oxford.

Pierce, K. L. 1979. History and dynamics of glaciation in the Northern Yellowstone Park area. *US Geological Survey Professional Paper* **729-F**.

Price, R. J. 1966. Eskers near the Casement glacier, Glacier Bay, Alaska. *Geografiska Annaler* **48A**, 111–125.

Price, R. J. 1969. Moraines, Sandur, Kames and Eskers near Breiðamerkurjökull, Iceland. *Transactions of the Institute of British Geographers* **46**, 17–43.

Price, R. J. 1970. Moraines at Fjallsjökull. *Journal of Arctic and Alpine Research* **2**, 27–42.

Price, R. J. 1971. The development and destruction of a sandur, Breiðamerkurjökull, Iceland. *Journal of Arctic and Alpine Research* **3**, 225–237.

Price, R. J. 1973. *Glacial and Fluvioglacial Landforms*. Oliver & Boyd, Edinburgh.

Rains, R. B. & Shaw, J. 1981. Some mechanisms of controlled moraine development, Antarctica. *Journal of Glaciology* **27**, 113–128.

Rose, J. 1989. Glacier stress patterns and sediment transfer associated with the formation of superimposed flutes. *Sedimentary Geology* **62**, 151–176.

Rose, J. & Letzer, J. M. 1977. Superimposed drumlins. *Journal of Glaciology* **18**, 471–480.

Russell, A. J. 1994. Subglacial jökulhlaup deposition, Jotunheimen, Norway. *Sedimentary Geology* **91**, 131–144.

Shakesby, R. A. 1989. Variability in Neoglacial moraine morphology and composition, Storbreen Jotunheim Norway: within-moraine patterns and the implications. *Geografiska Annaler* **71A**, 17–29.

Sharp, M. 1984. Annual moraine ridges at Skálafellsjökull, South-east Iceland. *Journal of Glaciology* **30**, 82–93.

Sharp, M. 1985. 'Crevasse-fill' ridges—a landform type characteristic of surging glaciers? *Geografiska Annaler* **67A**, 213–220.

Shaw, J. 1983. Drumlin formation related to inverted meltwater erosional marks. *Journal of Glaciology* **29**, 461–479.

Shaw, J. 1989. Drumlins, subglacial meltwater floods and ocean responses. *Geology* **17**, 853–856.

Shaw, J. & Kvill, D. 1984. A glaciofluvial origin for drumlins of Livingstone lake area, Saskatchewan. *Canadian Journal of Earth Sciences* **21**, 1442–1459.

Shaw, J. & Sharpe, D. R. 1987. Drumlin formation by subglacial meltwater erosion. *Canadian Journal of Earth Science* **24**, 2316–2322.

Shaw, J., Kvill, D. & Rains, B. 1989. Drumlins and catastrophic subglacial floods. *Sedimentary Geology* **62**, 177–202.

Shreve, R. L. 1985. Esker characteristics in terms of glacier physics, Katahdin esker system, Maine. *Geological Society of America Bulletin* **96**, 639–646.

Small, R. J. 1983. Lateral moraines of glacier de Tsidjiore Nouve: form, development, and implications. *Journal of Glaciology* **29**, 250–259.

Small, R. J., Beecroft, I. R. & Stirling, D. M. 1984. Rates of deposition on lateral moraine embankments, Glacier de Tsidjiore Nouve, Valais, Switzerland. *Journal of Glaciology* **30**, 275–281.

Souchez, R. A. 1967. The formation of shear moraines: an example from South Victoria Land, Antarctica. *Journal of Glaciology* **6**, 837–843.

Syverson, K. M., Gaffield, S. J. & Mickelson, D. M. 1994. Comparison of esker morphology and sedimentology with former ice-surface topography, Burroughs Glacier, Alaska. *Geological Society of America Bulletin* **106**, 1130–1142.

Thomas, G. S. P., Connaughton, M. & Dackombe, R. V. 1985. Facies variation in a Late Pleistocene supraglacial outwash sandur from the Isle of Man. *Geological Journal* **20**, 193–213.

Warren, W. P. & Ashley, G. M. 1994. Origins of the ice-contact stratified ridges (eskers) of Ireland. *Journal of Sedimentary Research* **A64**, 433–449.

Weertman, J. 1961. Mechanism for the formation of inner moraines found near the edge of cold ice caps and ice sheets. *Journal of Glaciology* **3**, 965–978.

Worsley, P. 1974. Recent 'annual' moraine ridges at Austre Okstindbreen, Okstidan, north Norway. *Journal of Glaciology* **13**, 265–277.

10
Glacial Sedimentation in Water

Where glaciers terminate in water the range of sediments produced is very different to that deposited on land. Where the glacier is **grounded** (i.e. in contact with its bed) subglacial deposition either by lodgement or subglacial melt-out occurs. At the ice margin the glacier may either remain grounded or become detached from its bed and float. Either way, the processes of deposition at the ice margin are very different from those on land. The glacier simply acts as a source of debris: the depositional processes are controlled not by the direct agency of glacier ice but through the processes of sedimentation within the water body. Tills may not be produced in this environment since disaggregation of the debris occurs within the water column (see Section 8.1), although diamictons (deposits which look like till) may form. Two types of environment exist: lacustrine and marine.

10.1 SEDIMENTATION IN LACUSTRINE ENVIRONMENTS

Lacustrine ice margins develop in a variety of different situations: glaciers may dam lakes; lakes may develop in front of a glacier due to the melting of stagnant ice beneath the proglacial surface; lakes may be dammed by moraines in front of a glacier or a glacier may simply drain into a rock basin. Some of the large mid-latitude ice sheets of the last glacial cycle terminated in huge proglacial lakes formed in part by the isostatic depression of the crust by the ice sheet. Ice margins which terminate in lakes are typically steep, often near-vertical, due to iceberg calving and have a linear snout plan form.

The pattern of sedimentation within glacial lakes is controlled by the density stratification that develops within the water body. Water density is controlled primarily by temperature and secondarily by salinity and the suspended sediment content. In the summer most lakes develop a strong water stratification in which a surface layer of warm, and therefore less dense, water (**epilimnion**) sits above a lower body of cold denser water (**hypolimnion**) (Figure 10.1). In the autumn this surface layer is cooled rapidly, becomes denser and therefore sinks to the bottom of the lake, causing the water body as a whole to mix. This stratification controls the resulting processes of sedimentation. Sedimentation within a glaciolacustrine lake may occur through any or all of the following eight processes:

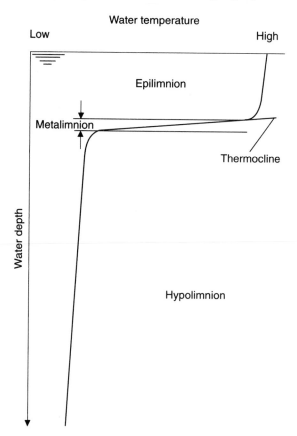

Figure 10.1   *The thermal stratification of lake waters. [Modified from: Drewry (1986) Glacial Geologic Processes, Arnold, Fig. 11.3, p. 169]*

1. **Deposition from meltwater flows.** The manner in which meltwater enters into a lake depends upon the density of the meltwater relative to that of the lake. If there is a significant difference in density, the sediment-laden meltwater will maintain its integrity as a **plume**. This plume may enter the lake in one of three ways, which are controlled by the density stratification within it: (1) as an **underflow** where the sediment plume is denser than the lake water, and therefore sinks to the base of the lake and flows as a sediment-laden body known as a **turbidity current**; (2) as an **interflow** where the sediment plume is of similar density to the surface water (epilimnion) but less dense than the basal water (hypolimnion) and the plume enters at intermediate depth; and (3) as an **over-**

flow where the sediment is less dense than the surface lake water and therefore rises to the surface. Meltwater introduced as an overflow or interflow
tends to produce a delta. A delta consists of three structural components:
topsets, **foresets** and **bottom sets** (Figure 10.2). The bedload of the meltwater

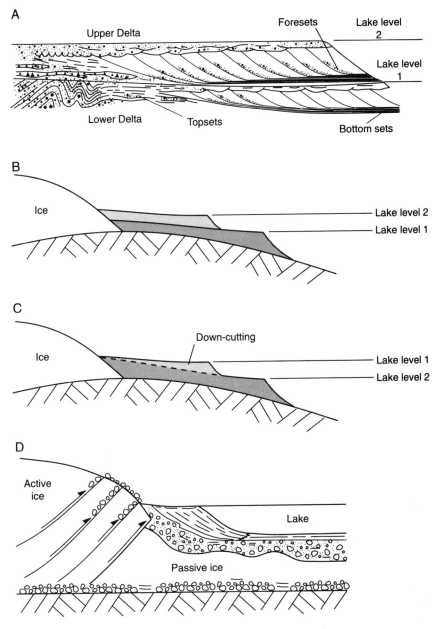

Figure 10.2   *Sedimentological components of deltas.* **A:** *Delta components, topsets,
foresets and bottom sets. Note the two superimposed deltas and that the lower one has
been deformed by a glacial advance.* **B:** *Delta response to rising lake levels.* **C:** *Delta
response to falling lake levels.* **D:** *Formation of a supraglacial delta*

flow is rapidly deposited close to the point of entry as the velocity is checked by the standing lake water. Deposition of coarse bedload gives rise to the topsets of the delta. This coarse bedload mixes with finer material settling from the sediment plume to give rise to the foresets of the delta which are formed as this material avalanches and flows beyond the edge of the topsets into the lake basin. Beyond the foresets the bottom sets are formed from fine material settling from suspension and that carried down the delta slope by small flows and avalanches.

In lakes where there is a pronounced underflow, the sediment-laden meltwater descends rapidly into the lake basin as a series of turbidity currents. These currents may inhibit the development of large deltas close to the lake margin as they carry much of the sediment into deep water. Persistent underflow eventually develops a series of subaqueous fans which may form in front of a small delta at the stream mouth. These fans consists of a system of channels, often with lateral levées, which cut into the delta front and feed a series of broad lobes reaching out towards the centre of the lake basin.

2. **Direct deposition from the glacier front.** Material may simply be dumped into the water body from the glacier front. The degree of disaggregation that occurs will depend on the water depth and upon the current activity within the lake. This will result in irregular-shaped deposits of diamicton.

3. **'Rain-out' from icebergs.** Calving icebergs may float debris out into the body of a lake where it may be released as the iceberg melts (Figure 10.3). The deposits produced vary, depending upon the density of the debris concentration within

Figure 10.3   *Icebergs on the proglacial lake Jökulsarlon in front of Breidamerkurjökull, Iceland. [Photograph: M. R. Bennett]*

the bergs and the number of bergs present. Low concentrations of debris and bergs may result in the deposition of occasional **dropstones**, out-sized clasts set in fine-grained sediments (Figure 10.4), while high concentrations may result in thick, laterally extensive deposits of diamicton. Irregular-shaped packages of debris may be associated with dropstones formed by the dumping of packages of sediment from the berg either by capsize or basal melting.

Figure 10.4   *Dropstones in glaciolacustrine deposits in Patagonia. [Photograph: P. Doyle]*

4. **Settling from suspension.** Sediment within the main body of lake water will gradually settle to form thin layers of mud and clay which drape other sediments. This type of sedimentation dominates over much of the lake floor away from its margins.
5. **Re-sedimentation by gravity flows.** Sediment within a lake may become unstable on steep slopes, particularly where rapid deposition occurs. Slumping or flow of this sediment may give rise to a range of diamictons, the properties of which depend on the fluidity of the flow. The greater the flow mobility, the greater the sorting and fabric development within the diamicton produced by the gravity flows.
6. **Current reworking.** Currents within the lake may rework and sort sediment which has already been deposited.
7. **Shoreline sedimentation.** The action of waves on lake shorelines may modify material collected here. The importance of wave action is limited by the size of the lake and the presence of a seasonal ice cover. The addition of hillside debris

from surrounding slopes may also be important in building up shoreline deposits. Non-glacial streams entering the lake may build small deltas.

8. **Biological sedimentation.** This is relatively unimportant as the biological component of glacial lakes is small due to the long freezing season, the ephemeral and unstable nature of many glacial lakes, and the high sediment load. However, the importance of biological sedimentation may increase with the size of the lake.

The distribution of these eight processes within a lacustrine environment will determine the facies pattern that develops. In small lakes this will reflect the dominance of, and fluctuation between, underflow, interflow and overflow conditions. Despite this variation there is in general a fall off in flow competence away from the point of water inflow within the lake. This results in a broad size grading within glaciolacustrine environments: course deposits occur near the point of water inflow or near the glacier while fine-grained deposits are dominant in the lake centre. In fact, one can identify two broad facies within a lake: (1) the basin-margin facies, and (2) the lake-floor facies.

The basin-margin environment and the associated sedimentary facies is dominated by the inflow of water into the lake. Figure 10.5 shows two conceptual facies models for two different lacustrine environments; one in which underflow is occurring and one in which overflow or interflow dominates. The principal difference between these two facies is the morphology of the delta which develops and the steepness of the gradient with which particle size decreases into the basin. Where overflow or interflow dominates there is a rapid decrease in particle size away from the lake margin and large deltas are common (Figure 10.5A). In contrast, where significant underflow occurs a higher proportion of coarse sediment is carried further into the lake basin, resulting in a smaller delta dissected by chan-

---

Figure 10.5 *Distribution of processes within glacial lakes.* **A:** *Overflow and interflow dominant. Note 1: Coarse deposition of bedload as topsets. Note 2: Mixture of bedload and coarse sediment from overflow plume avalanches and slumps down the delta front to form foresets. Note 3: Laminated and often graded bedding formed from settling out of suspended sediment and from turbidity currents; occasional diamicton units associated with slumped sediment. Note 4: Laminated rhythmitic sediments, plus occasional dropstones. Note 5: Homogeneous, massive or weakly graded silt and clay deposits. Note 6: Thick deposits of diamicton formed by 'rain-out' of debris from icebergs. Reworking of this diamicton may occur by current action and by sediment gravity flow.* **B:** *Underflow dominant. Note 1: Coarse deposition of bedload as topsets. Note 2: Delta front heavily channelled by erosive turbidity currents which carry the coarse fraction into the lake basin; delta growth is inhibited. Note 3: Graded beds with cross-laminations produced by ripples occur; layers of fine clay may be draped over the ripples to give draped laminations; occasional diamicton units associated with sediment slumping may be present. Note 4: Laminated rhythmitic sediments, plus occasional dropstones. Note 5: Thick deposits of diamicton formed by 'rain-out' of debris from icebergs; reworking of this diamicton may occur by current action and by sediment gravity flow*

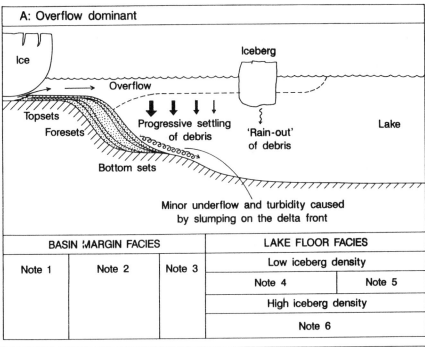

A: Overflow dominant

Ice

Iceberg

Overflow

Topsets

Foresets

Progressive settling
of debris

'Rain-out'
of debris

Lake

Bottom sets

Minor underflow and turbidity caused
by slumping on the delta front

BASIN MARGIN FACIES			LAKE FLOOR FACIES	
			Low iceberg density	
Note 1	Note 2	Note 3	Note 4	Note 5
			High iceberg density	
			Note 6	

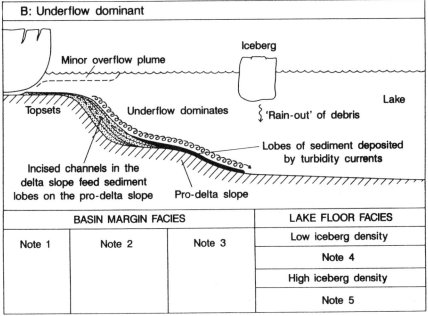

B: Underflow dominant

Iceberg

Minor overflow plume

Lake

Topsets

Underflow dominates

'Rain-out' of debris

Lobes of sediment deposited
by turbidity currents

Incised channels in the
delta slope feed sediment
lobes on the pro-delta slope

Pro-delta slope

BASIN MARGIN FACIES			LAKE FLOOR FACIES
			Low iceberg density
Note 1	Note 2	Note 3	Note 4
			High iceberg density
			Note 5

nels that feed sediment lobes which broaden and merge toward the basin centre (Figure 10.5B).

The nature of the processes of sedimentation away from the lake margin depends upon the amount of ice-rafted debris being introduced into the system. In part this depends on the size of the ice margin in contact with the lake and the size of the lake itself. The greater the length of ice margin in contact with the lake, the greater the potential for iceberg calving. One can identify a broad continuum from those lakes in which the input of ice-rafted debris is small, and therefore sedimentation is dominated by settling from suspension, and those in which the ice-rafted component is high.

If the ice-rafted component is small then rhythmic fine-grained sediments produced by settling of material out of suspension dominate. This process of settling gives rise to fine laminations of silt and clay (Figure 10.6). This process is interrupted by the deposition of coarser laminations from material introduced by turbidity currents, generated either by underflow or by slumping on the delta front or lake margin. This forms a rhythmic sediment in which there are coarse- and fine-grained laminations. Sedimentation by turbidity currents gives coarser-grained laminations. When or where underflow is not active, for example during winter months when the inflow or discharge of water into the lake is low, settling from suspension will occur. These laminations typically grade in size from silt and clay particles at the base to fine clay at the top. They are usually terminated by a sharp contact formed by the influx of a new underflow of coarse material. The relative thickness and importance of these two components changes with distance away

Figure 10.6 *Laminated lake sediments. [Photograph: S. Lewis]*

from the sediment source or point of water inflow. Close to the delta the coarse sand/silt layer, formed by underflow, will be thick. In the basin centre the coarse layers will be much thinner and may be absent. In such cases a continuous deposit of homogeneous fine clay may result. This laminated or rhythmic sediment may have an annual pattern. Underflow and therefore coarse lamination dominate in the summer when the influx of meltwater to a lake is high, while suspension settling and fine lamination tend to dominate in the winter when the influx of turbid flows is small or absent. Where an annual cycle can be identified within the rhythmic sediments they are known as **varves**. It is important to emphasise, however, that not all rhythmic lake sediments necessarily contain an annual signature. These rhythmic deposits are easily disturbed or deformed during or after deposition and may show a variety of deformation structures.

In contrast, if the ice-rafted debris component is large, sedimentation is dominated by coarse debris melting from icebergs. In this case the deposition of diamictons and resedimentation by gravity flows will dominate. In some of the large glaciolacustrine lakes along the margins of the former mid-latitude ice sheets, large extensive deposits of diamictons have been deposited in this way (Box 10.1).

The types of facies patterns or facies architecture that occur are largely controlled by the geometry of the lake basin. For example, the number of meltwater streams which feed a glacial lake, the presence and geometry of any ice margin, and the presence of buried ice, all determine the facies pattern within a given glacial lake. The long-term stability of the lake and its evolution are also very important. For example, a supraglacial lake formed on dead ice is highly unstable and constantly prone to change as the ice beneath melts. As a consequence of the range of possible variables, it is hard to predict the facies patterns that might occur within glaciolacustrine environments.

10.2 SEDIMENTATION IN MARINE ENVIRONMENTS

Glaciers may terminate in the sea either within the confines of a drowned glacial valley, a **fjord**, or in the open sea. During the Cenozoic Ice Age many of the mid-latitude ice sheets extended towards the edge of their continental shelves, as did the Antarctic ice sheet. In recent years there has been an increased emphasis on the importance of glaciomarine sedimentation and several glacial sedimentary sequences traditionally interpreted as being deposited by grounded ice have been reinterpreted.

One controversial reinterpretation has involved the sediments of the Irish Sea and its coastal periphery. It has been argued that the last British ice sheet may have decayed rapidly from the Irish Sea due to rapid calving. The argument is that at the close of the last glacial cycle the British Ice Sheet was at first slow to decay. Isostatic depression of the crust around this Ice Sheet, coupled with rising eustatic sea levels due to the decay of the North American ice sheet, would have produced very high relative sea-levels. These high sea-levels may have uncoupled the British ice sheet from its bed, increasing the importance of glaciomarine sedimentation

BOX 10.1: GLACIAL TILL OR ICE-RAFTED DIAMICTONS?

Along the shore of Lake Ontario to the east of Toronto, in Canada, a sequence of glacial sediments is exposed at Scarborough Bluffs. This sequence is dominated by large units of fine-grained diamicton separated by bedded and laminated sand and mud units. Traditionally, this sequence has been interpreted as the product of multiple ice advances by a grounded ice sheet. Between each advance, lakes formed to give the sandier sediments. Eyles and Eyles (1983) challenged this interpretation on the basis of detailed sedimentological logging. They made the following observations and inferences:

- *Observation:* there were no tectonic structures in the sand and mud units.
  Inference: this is inconsistent with the passage of a grounded ice sheet over the sediment.
- *Observation:* thin stringers of silt and sand are found within the diamictons.
  Inference: water currents were active during the deposition of the diamictons.
- *Observation:* diamicton units show evidence of having flowed into and accumulated preferentially within topographic lows.
  Inference: gravity was important in the sedimentation of the diamicton units.
- *Observation:* transitional and interbedded contacts occurred between the sand and the diamicton units.
  Inference: there was a gradual transition from the deposition of the sand to diamicton units. This is inconsistent with an ice advance, which would tend to give erosional, non-conformable contacts.
- *Observation:* load structures are present at the contact of the diamicton and sand units.
  Inference: the diamicton was deposited onto saturated sands and muds.
- *Observation:* the sand units contain structures typical of deltas.
  Inference: the sand units were deposited in deltas.

On the basis of these observations and inferences Eyles and Eyles (1983) rejected the traditional model of multiple ice advances. The sediment sequence at Scarborough Bluffs was reinterpreted as being glaciolacustrine, formed by the repeated progradation of deltas over glaciolacustrine diamictons, deposited by both the 'rain-out' of ice-rafted debris and by sediment gravity flows. This interpretation has significant implications for the glacial stratigraphy of the area, in particular for the number of ice advances recognised within the region.

This work represented a significant step forward in glacial geology and opened the way for the reinterpretation of many diamicton sequences traditionally viewed simply as the product of grounded glaciers. It also illustrates the process of sediment interpretation, which involves three steps: (1) careful observation/recording; (2) inferences being made from each observation; and (3) each inference then being used to examine all the possible explanations or models.

Source: Eyles, C. H. & Eyles, N. 1983. Sedimentation in a large lake: a reinterpretation of the late Pleistocene stratigraphy at Scarborough Bluffs, Ontario, Canada. *Geology* **11**, 146–152.

and accelerating deglaciation via rapid glacier calving. At present there is considerable controversy over the interpretation of many of the sedimentary sequences exposed in sea cliffs along the margins of the Irish Sea. This model of deglaciation has also been suggested for several other Cenozoic ice sheets and has led to the reinterpretation of other sedimentary sequences as being of glaciomarine origin.

Glaciomarine sedimentation involves a similar range of processes to sedimentation in glacial lakes these are:

1. **Direct deposition from the glacier front.** The dumping or release by melting of supraglacial and englacial debris at the ice margin may be an important process. The rate of sedimentation will depend on: (1) the volume of ice that is melted; (2) the forward velocity of the ice; and (3) its debris content.
2. **'Rain-out' from icebergs and seasonal sea ice.** The heavily crevassed and dynamic nature of most marine margins means that iceberg calving is not only much more rapid but is in general a much more violent process than in lacustrine environments. In particular, the presence of tidal variation causes flexure of the margin, which accelerates iceberg calving. The waves generated by the calving of an iceberg may have a significant effect on adjacent beaches, particularly in relatively enclosed fjords. Supraglacial debris is often shaken from the upper surface of the berg as it calves or as it subsequently capsizes due to differential melting. Melting of icebergs will ultimately release subglacial and englacial debris. The rate of sedimentation at any one point in front of the ice margin depends on: (1) the concentration of debris within the glacier and therefore the iceberg; (2) the residence time of an iceberg in the area; (3) the rate at which calving takes places; (4) the rate of iceberg melting; and (5) the wave climate, which influences the frequency with which icebergs may capsize. The greater the englacial debris content of the glacier, the greater the sediment an iceberg calved from it will contain. Warm-based glaciers have a relatively thin basal debris layer, whereas glaciers with cold or mixed regimes may have a greater thickness of basal debris due to freezing-on and due to thickening caused by basal folding and thrusting. Icebergs from cold and mixed regime glaciers are likely, therefore, not only to contain a greater volume of debris, but this debris will be distributed throughout a greater part of the berg and therefore sediment will be released over a longer period. It is important to note that icebergs from ice shelves will contain relatively little debris within them, since most of the debris melts from the base of the ice shelf prior to calving. The sediments produced by 'rain-out' vary from occasional dropstones to dump structures and large deposits of diamicton. These diamicton deposits are sometimes incorrectly referred to as **waterlain tills**, but since they have been disaggregated by the water column they are not tills in the strict sense of the definition (see Section 8.1). Diamictons produced by the 'rain-out'

of ice-rafted debris may possess crude **graded bedding** (size sorting) if the rate of debris supply is periodic or pulsed. When a pulse of debris is released into the water column it will become sorted as it settles; the larger clasts will settle first, followed by the finer fraction. In this way a crude graded bed may form. The next pulse of debris will form a new graded unit. If the water column is too shallow or the supply or 'rain-out' of debris continuous then no graded units will form.

Seasonal sea ice may also transport debris within this environment. Debris is entrained by the sea ice through: (1) avalanches, which carry debris-laden snow out on to the sea ice; (2) rockfalls and mudflows, which may carry debris on to the ice; (3) stream flow, which in the spring may deliver debris to the shore before sea ice has broken up, depositing sediment on the ice; (4) bottom freezing, where debris may be incorporated by freezing-on from the sea-floor as the ice rests in shallow water; (5) sediment capture, where suspended sediment may be frozen into sea ice as it grows; and (6) aeolian deposition. As the sea ice breaks up it may raft sediment, depositing it as the ice melts.

3. **Deposition from meltwater flows.** The influx of subglacial meltwater is an important process. This sediment-laden water is fresh and is therefore usually less dense than sea-water. Consequently it will enter as an overflow plume. It rises quickly from the base of the glacier where it spreads out over the sea-water surface. Deposition from this plume is rapid and the meltwater portal or exit is usually marked by a fan of sand and gravel. Underflow may occur occasionally where the fresh water is particularly dense due to very high sediment concentrations. The importance of meltwater is affected by basal thermal regime, being more important for warm-based than for cold-based glaciers (see Section 3.4).

4. **Settling from suspension.** Suspended sediment introduced into the sea will gradually settle out. This is accelerated in sea-water due to a number of processes: (1) **flocculation**—fine clay particles, which normally carry small electrical charges which repulse one another, attract each other in sea-water because salt neutralises the electrical charges; (2) **agglomeration**—particles may be attached to one other by organic matter; and (3) **pellitisation**—some planktons and another small organisms ingest fine sediment and bind it into large faecal pellets which then sink rapidly.

5. **Subaqueous re-sedimentation by gravity flows.** Sediment may become unstable on steep slopes. Slumping or flow of this sediment may give rise to a range of diamictons. This process may also generate turbidity currents which further redistribute sediment.

6. **Subaerial rockfall and mass flow.** In fjords material may be deposited by rockfall and mass flow directly from the valley sides into the water body. Subaqueous talus cones may develop.

7. **Re-mobilisation by iceberg scour.** The keels of large bergs may ground in shallow water. This may remobilise sediment, returning it to suspension.

8. **Current reworking.** Sediment reworking by wave-induced currents close to the shore and by tides, particularly in fjords, is an important process in redistributing sediments.

9. **Shoreline sedimentation.** The action of waves on shorelines may modify material collected here. The addition of material from surrounding slopes may also be important in building up shoreline deposits. Streams entering the lake may build small deltas.
10. **Biological sedimentation.** This is an important component of the marine environment. The skeletal remains of micro-organisms such as diatoms, Foraminifera and Radiolaria may add to the sediment in these environments, while larger organisms may be present, depending on the proximity of the glacier and the nature of the environment. Organisms also have an important role in mixing (**bioturbating**) sediment that has already been deposited.
11. **Coriolis force.** Sedimentation in fjords is partly controlled by the effect of the Earth's rotation on the water body. Sediment plumes are commonly deflected towards the right-hand side of a fjord in the northern hemisphere, and to the left-hand side in the southern hemisphere. This may cause asymmetry in sediment accumulation on the fjord floor. The importance of this increases towards the poles.

The distribution of these processes within the context of a glacial fjord is illustrated in Figure 10.7. In general, the distribution of glaciomarine environments and therefore the sedimentary facies associated with them can be seen as a sequence of environments away from the ice front: (1) the ice contact, (2) the proximal zone (inner shelf or inner fjord), and (3) the distal zone (outer shelf or outer fjord).

The **ice-contact environment** comprises the subglacial environment that will be progressively exposed on the sea-floor during glacier retreat and the immediate proglacial area. Subglacial processes will be the same as those for glaciers which terminate on land. It has been frequently noted that greater thicknesses of tills are produced by marine-based glaciers. For example, marine-based glaciers may typically deposit a layer of till between 5 and 20 m thick, while most tills deposited by terrestrial glaciers are only 1–2 m thick. This is probably due to the availability of soft deformable sediment beneath the glacier. A glacier advancing over saturated marine muds will remould and mix them with subglacial debris through the process of subglacial deformation. Consequently, large thicknesses of till result. On land, by contrast, a glacier will predominantly override coarse proglacial fluvial sands and gravels, which will be better drained and therefore less easily deformed.

Excepting cold-based glaciers, deposition at the ice front is dominated by the outflow of meltwater, which rises to the surface to form a plume. The flow velocity of this plume is checked dramatically as it enters the body of water and consequently, unlike terrestrial environments, most of the fluvial sediment is deposited close to the ice margin, where submarine fans develop. These fans are known as **grounding line fans**. These fans may be associated with **subaqueous push moraines** and **moraine banks** if the glacier is advancing or if seasonal ice-marginal fluctuations occur (see Section 11.1). If the ice front is stationary these fans can grow rapidly and may emerge from the water to form **ice-contact deltas**. The fans grow by the direct addition of material from the meltwater flow and are extended laterally, at the front of the delta, by gravity flows. The height of these fans or

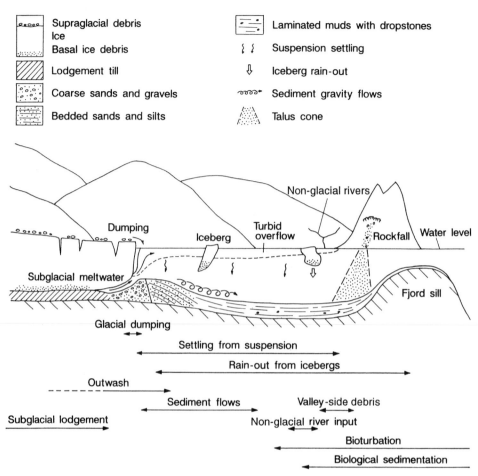

Figure 10.7   *Sediment sources and the distribution of processes within a glacially influenced fjord*

deltas may be increased by pushing. As ice pushes, perhaps seasonally, into the delta it is buckled and compressed. This model only holds if the grounding line coincides with the ice front. If the ice front is partially floating, as will be the case for an ice shelf, then a subglacial delta or fan may form.

Beyond this ice-contact environment the **proximal zone** (i.e. the inner shelf or inner fjord) is dominated by sedimentation from suspension, which builds up large thicknesses of sediment. The proximal zone can be divided into an inner zone where the rate of sedimentation is so fast that it inhibits benthic life (bottom-dwellers), and an outer zone where benthic life can exist. This outer zone will be defined by the presence of bioturbation within the sediment, caused by burrowing organisms. There is usually a very rapid decrease in particle size away from the glacier within this zone (Box 10.2). The rate of sedimentation also falls rapidly with distance from the ice margin (Figure 10.8). Sediment accumulation in the inner part

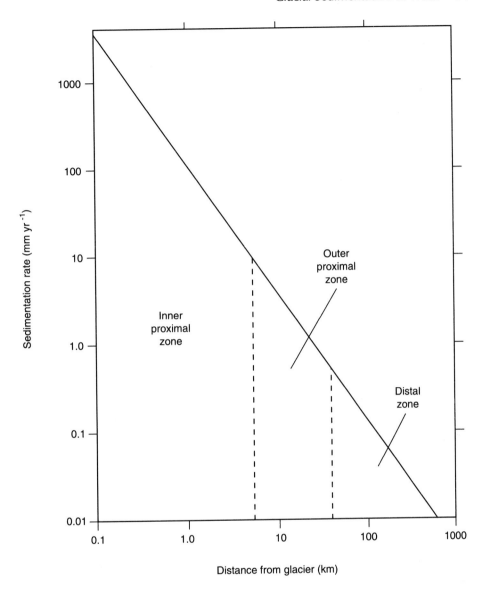

Figure 10.8 *The relationship between the rate of sedimentation and distance away from the glacier in glaciomarine environments. [Reproduced with permission from: Boulton (1990) In: Dowdeswell, J. A. & Scourse, J. D. (Eds) Glacimarine Environments: Processes and Sediments, Geological Society Special Publication No. 53, Fig. 7A, p. 23]*

of the proximal zone may give rise to considerable thicknesses of sea-floor sediment. This sediment may be laminated with a rhythmitic form. The laminations and rhythmites form in a number of different ways. For example, tidal currents can produce two graded couplets each day. Sand is predominantly deposited from the

BOX 10.2: SEDIMENTATION WITHIN KONGSFJORD

The adjacent diagram shows the results obtained from observations made in Kongsfjord, Svalbard during the summer of 1973 (Boulton 1990). The upper part of the diagram shows the suspended sediment content at the water surface of the fjord during a two-hour period either side of high tide on 23 July 1973. The sediment content is indicated by the contours. Most suspended sediment is concentrated in two sediment plumes associated with the principal meltwater portals of the Kongsvegen glacier. The lower part of the diagram shows a section down the centre of the fjord. The suspended sediment concentration is shown along with the grain-size distributions for samples of suspended sediment on the surface. These grain-size distributions show a rapid decrease in particle size away from the glacier as the coarse sediment settles out. A similar trend is found with the samples taken from sediment traps on the fjord floor. The first three grain-size distributions show a rapid fall in grain size away from the ice margin, although this trend is reversed in the fourth sample, furthest away from the glacier, which reflects the introduction of ice-rafted debris from icebergs stranded at the fjord entrance. Note the similarity in size distribution between the sediment sampled from the iceberg and the sea-floor sample beneath the iceberg.

Source: Boulton, G. S. 1990. Sedimentary and sea level changes during glacial cycles and their control on glacimarine facies architecture. In: Dowdeswell, J. A. & Scourse, J. D. (Eds) *Glacimarine Environments: Processes and Sediments*, Geological Society Special Publication No. 53, 15–52. [Diagram modified from: Boulton (1990) In: Dowdeswell, J. A. & Scourse, J. D. (Eds) *Glacimarine Environments: Processes and Sediments*, Geological Society Special Publication, Fig. 3, p.18]

plume at low tide, while finer grained sediment dominates at high tide. Diurnal variation in discharge may also affect the sediment plume and cause rhythmic lamination to develop. Two types of rhythmically laminated sediment are commonly identified, based on the size fractions involved. **Cyclopsams** consist of graded sand–mud couplets, while **cyclopels** consist of graded silt–mud couplets.

The high rates of sedimentation in this inner proximal zone result in a sedimentary pile that frequently becomes unstable and is redistributed by gravity flows which may feed turbidity currents. In the inner part of the proximal zone suspension settling may dominate over 'rain-out' from icebergs, while in the outer part of the zone the reverse may be true (Figure 10.9). This may re-introduce coarse particles to the sea-floor in the outer part of the proximal zone (Box 10.3). The action of currents removing finer particles (**winnowing**) may also cause the surface sediment to become slightly coarser in this area. Streams or rivers entering the fjord within this zone may develop deltas and provide a source of coarser sediment input. Coarse sediment in fjords may also be derived from the valley sides where subaqueous talus cones may develop.

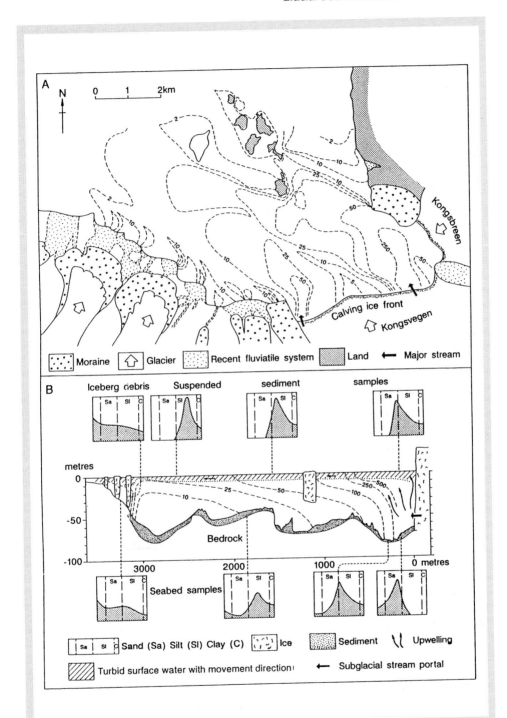

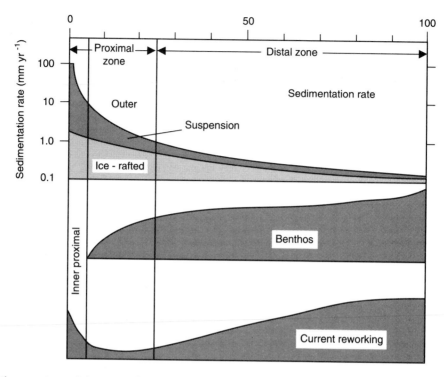

Figure 10.9   *Schematic diagram showing the changes in the process of sedimentation with distance from the glacier. [Reproduced with permission from: Boulton (1990) In: Dowdeswell, J. A. & Scourse, J. D. (Eds) Glacimarine Environments: Processes and Sediments, Geological Society Special Publication No. 53, Fig. 7B, p. 23]*

The **distal sedimentary environment** (i.e. the outer shelf or outer fjord) contrasts strongly with that in the proximal zone. Suspended sediment concentrations in the water column are significantly less than in the inner proximal zone. Increased current activity, due to upwelling of deep water along continental margins or at the mouths of fjords, ensures that there are usually strong sea-floor currents (Figure 10.9). As a consequence, sediment reworking dominates in this environment. Reworking of sediment by currents tends to remove finer particles and concentrate coarser sediment into lag horizons. Due to the current activity, nutrient supply is good and this results in high biological productivity (Figure 10.9). Benthonic and planktonic floras and faunas often contribute significantly to sedimentation. The proportion of sediment from icebergs is relatively small, although dropstones may be found.

The pattern of facies or the **facies architecture** that develops in glaciomarine environments can be very complex. Unlike some terrestrial glacial environments, the preservation potential for complex glaciomarine facies assemblages can be quite good, especially on gently subsiding continental shelves, like many of those along the margin of the North Atlantic. Figure 10.10 shows a hypothetical facies sequence for a glacier advancing and retreating across a continental shelf. The

BOX 10.3: THE INTERPRETATION OF PRECAMBRIAN TILLITES IN SCOTLAND: GLACIOMARINE?

The 15–20 km thick upper Precambrian Dalradian Supergroup in Scotland and Ireland contains multiple diamictites, known as the Port Askaig Formation. A total of 47 diamictite horizons have been recognised, separated by interbeds of siltstone, sandstone, conglomerate and dolomite. Spencer (1971) interpreted these diamictites as tillites deposited by at least 17 major glacial advances of a grounded ice sheet. These glacial advances were separated by the deposition of sands and silts in shallow marine and emergent conditions. Large networks of wedge-like structures, illustrated below, were interpreted as periglacial frost wedges.

The Port Askaig Formation was re-examined by Eyles and Eyles (1983), who proposed a glaciomarine interpretation (see also Eyles 1988). They argued that the diamictites contain no direct evidence of deposition beneath a glacier. In this interpretation the sequence was rarely if ever emergent from the sea and the wedge-like structures are interpreted as the product of soft sediment deformation caused by loading, via deposition, of saturated marine sediments. Spencer (1985) responded to the work of Eyles and Eyles (1983), emphasising the importance of the wedge-like structures which he continued to believe were fossil ice wedges. This discussion has yet to be completely resolved, although the balance of evidence currently favours a glaciomarine interpretation. This debate is of considerable importance in attempts not only to reconstruct the

palaeogeography of this region during the Precambrian but also to understanding this episode of Precambrian glaciation.

Sources: Eyles, C. H. 1988. Glacially- and tidally-influenced shallow marine sedimentation of the Late Precambrian Port Askaig Formation, Scotland. *Palaeogeography, Palaeoclimatology, Palaeoecology* **68**, 1–25. Eyles, C. H. & Eyles, N. 1983. Glaciomarine model for upper Precambrian diamictites of the Port Askaig Formation, Scotland. *Geology* **11**, 692–696. Spencer, A. M. 1971. *Late Precambrian Glaciation in Scotland*, Geological Society of London Memoir **6**. Spencer, A. M. 1985. Mechanisms and environment of deposition of Late Precambrian geosynclinal tillites Scotland and East Greenland. *Palaeogeography, Palaeoclimatology, Palaeoecology* **51**, 143–157. [Photograph: N. Eyles]

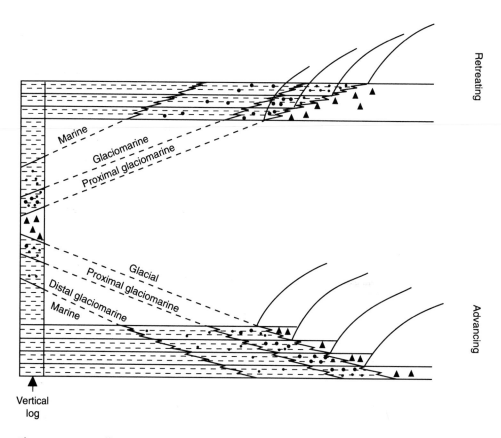

Figure 10.10 *Illustration of Walther's principle with reference to glaciomarine environments and sedimentary facies during a glacial cycle in which a grounded glacier advances and decays across a continental shelf*

sequence of glaciomarine environments in front of the glacier are superimposed in a vertical log as the glacier advances and retreats, as explained by Walther's principle (see Section 8.1.7). As the glacier advances there is a gradual increase in the amount of ice-rafted debris, which is followed by the deposition of a lodgement or deformation till as the area is covered by the advancing ice. Deglaciation is first marked by the development of a grounding line fan and subsequently by a gradual decrease in the amount of ice-rafted debris as the glacier retreats and the glacial influence falls. A hypothetical vertical log produced by such a glacial cycle is shown in Figure 10.11. In practice, the pattern of sedimentary facies associated with the advance and retreat of a glacier across a continental shelf may be much more complex.

Within fjords the facies architecture depends on the dynamics and behaviour of the glacier and may therefore be even more complex than that found on the continental shelf. Work on fjord glaciers in Alaska has identified five types of glacier behaviour, each associated with its own pattern of glaciomarine facies.

1. **Association I: rapid deep-water retreat.** In this case the glacier is retreating rapidly in deep water by iceberg calving. In the ice proximal zone the sediment facies consist of reworked subglacial till and glaciofluvial sands and gravels exposed as the glacier retreats. These are associated with supraglacial debris dumped from the ice margin. The proximal and distal glaciomarine zones contain large amounts of ice-rafted debris.

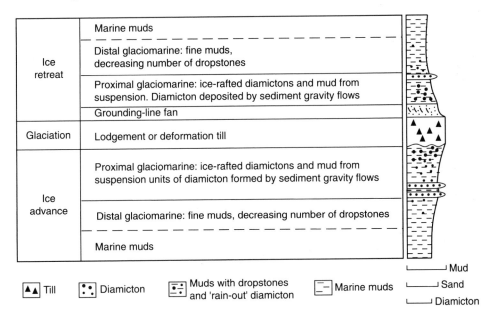

Figure 10.11 Hypothetical vertical log deposited on the continental shelf during a single glacial cycle in which a grounded glacier advances and decays across a continental shelf. [Modified from: Hambrey (1994) Glacial Environments, University College London Press, Fig. 8.20, p. 260]

2. **Association II: stabilised or slowly retreating ice margin.** In this case the ice recession has been retarded or stopped by a fjord constriction or pinning point (see Section 3.6). Calving continues, however, and thick deposits of ice-rafted sediment accumulate. The ice margin is marked by the deposition of coarse-grained sediment in a grounding line fan, moraine bank or ice-contact delta.

3. **Association III: slow retreat in shallow water.** In this case the glacier is either retreating or slowly advancing in shallow water. Calving is severely reduced and the amount of ice-rafted sediment is therefore relatively small. At the ice margin an ice-contact delta may develop.

4. **Association IV: proximal terrestrial glacier.** In this case the glacier is terrestrial and produces a large outwash delta which progrades into the fjord. The resulting facies consists of coarse-grained sediment on the delta top, while the delta front comprises sand and gravel which intertongue with marine muds deposited from suspension. Little ice-rafted debris is present since only small icebergs are introduced into the fjord via meltwater streams. Sand and silt rhythmites may be deposited on the fjord floor due to the interplay of sedimentation from suspension and turbidity currents generated from the delta slope.

5. **Association V: distant terrestrial glacier.** The glacier is distant and facies comprise tidal-flat muds and braided stream gravels.

These facies associations can be combined in a variety of different ways depending on: (1) the morphology of the fjord; and (2) the behaviour of an individual glacier. Fjord morphology controls the location and frequency of pinning points and the depth of water in which calving will occur. The behaviour of the individual glacier is also important. Some fjord glaciers are inherently unstable and prone to cyclic episodes of advance and decay driven not by climate but by decoupling of the glacier from the stabilising effect of moraine banks or pinning points (Box 11.3). In practice, therefore, the facies architecture within fjords may be very complex, reflecting a complex glacial history.

Figure 10.12 illustrates a relatively simple facies pattern produced by the retreat of a glacier from deep water on the left of the diagram onto the landsurface on the right. In this model the five facies associations recognised above occur sequentially as the glacier retreats from left to right. To the left of the bedrock lip of this hypothetical fjord the ice margin calves into deep water and facies Association I is deposited (Figure 10.12). Continued retreat brings the ice margin into contact with the fjord lip, along which it becomes stabilised and a moraine bank develops. Further retreat will uncouple the glacier terminus from the fjord lip and it will retreat up the fjord. Glacier calving now occurs in the shallow water behind the fjord lip where it is reduced, thereby decreasing the output of ice-rafted debris. An ice-contact delta begins to form along the ice margin as the glacier stabilises in the shallow water at the head of the fjord. With continued glacial retreat and the growth of the delta, the ice margin becomes land-based and calving ceases. Continued glacier retreat removes the glacial influence from the fjord.

This is a highly simplified model and the level of behavioural complexity that can be added is considerable, particularly since fjord glaciers are very sensitive to

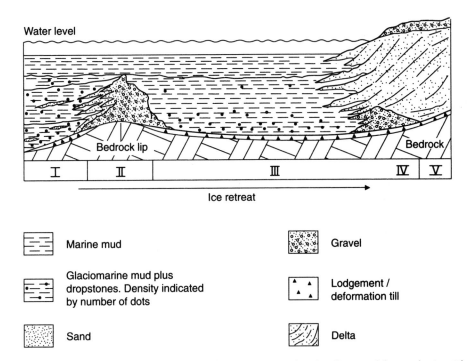

Water level

I   II   III   IV   V

Ice retreat

	Marine mud		Gravel
Glaciomarine mud plus dropstones. Density indicated by number of dots		Lodgement / deformation till
Sand		Delta

Figure 10.12   *Hypothetical facies architecture in a fjord influenced by a glacier. The facies pattern results from the retreat of a calving glacier from left to right. See text for facies descriptions. [Modified from: Hambrey (1994) Glacial Environments, University College London Press, Fig. 7.20, p. 219]*

mass balance changes and can retreat and advance dramatically (see Section 3.6). They are therefore highly dynamic and consequently may generate very complex facies patterns.

10.3 DISTINGUISHING GLACIOLACUSTRINE AND GLACIOMARINE DIAMICTONS FROM GLACIAL TILLS

Distinguishing the diamictons produced in glaciolacustrine and glaciomarine environments from those formed in other settings, particularly from glacial tills, is difficult on the basis of internal sediment properties alone (see Section 8.1.7). In general, waterlain diamictons may be recognised by the presence of stratification and graded bedding, although this is not always present. Waterlain diamictons tend not to have a strong particle fabric and that which is present usually reflects either local slopes or the direction of sediment flow (Figure 8.8). Waterlain diamictons may be found in association with shells and other biogenic material, although reworked fossils may occur in subglacial till, particularly in deformation tills that have assimilated marine sediment (Box 8.4). On the Antarctic continental shelf, ice-rafted diamictons have been separated from both subglacial till and current-winnowed diamictons on the basis of particle size and sorting, but these relationships have yet to be established for diamictons elsewhere. Clast shape within glaciolacustrine and glaciomarine diamictons simply records the transport history of the sediment while it was in contact with the glacier (Figure 7.7). Striated and faceted clasts may be present, depending upon the amount of basal debris within the icebergs from which the debris was deposited.

Glaciolacustrine and glaciomarine sediments can, however, be recognised from other glacial environments if the sedimentary facies or setting of the diamictons is examined (i.e. the external sediment properties). Facies analysis is the most important tool in recognising these glacial environments. In particular, the contact between the sediment units in glaciomarine and glaciolacustrine environments is highly diagnostic. In contrast to most other glacial environments, where contacts are usually unconformable, the sedimentary units in these environments frequently onlap one another. More importantly, the contacts between the units frequently contain deformation structures associated with sediment loading. When a relatively heavy sediment, such as a diamicton, is deposited onto a finer-grained and saturated sediment the contact between the two sediments is deformed by the weight of the overlying diamicton. The deformation of the contact gives rise to **load structures**. Other structures associated with the escape of water from these sediments as they accumulate (**dewatering structures**) are also common. The interpretation of these sedimentary structures is often crucial to the recognition of glaciomarine environments (Box 10.3). The association of laminated and graded sediments with the diamictons is also important, although laminated sediments may also be generated by tectonic deformation (Box 10.4).

In general, glaciolacustrine and glaciomarine sediments can be recognised from one another on the basis of the environmental setting in which the sediments are found and by the level and nature of the biogenic material present.

10.3 SUMMARY

Glaciers that terminate in water produce a very different assemblage of sediments to those deposited on land. The glacier simply acts as a source of debris and the depositional processes are controlled not by the direct agency of glacier ice but through the processes of sedimentation within a water body. Tills may not be produced in this environment since disaggregation of the debris occurs within the water column, although diamictons occur frequently. Sedimentation occurs via eight main processes: (1) deposition from meltwater flows or plumes; (2) direct deposition from the glacier margin; (3) the 'rain-out' of debris from icebergs; (4) settling of particles from suspension; (5) resedimentation by sediment gravity flows; (6) resedimentation by current reworking; (7) shoreline sedimentation; and (8) biological sedimentation. Within glacial lakes the facies architecture that results depends on the geometry of the lake, but in general two facies can be identified: a basin-margin facies and a lake-floor facies. The facies architecture within glaciomarine environments is more complex and varies with distance from the ice margin and with the character of the glaciomarine margin.

BOX 10.4: DISTINGUISHING SUBGLACIAL GLACIOTECTONICS FROM GLACIOMARINE SEDIMENTS

In recent years there has been considerable controversy over the interpretation of some glacial sequences. A glaciomarine interpretation has been advocated for several sequences traditionally interpreted as the product of subglacial deposition and deformation. In particular, sediments known as the North Sea or Contorted Drifts, exposed on the coast of Norfolk, England, have attracted considerable controversy. The North Sea Drifts are some of the most complex glacial sediments found in Britain and consist of till units, above deformed interglacial sediments. The tills commonly contain laminations, chalk clasts, pods of sand and folds. Large chalk rafts several hundred metres long are also found exposed in the coastal cliffs. Eyles *et al.* (1989) have interpreted these deposits as glaciomarine diamictons formed by the 'rain-out' of ice-rafted debris and by sediment gravity flow. The folding present within the diamictons and within the interglacial sediments beneath is interpreted in this model as the product of large-scale slumping. This explanation is also proposed for the giant chalk rafts.

In contrast, Hart and Boulton (1991) argue that this sequence represents a classic example of subglacially deformed sediments. They interpret the laminations as the product of intense folding. As Figure 8.5 shows, attenuation of folds may generate tectonic laminations and boudins. In this case the boudins consist of pods of sand broken from the nose of folds by attenuation. A tectonic explanation is consistent with the interpretation of the large chalk rafts as the product of large-scale tectonic thrusting. In order to help distinguish between glaciotectonic and glaciomarine sequences, Hart and Roberts (1994) have developed a number of criteria, some of which are listed in the table below.

	Glaciomarine	Glaciotectonic
Sedimentary units	Laterally continuous, onlapping relationships	Laterally discontinuous, tectonic boundaries
Basal boundary	Sedimentary	Décollement surface
Laminations	Graded	Non-graded
Shells	Common, *in situ*	Rare, not *in situ*
Folds	Gravitational flow folds, restricted to local areas, orientated downslope	Tectonic folds, deformation throughout, orientated in the direction of ice flow
Boudins	Rare	Common
Lonestones	Dropstones	Sinking clasts
Fabric	Variable, if present will reflect local slopes or flow directions	Variable, but may be well developed in the direction of shear

A compromise between these two alternative explanations has been recently proposed by Lunkka (1994) for the contorted Drifts of Norfolk. He suggested that the sediments within this area were deposited both in a proglacial lake and by subglacial deformation. Fluctuations in the size of the lake allowed ice to advance and deform sediment. Consequently, both waterlain and subglacial deposits can be found.

Sources: Eyles, N., Eyles, C. H. & McCabe, A. M. 1989. Sedimentation in an ice-contact subaqueous setting: the Mid-Pleistocene 'North Sea Drifts' of Norfolk, UK. *Quaternary Science Reviews* **8**, 57–74. Hart, J. K. & Boulton, G. S. 1991. The interrelation of glaciotectonic and glaciodepositional processes within the glacial environment. *Quaternary Science Reviews* **10**, 335–350. Hart, J. K. & Roberts, D. H. 1994. Criteria to distinguish between subglacial glacio-tectonic and glaciomarine sedimentation, I. Deformation styles and sedimentology. *Sedimentary Geology* **91**, 191–213. Lunkka, J. P. 1994. Sedimentation and lithostratigraphy of the North Sea Drift and Lowestoft Till Formations in the coastal cliffs of northeast Norfolk, England. *Journal of Quaternary Science* **9**, 209–233. [Table modified from: Hart & Roberts (1994) *Sedimentary Geology* **91**, Table 3, p. 211]

10.4 SUGGESTED READING

An excellent review of glaciolacustrine sedimentation is provided by Drewry (1986). Sedimentary processes in glacial lakes are discussed by Smith (1978, 1981) and Smith *et al.* (1982), while Shaw (1975) and Benn (1989) examine the sedimentology of former ice-dammed lakes. Gustavson *et al.* (1975) and Cohen (1979) discuss the sedimentology of glaciolacustrine deltas. Ashley (1989) provides a useful overview of glaciolacustrine sediments. Eyles & Eyles (1983a) discuss the interpretation of diamictons deposited by 'rain-out' from icebergs, while Eyles (1987) describes a series of diamictons deposited in a glacial lake by sediment gravity flows.

The volume edited by Dowdeswell & Scourse (1990) contains several good papers on glaciomarine sedimentation. Excellent reviews on the processes and products of glaciomarine sedimentation are provided by Elverhøi (1984), Eyles *et al.* (1985), Dowdeswell (1987), Powell & Molnia (1989), Syvitiski (1989), Boulton (1990), and Hart & Roberts (1994). Sedimentology of grounding-line fans are discussed by Powell (1990), while the sedimentology of an ice-contact delta is examined by McCabe & Eyles (1988). Specific characteristics of glaciomarine diamictons are considered by Anderson *et al.* (1980), Domack *et al.* (1980), Mackiewicz *et al.* (1984), Domack & Lawson (1985), Dowdeswell (1986), and Cowan & Powell (1990). Examples of deformation structures produced by loading within glaciomarine sediments are examined in Eyles & Clark (1985). The facies architecture of fjord glaciers is examined by Powell (1981, 1984), while Hambrey *et al.* (1992) and Boulton (1990) consider the sedimentary architecture of continental shelves (see also Hambrey 1994). The interpretation of the glacial sediments in the Irish Sea is covered by Eyles *et al.* (1989b). The discussion over the origin of the Port Askaig

Formation is covered in Spencer (1971, 1985), Eyles & Eyles (1983b), Eyles & Clark (1985), and Eyles (1988). The controversy over the interpretation of the glacial sediments of the Norfolk coast in England is dealt within in Eyles *et al.* (1989); Hart & Boulton (1991), and Lunkka (1994). Other good examples of glaciomarine sediment studies include Visser *et al.* (1987), Eyles & Lagoe (1990), and Barrett & Hambrey (1992).

Anderson, J. B., Kurtz, D. D., Domack, E. W. & Balshaw, K. M. 1980. Glacial and glacial marine sediments of the Antarctic continental shelf. *Journal of Geology* **88**, 399–414.

Ashley, G. M. 1989. Classification of glaciolacustrine sediments. In: Goldthwait, R. P. & Matsch, C. L. (Eds) *Genetic Classification of Glacigenic Deposits.* Balkema, Rotterdam, 243–260.

Barrett, P. J. & Hambrey, M. J. 1992. Plio-Pleistocene sedimentation in Ferrar fjord, Antarctica. *Sedimentology* **39**, 109–123.

Benn, D. I. 1989. Controls on the sedimentation in a late Devensian ice-dammed lake, Achnasheen, Scotland. *Boreas* **18**, 31–42.

Boulton, G. S. 1990. Sedimentary and sea level changes during glacial cycles and their control on glacimarine facies architecture. In: Dowdeswell, J. A. & Scourse, J. D. (Eds) *Glacimarine Environments: Processes and Sediments.* Geological Society Special Publication No. 53, 15–52.

Cohen, J. M. 1979. Deltaic sedimentation in glacial Lake Blessington, County Wicklow, Ireland. In: Schlüchter, Ch. (Ed.) *Moraines and Varves.* Balkema, Rotterdam, 357–367.

Cowan, E. A. & Powell, R. D. 1990. Suspended sediment transport and deposition of cyclically interlaminated sediment in a temperate glacial fjord, Alaska, USA. In: Dowdeswell, J. A. & Scourse, J. D. (Eds) *Glacimarine Environments: Processes and Sediments.* Geological Society Special Publication No. 53, 75–89.

Domack, E. W. & Lawson, D. E. 1985. Pebble fabric in an ice-rafted diamicton. *Journal of Geology* **93**, 577–591.

Domack, E. W., Anderson, J. B. & Kurtz, D. D. 1980. Clast shape as an indicator of transport and depositional mechanisms in glacial marine sediments: George V continental shelf, Antarctica. *Journal of Sedimentary Petrology* **50**, 813–820.

Dowdeswell, J. A. 1986. Distribution and character of sediments in a tidewater glacier, southern Baffin Island, NWT, Canada. *Arctic and Alpine Research* **18**, 778–781.

Dowdeswell, J. A. 1987. Processes of glacimarine sedimentation. *Progress in Physical Geography* **11**, 52–90.

Dowdeswell, J. A. & Scourse, J. D. 1990. *Glacimarine Environments: Processes and Sediments.* Geological Society Special Publication No. 53, Bath.

Drewry, D. 1986. *Glacial Geologic Processes.* Arnold, London.

Elverhøi, A. 1984. Glacigenic and associated marine sediments in the Weddell Sea, fjords of Spitsbergen and the Barents Sea: a review. *Marine Geology* **57**, 53–88.

Eyles, C. H. 1988. Glacially- and tidally-influenced shallow marine sedimentation of the Late Precambrian Port Askaig Formation, Scotland. *Palaeogeography, Palaeoclimatology, Palaeoecology* **68**, 1–25.

Eyles, C. H. & Eyles, N. 1983a. Sedimentation in a large lake: a reinterpretation of the late Pleistocene stratigraphy at Scarborough Bluffs, Ontario, Canada. *Geology* **11**, 146–152.

Eyles, C. H. & Eyles, N. 1983b. Glaciomarine model for upper Precambrian diamictites of the Port Askaig Formation, Scotland. *Geology* **11**, 692–696.

Eyles, C. H. & Lagoe, M. B. 1990. Sedimentation patterns and facies geometries on a temperate glacially-influenced continental shelf: the Yakataga Formation, Middleton Island, Alaska. In: Dowdeswell, J. A. & Scourse, J. D. (Eds) *Glacimarine Environments: Processes and Sediments.* Geological Society Special Publication No. 53, 363–386.

Eyles, C. H., Eyles, N. & Miall, A. D. 1985. Models of glaciomarine sedimentation and their application to the interpretation of ancient glacial sequences. *Palaeogeography, Palaeoclimatology, Palaeoecology* **51**, 15–84.

Eyles, N. 1987. Late Pleistocene debris-flow deposits in large glacial lakes in British Columbia and Alaska. *Sedimentary Geology* **53**, 33–71.

Eyles, N. & Clark, B. M. 1985. Gravity-induced soft-sediment deformations in glaciomarine sequences of the Upper Proterozoic Port Askaig Formation, Scotland. *Sedimentology* **32**, 789–814.

Eyles, N. & McCabe, A. M. 1989a. Glaciomarine facies within subglacial tunnel valleys: the sedimentary record of glacioisostatic downwarping in the Irish Sea Basin. *Sedimentology* **36**, 431–448.

Eyles, N. & McCabe, A. D. 1989b. The Late Devensian (< 22, 000 BP) Irish Sea Basin: the sedimentary record of a collapsed ice sheet margin. *Quaternary Science Review* **8**, 307–351.

Eyles, N., Eyles, C. H. & McCabe, A.M. 1989. Sedimentation in an ice-contact subaqueous setting: the Mid-Pleistocene 'North Sea Drifts' of Norfolk, UK. *Quaternary Science Reviews* **8**, 57–74.

Gustavson, T. C., Ashley, G. M. & Boothroyd, J. C. 1975. Depositional sequences in glaciolacustrine deltas. In: Jopling, A. V. & McDonald, B. C. (Eds) *Glaciofluvial and Glaciolacustrine Sedimentation.* The Society of Economic Paleontologists and Mineralogists, Special Publication 23, Tulsa, 264–280.

Hambrey, M. J. 1994. *Glacial Environments.* University College London Press, London.

Hambrey, M. J., Barrett, P. J., Ehrmann, W. U. & Larsen, B. 1992. Cenozoic sedimentary processes on the Antarctic continental shelf: the record from deep drilling. *Zeitschrift für Geomorphologie* **86**, 77–103.

Hart, J. K. & Boulton, G. S. 1991. The interrelation of glaciotectonic and glaciodepositional processes within the glacial environment. *Quaternary Science Reviews* **10**, 335–350.

Hart, J. K. & Roberts, D. H. 1994. Criteria to distinguish between subglacial glaciotectonic and glaciomarine sedimentation, I. Deformation styles and sedimentology. *Sedimentary Geology* **91**, 191–213.

Lunkka, J. P. 1994. Sedimentation and lithostratigraphy of the North Sea Drift and Lowestoft Till Formations in the coastal cliffs of northeast Norfolk, England. *Journal of Quaternary Science* **9**, 209–233.

Mackiewicz, N. E., Powell, R. D., Carlson, P. R. & Molina, B. F. 1984. Interlaminated ice-proximal glacimarine sediments in Muir Inlet, Alaska. *Marine Geology* **57**, 113–147.

McCabe, A. M. & Eyles, N. 1988. Sedimentology of an ice-contact glaciomarine delta, Carey Valley, Northern Ireland. *Sedimentary Geology* **59**, 1–14.

Powell, R. D. 1981. A model for sedimentation by tidewater glaciers. *Annals of Glaciology* **2**, 129–134.

Powell, R. D. 1984. Glacimarine processes and inductive lithofacies modelling of ice shelf and tidewater glacier sediments based on Quaternary examples. *Marine Geology* **57**, 1–52.

Powell, R. D. 1990. Glacimarine processes at grounding-line fans and their growth to ice-contact deltas. In: Dowdeswell, J. A. & Scourse, J. D. (Eds) *Glacimarine Environments: Processes and Sediments.* Geological Society Special Publication No. 53, 53–73.

Powell, R. D. & Molnia, B. F. 1989. Glacimarine sedimentary processes, facies and morphology of the south-southeast Alaska shelf and fjords. *Marine Geology* **85**, 359–390.

Shaw, J. 1975. Sedimentary successions in Pleistocene ice-marginal lakes. In: Jopling, A. V. & McDonald, B. C. (Eds) *Glaciofluvial and Glaciolacustrine Sedimentation.* The Society of Economic Paleontologists and Mineralogists, Special Publication 23, Tulsa, 281–303.

Smith, N. D. 1978. Sedimentation processes and patterns in a glacier-fed lake with low sediment input. *Canadian Journal of Earth Science* **15**, 741–756.

Smith, N. D. 1981. The effect of changing sediment supply on sedimentation in a glacier-fed lake. *Arctic and Alpine Research* **13**, 75–82.

Smith, N. D., Venol, M. A. & Kennedy, S. K. 1982. Comparison of sedimentation regimes in four glacier-fed lakes in western Alberta. In: Davidson-Arnott, R., Nickling, W. & Fahey, B. D. (Eds) *Research in Glacial, Glacio-fluvial and Glacio-lacustrine Systems.* Proceedings of the 6th Guelp Symposium on Geomorphology 1980, Geo Books, Norwich, 203–238.

Spencer, A. M. 1971. Late Precambrian glaciation in Scotland. *Geological Society of London Memoir 6*.

Spencer, A. M. 1985. Mechanisms and environment of deposition of Late Precambrian geosynclinal tillites Scotland and East Greenland. *Palaeogeography, Palaeoclimatology, Palaeoecology* **51**, 143–157.

Syvitiski, J. P. M. 1989. On the deposition of sediment within glacier-influenced fjords: oceanographic controls. *Marine Geology* **85**, 301–329.

Visser, J. N. J., Loock, J. C. & Colliston, W. P. 1987. Subaqueous outwash fan and esker sandstones in the Permo-Carboniferous Dwyka Formation of South Africa. *Journal of Sedimentary Petrology* **57**, 467–478.

11
Landforms of Glacial Deposition in Water

When deposition occurs from a glacier into water a different landform assemblage develops. Despite this, some of the processes and landforms which result are genetically similar to those found on land. For example, push moraines form in glaciolacustrine and glaciomarine environments in the same way as those on land, while outwash fans develop in all environments. In this chapter we shall first examine the landforms associated with glaciolacustrine environments before tackling those which form along ice margins terminating in the sea.

11.1 GLACIOLACUSTRINE LANDFORMS

There is a wide variety of types of glacial lake. One can identify two broad types: (1) ice-marginal, and (2) supraglacial. Ice-marginal lakes may form in front of glaciers or when ice dams water in a valley or against a hillside (Figure 4.9). Supraglacial lakes may occur either where an ice-dammed lake expands over an ice margin (Figure 11.1) or in areas of ice-cored moraines. The geomorphological products of these two categories of lake are different and will be considered in turn.

In ice-marginal lakes there are two types of glacier–lake interface (Figure 11.2): those at which large fans or deltas form, separating the ice margin from the lake; and those at which the ice front is actively calving. At calving margins small fans and moraine banks form. The type of landforms which develop along the ice margin depends on whether it is stationary and upon the amount of meltwater discharge. The development of a large delta is, for example, favoured by a stable ice margin and high meltwater/sediment discharge. The landforms associated with these different lake margins and with lakes in general are discussed below.

Figure 11.1   *Supraglacial lake submerging part of an ice margin in South Greenland. Note the debris bands on the glacier surface formed by ice-marginal thrusting [Photograph: N. F. Glasser]*

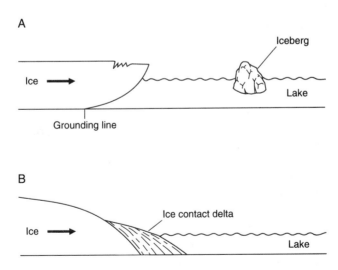

Figure 11.2   *Types of glacier–lake interface. **A:** Calving margin. **B:** Ice-contact delta type margin*

1. **Glaciolacustrine landforms at non-calving ice margins.** At a stationary ice margin with high meltwater discharge an ice-contact delta may develop. Where sediment is delivered from a single or narrowly confined group of channels, delta fronts are arcuate in plan form. In contrast, where meltwater streams switch from one side of a valley to another the delta front may be more straight. The ice-contact face of the delta may be deformed and raised by post-depositional ice pushing. This face may also be subject to collapse and slumping when the ice support is removed.

   A **delta moraine** may develop where an ice front remains stationary for a considerable period of time. A delta moraine consists of a series of deltas or fans which coalesce along an ice margin to form a continuous ridge. The moraine is not formed from one but from many fans and deltas. Probably the longest delta moraines exposed on land are the Salpausselkä Moraine in Finland which are over 600 km long and formed in the Baltic ice lake (Box 11.1).

BOX 11.1: THE DELTA MORAINES OF THE BALTIC ICE LAKE

At the close of the last glacial cycle a very large lake lay along the southern margin of the Scandinavian Ice Sheet, covering much of what is the Baltic Sea today. Along the southern margin of the ice sheet large delta moraines accumulated. These are best developed in southern Finland where the Salpausselkä moraines occur. These moraines extend for over 600 km and are composed for the most part of outwash sediments. They formed along a stationary ice margin. Fyfe (1990) studied the morphology and sedimentology of one of the three Salpausselkä moraines. She found that it was composed along its length of three geomorphological components:

1. Large individual deltas with braided tops, ice-contact deltas, which built up to the water level and were produced in relatively shallow water where the meltwater outflow was concentrated into a small number of concentrated outflows or portals.
2. Low narrow fans, grounding-line fans, which coalesce along the former ice margin and were formed in deeper water where the meltwater outflow was more distributed along the ice margin.
3. Small laterally overlapping subaqueous fans formed where the location of the meltwater portals was unstable due to calving into deep water. Consequently a large number of small fans developed along the margin.

The water depth and the nature of the subglacial drainage system—concentrated, distributed, stable or unstable—controlled the nature of the outwash accumulation which formed along the former ice margin and therefore controlled the morphology of the delta moraine formed.

Source: Fyfe, G. J. 1990. The effect of water depth on ice-proximal glaciolacustrine sedimentation: Salpausselkä I, southern Finland. *Boreas* **19**, 147–164.

2. **Glaciolacustrine landforms at calving ice margins.** Moraine ridges may develop at the grounding line. This is the line at which the ice begins to float and loses contact with its bed. These moraines are frequently referred to as **De Geer moraines** (washboard or cross-valley moraines). They are low (< 5 m), often asymmetric ridges, which are either straight or slightly concave in plan form. This plan form reflects the linear or concave morphology typical of a calving glacier margin. They appear to form at the grounding line, which may be some distance behind the ice margin. There is considerable confusion surrounding the nomenclature and formation of these ridges. Most simply form as seasonal push moraines, formed by winter or periodic readvances of the grounding line or ice margin in the same way as terrestrial push moraines. It has also been suggested that the calving of large icebergs may also be important in the formation of these moraines. Debris accumulates along sub-lacustrine margins by melt-out from beneath a floating ice margin and then is pushed up by the calving of icebergs (Figure 11.3A). The internal structure of some larger ridges or moraine banks suggests that they may also form by the accumulation of subglacial or englacial debris. Debris is transported either in

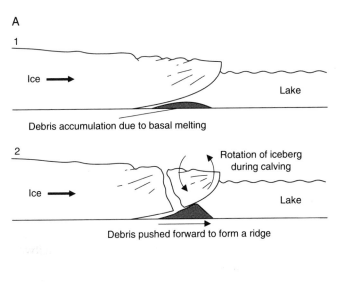

A

1

Ice ⟶

Lake

Debris accumulation due to basal melting

2

Rotation of iceberg during calving

Ice ⟶

Lake

Debris pushed forward to form a ridge

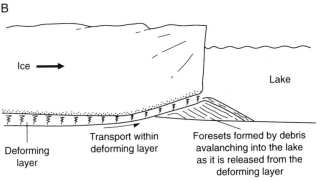

B

Ice ⟶

Lake

Deforming layer

Transport within deforming layer

Foresets formed by debris avalanching into the lake as it is released from the deforming layer

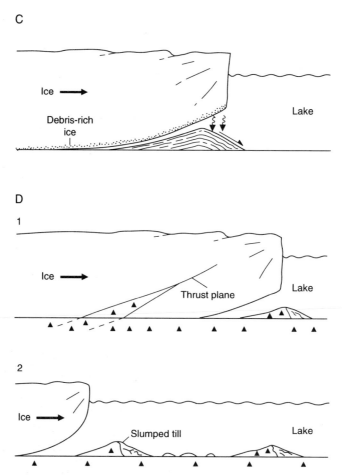

Figure 11.3 *Possible modes of formation for De Geer moraines and other similar transverse sub-lacustrine moraine banks.* **A:** *Formation during iceberg calving.* **B:** *Formation by the release of debris from a deforming layer.* **C:** *Formation by basal melting and debris accumulation.* **D:** *Formation at the base of englacial thrusts*

basal ice or as a deforming layer of subglacial sediment to the grounding line where it avalanches or flows forward into the lake to form a layer or foreset (Figure 11.3B and C). The bank or moraine grows by the addition of these debris foresets. Alternative explanations for De Geer moraines include the formation of ridges by the squeezing of saturated till into basal crevasses behind a calving ice margin. The production of the necessary basal crevasses is usually seen as a function of local glacier surges. Recent work has also recognised the potential of glacial thrusting to produce small linear ridges; significantly these do not form at the ice margin (Figure 11.3D).

Small sub-lacustrine fans form at steep calving ice margins wherever a meltwater portal exists. These small fans typically have steep gradients since deposition of the coarse fraction occurs rapidly as the meltwater enters the standing water body. The position of meltwater portals is relatively unstable due to calving and the point of fan formation may move laterally along an ice margin. In this way a series of small coalescing fans may form along the margin. Their steep ice-contact faces may merge to give a ridge-like form. In some cases these coalesced fans may be mistaken for cross-valley or De Geer moraines. Small fans may also develop at the sides of a lake, where water is channelled along the ice margin. The sediment of these fans, being derived from the valley sides and lateral moraines of the ice margin, is typically much coarser than the sediments which form in the centre of the lake from subglacial meltwater debris.

3. **Glaciolacustrine landforms not associated with an ice margin.** Where lake levels are stationary for some time, shorelines or strandlines may develop. Fossil shorelines can be identified on hillsides as horizontal lines with a faint terrace-like form (Figure 11.4). The shoreline terrace may be purely depositional in character or may also be eroded into the hillside. Depositional elements involve the reworking of slope debris by wave action moving down the hillside towards the lake. The amount of sediment modification will largely depend on the size of the lake, which controls the available fetch and therefore the size of waves upon it. Waves generated by the calving of icebergs are often of greater magnitude and therefore more significant than wind-generated waves. Small berms produced by wave action are commonly superimposed upon the beach

Figure 11.4 *Glacial lake shorelines formed by a former ice-dammed lake at Heinabergsjökull, Iceland. The shorelines, visible as horizontal terraces on the right-hand side of the picture, terminate abruptly on the left. This is the ice-marginal position of the glacier which blocked the valley to form an ice-dammed lake during the Little Ice Age. [Photograph: M. R. Bennett]*

terrace. These features are transitory and unlikely to be preserved. Sediment on the shoreline is often deformed or pushed up by lake ice and a variety of pushed ridges may form. Some shorelines are formed not only of depositional terraces but have also been eroded into bedrock or slope debris. Given that most ice-dammed lakes are transitory features and that wave action upon them is likely to be limited, it seems likely that erosion is the result of the action of freeze–thaw weathering and of icebergs rather than being associated with wave action. It has been suggested that lake shorelines may be an ideal location for intense freeze–thaw activity (Box 11.2). The size of a shoreline terrace may be increased where small streams enter the lake and deposit sediment in small fans or deltas. The lake floor may also be subject to scour by grounded icebergs, especially towards the lake margins. The keel or bottom of an iceberg may drag along the lake floor to form linear **plough marks** (Figure 11.5).

Figure 11.5 *Iceberg scour marks from the bed of the proglacial lake in front of Heinabergsjökull in south-east Iceland. The rocking or bouncing of icebergs by the passage of waves is responsible for the transverse tread visible in some of the scour marks.*
[Photographs: M. R. Bennett]

BOX 11.2: THE PROCESS OF SHORELINE FORMATION

Matthews *et al.* (1986) studied the rate of formation of lake shorelines in the Jotunheimen mountains of southern Norway. An outlet glacier of the Smörstabbreen ice cap, Böverbreen, dammed a small ice-dammed lake during the Little Ice Age. A shoreline was cut into metamorphic bedrock during the short life of the ice-dammed lake. This shoreline is up to 5.3 m wide. Lichenometric dating has been used to establish the length of time that the ice-dammed lake existed. Lichenometric dating relies upon the fact that certain lichens colonise moraines immediately on deglaciation and grow in size in a consistent fashion. Consequently, if one measures the size of lichens on a moraine and compares their size with a lichen growth curve (a curve of the increase in lichen size with time) then one can produce age estimates for deglaciation of the moraine. By dating moraines in this way they estimated that the shorelines were formed by the ice margin which dammed the lake over a time span of between 75 and 125 years. Given the width of the rock-cut shorelines, this implies that the average rate of erosion must have been between 26 and 44 mm a year. Matthews *et al.* (1986) argue that such rapid rates of erosion are only possible if intense freeze–thaw weathering operated. They suggest that a shoreline is an ideal location for intense frost weathering and that the removal of debris by a seasonal lake would have assisted this procedure.

Source: Matthews, J. A., Dawson, A. G. & Shakesby, A. 1986. Lake shoreline development, frost weathering and rock platform erosion in an alpine periglacial environment, Jotunheimen, southern Norway. *Boreas* **15**, 33–50.

Supraglacial lakes contain a similar assemblage of landforms (fans, deltas and shorelines), but due to the presence of buried ice these landforms are unlikely to be preserved on deglaciation. The melt-out of the buried ice will produce a kame and kettle topography in which the only evidence of a lake may be the presence of glaciolacustrine sediments within the kames.

11.2 GLACIOMARINE LANDFORMS

A similar assemblage of landforms to those in glaciolacustrine environments may develop at marine ice margins (Figure 11.6). In general, however, the glaciomarine landforms are often larger and may be found over a wider area, often extending to the edge of the continental shelf. The landforms typical of calving margins are discussed below (Figure 11.6).

Moraine banks may form against calving margins in a variety of different ways. They may form by the direct dumping of debris from the ice front, by basal squeezing or by pushing. Squeezing of saturated subglacial till and marine muds may occur at the margin of some calving glaciers. Sediment is squeezed out in front of

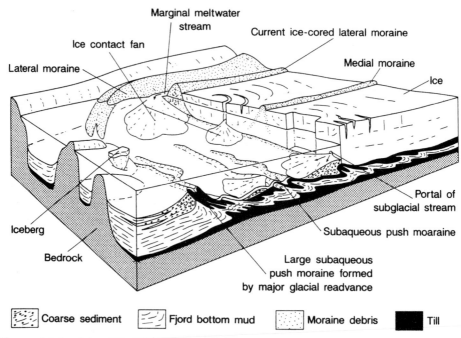

Figure 11.6   *Schematic block diagram showing the principal seabed features in the proglacial area of a calving fjord glacier. [Modified from: Boulton (1990) In: Dowdeswell, J. A. & Scourse, J. D. (Eds) Glacimarine Environments: Processes and Sediments, Geological Society Special Publication No. 53, Fig. 4, p. 19]*

the margin, particularly if the glacier has been advancing rapidly or surging and has subsequently settled into its bed following the surge. As sediment is squeezed out it may flow forward and fold to form a series of gently undulating ridges in front of the ice margin. These ridges are not produced by pushing but by sediment flow. The ridge produced in front of Seftrømbreen on to the Island of Coraholmen, described in Section 9.2.2, provides an excellent example of a moraine ridge formed in this way (Figure 9.28). Immediately behind the ice margin, sediment may be squeezed into basal crevasses.

Morainal banks may also form by pushing. Subaquatic push moraines are very similar in character to terrestrial push moraines (see Section 9.1.1), although they are generally formed from much finer-grained sediment and are frequently modified by slumping and sediment gravity flows. Moraines may also become buried by a blanket of mud deposited from suspension. They are often formed by seasonal readvances in a similar manner to push moraines on land. Glacier calving is dramatically reduced in winter if the sea freezes over, but as the glacier continues to flow its margin must advance, pushing up a moraine and buckling the sea ice in front of the ice margin. Like their terrestrial counterparts, subaquatic push moraines are often intimately associated with subaquatic outwash or grounding line fans; points against which the glacier may push efficiently. Large moraine banks may also tend to be associated with stable ice positions which are controlled, in the case of fjords or restricted marine margins, by the location of topographic pinning points (see Section 3.6). The formation of morainal banks is an important control on the glaciodynamics of ice margins, particularly in fjords or restricted sea-ways (Box 11.3).

The outflow of meltwater from an ice margin will lead to the formation of a **sub-aquatic outwash** or **grounding line fans**. These are very similar to those which develop in glaciolacustrine environments. Unlike their terrestrial counterparts, grounding line fans do not form along the whole of the glacier front but occur close to the specific points at which meltwater streams emerge from the glacier. On land, sediments spread out over a large area, constructing a fan along much of the ice margin. In water, sedimentation occurs very rapidly as the meltwater velocity is checked by the standing water into which it flows. Consequently grounding line fans are located close to the meltwater portal. The position of this portal may be locally unstable within a season, or from season to season, but will tend to emerge at roughly the same positions through time. There are strong contrasts between those which develop at the lateral margins and those which do not. Since sedimentation occurs very rapidly close to the meltwater portal, the fan has a steep gradient. As the ice front, against which the fan accumulates, is also steep little buried ice is usually incorporated into the ice-contact face of the fan. However, the ice-contact face frequently shows evidence of slumping, which follows as a consequence of the removal of support provided by the ice front. Grounding line fans can aggrade rapidly and at some temperate sites sedimentation rates of 1×10^6 m^3 per year are typical. Pushing against the ice-contact face of the fans is common and may increase the height of a fan dramatically. It is important to note that subaqueous outwash fans form at the grounding line, which may occur some way behind the ice margin if a shelving or floating margin exists.

The morphology of the fan and the nature of the sediments deposited within them appear to be controlled by the discharge of water from the glacier and therefore the morphology of the debris plume (Figure 11.7). At low water discharges the plume rises rapidly to the surface. Coarse tractional deposits are dumped close to the ice margin at the meltwater portal. Most of the fan is deposited on the distal slope of the fan by settling from the debris plume and by slumping or sliding debris down the distal face of the fan. This slumping gives rise to a series of foresets. At higher water and sediment discharges the plume remains in contact with the fan for longer and flattens or planes off the apex of the fan. Coarse sediment is also deposited at this point in sheets and scour and fill structures. Deposition by settling and by sediment gravity flows again builds up the distal part of the fan. At very high discharges deposition from the plume while it is in contact with the fan may deposit a barchinoid bar at the point at which the plume is detached from the fan (Figure 11.7).

The morphology of grounding line fans is also controlled by the stability of the ice margin and the rate of discharge. At advancing ice margins grounding line fans will be small and heavily tectonised. In contrast, at retreating ice margins the fan morphology will depend upon: (1) the rate of ice-marginal retreat; (2) the stability in the position of the meltwater portal; and (3) the rate of the sediment discharge.

BOX 11.3: MORAINE BANKS AND CYCLIC GLACIER FLUCTUATIONS

The behaviour of individual Alaskan tidewater or fjord glaciers has often been observed to be asynchronous with other tidewater glaciers and with glaciers on land. It has been suggested that tidewater glaciers may behave in a cyclic fashion due to the inherent instability of a calving glacier margin. The rate of glacier

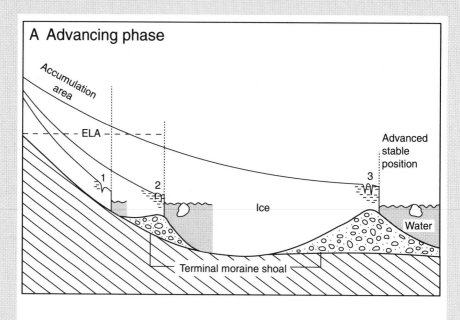

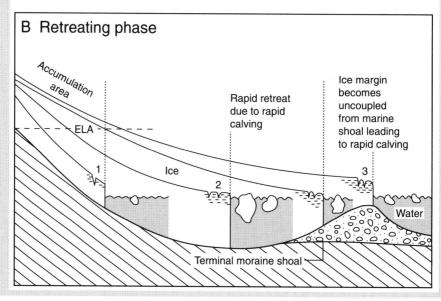

calving is strongly influenced by water depth. As a tidewater glacier advances down a fjord in equilibrium with its mass balance, a moraine shoal or moraine bank will form along its margin and will advance with the glacier. This bank will help stabilise the ice margin by reducing the water depth and therefore the rate of glacier calving. If this balance is disturbed and the ice margin retreats away from this bank into deep water, rapid calving and retreat may occur. This retreat may be triggered by a small rise in the equilibrium line of a glacier or by some non-climatic event such as a catastrophic glacier flood. Either way, once the glacier terminus is uncoupled from its moraine bank, calving can proceed rapidly in the deep water behind. Rapid calving may cause the glacier to retreat out of proportion with the magnitude of the event that triggered it and it may continue to retreat until the glacier reaches shallow water. At this point the position of the ice margin is likely to be out of equilibrium with its mass balance and it will start to advance again, forming a new moraine bank. The glacier, therefore, behaves in a cyclic fashion and is uncoupled from climate. The margin is not advancing and retreating due to climate but as a consequence of the inherent instability of a calving glacier margin. The presence of a moraine bank, therefore, is important to the stability of the ice margin.

Source: Mayo, L. R. 1988. Advance of Hubbard glacier and closure of Rusell fjord, Alaska—environmental effects and hazards in the Yakutat area. *United States Geological Survey Circular 1016*. [Diagram modified from: Warren (1992) *Progress in Physical Geography* **16**, Fig. 7, p. 265]

in the position of the meltwater portal; and (3) the rate of the sediment discharge. Given these variables the following morphologies may develop:

A. Advancing ice margin: in this situation a small fan would develop in front of the ice margin during the summer ablation season when calving is high. These fans will be overrun and tectonised each winter when the glacier advances more rapidly.

B. Stationary ice margin: in this situation a single large fan will develop. If the rate of sedimentation is high this fan may grow to sea-level and form an ice-contact delta.

C. Slowly retreating ice margins with a strong seasonal readvance: in this situation a series of fans, which may merge in the direction of retreat, form as the ice margin withdraws. Merged fans give an esker-like ridge, a **paraesker**, in front of the retreating meltwater portal. This ridge may be associated with closely spaced push moraines. If the position of the meltwater portal is not constant during retreat a series of closely spaced fans develop which are offset from one another in the direction of retreat to form a zigzag-type pattern.

D. Rapidly retreating ice margin with a seasonal readvance: in this situation widely spaced ice-marginal fans would develop in the direction of retreat, associated with well-spaced push moraines. These widely spaced fans often look like a string of beads (each fan being an individual bead) when viewed from above and have consequently been sometimes described as **beaded eskers**.

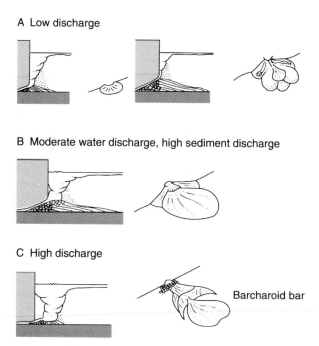

Figure 11.7   *Schematic representation of the variation in the discharge plume in front of a calving glacier in relation to water discharge, and the impact of this variation on the morphology of the grounding line fan produced.* **A:** *At low water discharges, the fan grows progressively with coarse traction deposits dumped at the meltwater portal and by deposition as a series of large foresets on the distal flank from settling, slumping/sliding and sediment gravity flows.* **B:** *At high sediment discharges, the base of the sediment plume remains in contact with the fan for longer. As a consequence, a series of sheet-like deposits of sand and gravel with crude scour and fill structures is deposited on the top of the fan. Settling and sediment gravity flow deposits make up the more distal slopes of the fan.* **C:** *At high water discharges the plume remains in contact with the fan for much longer and may deposit a migrating barchanoid bar along the line of detachment between the plume and the seabed. [Reproduced with permission from: Powell (1990) In: Dowdeswell, J. A. & Scourse, J. D. (Eds) Glacimarine Environments: Processes and Sediments, Geological Society Special Publication No. 53, Fig. 2, p. 58]*

E. Rapidly retreating ice margin with no seasonal readvance: in this situation fan and push moraine development would be minimal, and a broad subaqueous outwash plain or surface would develop if sediment discharge was high.

In general, subaquatic outwash fans (marine or lacustrine) may be distinguished from terrestrial outwash fans on the basis of the following:

A. Dead ice collapse features are common on the proximal flanks of terrestrial forms.
B. In marine forms, the coarse-grained material is concentrated at the point of ice

contact, with rapid distal reduction in grain size. This gradient in particle size is less pronounced in terrestrial forms.

C. The thickness of marine fans is limited only by the depth of the water body while the thickness of terrestrial fans is limited by the local base level and they therefore tend to be thin.

Another type of grounding line fan may develop, known as a **till delta**. These landforms are more common beneath ice shelves and broad glaciomarine margins than in the restricted space of a fjord. They are fans developed by the delivery of transported glacial debris to the grounding line, as opposed to fluvial sediment. Transport and deposition may occur in basal or englacial ice which melts out at the grounding line or alternatively through subglacial deformation. Either way, sediment is delivered to and deposited along the grounding line. As material is brought to the grounding line it avalanches and flows forward to form a series of foresets (Figure 11.8). If the glacier is advancing, the till delta will advance and the foresets will be capped by basal till. Bottomsets will be formed by settling from suspension. In contrast to grounding-line fans that develop through stream flow and which are focused on meltwater portals, till deltas form along the whole length of the grounding line. They may record a history of sea-level changes in the same way that lacustrine deltas may record the history of lake levels (Figures 10.2 and 11.8).

A genetically similar landform, a **trough mouth fan**, is found at the edge of continental shelves that have been glaciated. Most northern hemisphere continental shelves which have been glaciated show a simple geometry that consists of shallow banks, of 100–200 m depth, crossed by glacially eroded troughs which descend 300–400 m below sea-level. These troughs probably reflect the location of lines of fast-flowing ice or ice streams within ice sheets. At glacial maximum, ice was located at the mouths of these troughs on the edge of the continental shelf. The glacier would not be able to advance into the deep water beyond the edge of the shelf. Here large fans—trough mouth fans—prograde out over the continental slope, extending the shelf. Sediment is transported to the grounding line by: (1) subglacial deformation; (2) basal ice; and (3) meltwater. This sediment is then deposited as giant foresets down the continental slope by sediment gravity flows (Figure 11.8). These foresets are prone to slumping and further downslope movement. While the ice margin is located close to the edge of the continental shelf, progradation continues and steep foresets result. However, as the glacier retreats away from the edge of the continental shelf slumping and downslope movement of material occurs on the fan (Figure 11.8).

Other glaciomarine landforms include iceberg grounding structures which form wherever the keel of an iceberg makes contact with the sea-floor. **Plough marks** produced by icebergs are common wherever an iceberg runs aground, and

Figure 11.8 *The formation of till deltas and trough mouth fans. **A:** The formation of a till delta. Steps 1 to 3 show the response of the till delta to an increase in sea-level. **B:** Formation of a trough mouth fan. [Reproduced with permission from: Boulton (1990) In: Dowdeswell, J. A. & Scourse, J. D. (Eds) Glacimarine Environments: Processes and Sediments, Geological Society Special Publication No. 53, Fig. 6, p. 21]*

A

1

2

Sea-level
rise

3

B

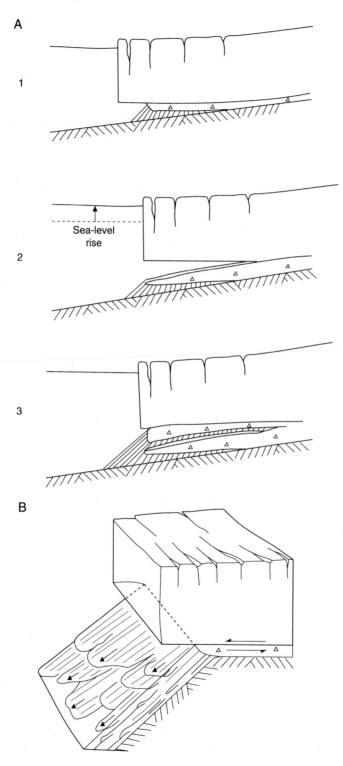

can be observed as both relict and contemporary features on many continental shelves today. The morphology of an individual plough mark depends upon: (1) the sedimentological characteristics of the sea-floor sediment; (2) the geometry of the iceberg keel; and (3) the motion of the iceberg. In stiff cohesive sediments ice keels may create irregular grooves flanked by low ridges. Blocks of sediment may be dislodged and pushed over the seabed in front of the keel. In less-cohesive soft sediments plough marks consist of more regular and continuous grooves which are rapidly modified by currents and wave action. The keels of icebergs may either be single- or multi-pronged. Multi-pronged bergs give rise to complex marks which consist of semiparallel troughs. The motion of the iceberg is also important and is controlled by the surface wind, currents and the interaction of one berg with another. For example, the rotation of an iceberg keel during ploughing may change the shape of a plough mark. Alternatively, unstable or 'wobbly' icebergs may form grooves with tread—ridges perpendicular to the groove orientation— known as 'sprag' or 'jigger' marks. Small bergs may also produce grooves with tread if they are lifted by waves. Ploughing of icebergs through glaciomarine sediment may also cause tectonic disturbance—faults and thrusts—in glaciomarine sediments. A grounded or stationary iceberg which settles on the seabed but does not plough forward may produce a gravity crater. This may form partly under the weight of the berg and partly by current scour around the keel.

Another glaciomarine landform is the striated boulder pavement, which develops as ice advances over marine sediment. Boulder lags develop on shallow continental shelf areas due to winnowing by waves and tidal currents of diamictons, produced by the 'rain-out' of debris from icebergs. The top surfaces of these boulder lags are then striated and planed as the glacier advances over them (Figure 11.9).

Finally it should be noted that beneath grounded ice, normal subglacial processes continue and subglacial landforms may develop. Flutes and drumlins may therefore be identified beneath marine margins provided the glacier was grounded. The morphology of these landforms may of course have been modified by current activity and by the deposition of grounding-line fans or the gentle 'rain-out' of glaciomarine sediment.

11.3 SUMMARY

A similar association of landforms—push moraines and outwash fans—is found at glaciolacustrine and glaciomarine ice margins as that which is found on land. The following landforms are typical of glaciolacustrine ice margins: ice-contact deltas, delta moraines, subaqueous outwash fans, push moraines (De Geer moraines), and shorelines. The following landforms may be associated with glaciomarine ice margins: outwash or grounding-line fans, ice-contact deltas, push moraines, till deltas, trough mouth fans, iceberg plough marks, and striated boulder pavements. At both glaciomarine and glaciolacustrine ice margins the actual assemblage of landforms and their detailed morphology is controlled by such variables as the mass balance, the

Figure 11.9   *Boulder pavements, South Shetland Islands. [Photograph: J. Hansom]*

geometry of the water body, the geometry and setting of the ice margin, and the basal thermal regime of the glacier and therefore the debris content of any icebergs present.

Table 11.1 summarises the inferences that can be made about the glacial environment in which each glaciolacustrine and glaciomarine landform is normally produced. It provides a key to the clues left about the former glacier which deposited the depositional landscape visible in areas once covered by Cenozoic glaciers.

Table 11.1   *Summary table of the principal glaciolacustrine and glaciomarine landforms and their significance for glacier reconstructions*

Glaciolacustrine

Ice-contact deltas	Morphology: mound with a flat or gently sloping top and a steep frontal slope. Often found in mid-valley locations with a steep ice-contact face. Indicative of: ice-marginal position in glacial lake.
Deltas	Morphology: mound with a flat or gently sloping top usually extending from a valley side into the former lake. Has a steep frontal slope and the mound top may merge laterally with lake shorelines if present. Indicative of: point of water inflow into a lake.
Delta moraines	Morphology: extensive mound or ridge composed of individual fans or deltas. Deltas may appear as large mounds with flat or gently sloping tops and steep frontal slopes. Fans may possess steep ice-contact faces and gently sloping distal

flanks. Morphologically very variable.
Indicative of: ice-marginal position in a glacial lake.
Morphology may provide some indication of water depth;
evidence suggests that fans tend to form in deep water while
deltas are more common in shallow water.

De Geer moraines	Morphology: low asymmetrical, often very straight, ridges transverse to the direction of ice flow. Tend to form distinct suites of ridges with a regular spacing. Indicative of: ice-marginal position in a glacial lake.
Lacustrine fans	Morphology: small steep-sided fan; may merge with other fans along a distinct line. Indicative of: ice-marginal position in a glacial lake and the location of meltwater portals.
Lake shorelines	Morphology: horizontal or near-horizontal bench, may be formed by either deposition or erosion. Tend to be paired across a valley side. Indicative of: lake water level.

Glaciomarine

Moraine banks	Morphology: irregular accumulation of coarse gravel and diamicton. Morphology poorly recorded. Indicative of: ice-marginal position.
Submarine push moraine	Morphology: low asymmetrical, often very straight, ridges transverse to the direction of ice flow. Tend to form distinct suites of ridges with a regular spacing. Indicative of: ice-marginal position and of seasonal readvances, which occur whenever winter calving is less than winter ice flow.
Grounding line fans	Morphology: steep-sided fan with a distinct ice-contact face, may merge with other fans along a distinct line. Indicative of: ice-marginal position.
Till deltas	Morphology: linear asymmetrical ridge with a steep frontal slope and gentle ice-contact face. Morphology poorly recorded. Indicative of: ice-marginal position.
Trough mouth fan	Morphology: fan-like body of sediment at the edge of the continental shelf. Indicative of: presence of ice margin close to the continental edge.
Plough marks	Morphology: linear furrows or depressions on the seabed. Indicative of: iceberg grounding.
Boulder pavements	Morphology: compact surface layer of packed boulders which have been levelled off and striated. Indicative of: grounding of an ice shelf or glacier in a glaciomarine environment.

11.4 SUGGESTED READING

A variety of glaciolacustrine landforms and landform assemblages are described in the following papers which cover both palaeo and contemporary ice-dammed lakes: Sissons (1977, 1978, 1982), Peacock & Cornish (1989), Benn (1989), Gore (1992), and Johnson & Kasper (1992). More specifically, glaciolacustrine deltas are described by Thomas (1984) and Clemmensen & Houmark-Nielsen (1981), while delta moraines are discussed by Fyfe (1990). Cross-valley, washboard or De Geer moraines are discussed in papers by Andrews (1963a,b), Andrews & Smithson (1966), Barnett & Holdsworth (1974), Zilliacus (1989), Beaudry & Prichonnet (1991), and Larsen *et al.* (1991). Shoreline development is discussed by Matthews *et al.* (1986) and iceberg plough marks on lacustrine shorelines are reported by Nichols (1953) and Bennett & Bullard (1991).

A wide range of glaciomarine landforms are described in the papers by Boulton (1986, 1990) and Powell & Molnia (1989). Stoker & Holmes (1991) describe a series of moraines from the Hebridean continental shelf. The formation of grounding-line fans is covered by Powell (1990), while McCabe & Eyles (1988) describe a glaciomarine ice-contact delta. The formation of till deltas is covered in the paper by King (1994). Nielsen (1991) describes the formation of boulder beaches formed along the margin of fjord by waves generated by glacier calving. Iceberg plough marks are discussed in detail in papers by Reimnitz & Barnes (1974), Belderson *et al.* (1973), Kovacs & Mellor (1974), and Dowdeswell *et al.* (1993). Boulder pavements are discussed in detail by both Hansom (1983) and Eyles (1988, 1994).

Andrews, J. T. 1963a. Cross-valley moraines of the Rimrock and Isortoq River valleys, Baffin Island, North West Territories. *Geographical Bulletin* **19**, 49–77.

Andrews, J. T. 1963b. Cross-valley moraines of north-central Baffin Island: a quantitative analysis. *Geographical Bulletin* **20**, 82–129.

Andrews, J. T. & Smithson, B. B. 1966. Till fabric of the cross-valley moraines of north-central Baffin Island, North West Territories, Canada. *Bulletin of the Geological Society of America* **77**, 271–290.

Barnett, D. M. & Holdsworth, G. 1974. Origin, morphology, and chronology of sublacus-trine moraines, Generator Lake Baffin Island, Northwest Territories, Canada. *Canadian Journal of Earth Science* **11**, 380–408.

Beaudry, L. M. & Prichonnet, G. 1991. Late Glacial De Geer moraines with glaciofluvial sediment in the Chapais area, Québec (Canada). *Boreas* **20**, 377–394.

Belderson, R. H., Kenyon, N. H. & Wilson, J. B. 1973. Iceberg plough marks in the northeast Atlantic. *Palaeogeography, Palaeoclimatology, Palaeoecology* **13**, 215–224.

Benn, D. I. 1989. Controls on sedimentation in a late Devensian ice-dammed lake, Achnasheen, Scotland. *Boreas* **18**, 31–42.

Bennett, M. R. & Bullard, J. E. 1991. Iceberg tool marks: an example from Heinabergsjökull, South East Iceland. *Journal of Glaciology* **37**, 181–183.

Boulton, G. S. 1986. Push-moraines and glacier-contact fans in marine and terrestrial environments. *Sedimentology* **33**, 677–698.

Boulton, G. S. 1990. Sedimentary and sea level changes during glacial cycles and their control on glacimarine facies architecture. In: Dowdeswell, J. A. & Scourse, J. D. (Eds) *Glacimarine Environments: Processes and Sediments.* Geological Society Special Publication No. 53, 15–52.

Clemmensen, L. B. & Houmark-Nielsen, M. 1981. Sedimentary features of a Weishselian glaciolacustrine delta. *Boreas* **10**, 229–245.

Dowdeswell, J. A., Villinger, H., Whittington, R. J. & Marienfield, P. 1993. Iceberg scouring in Scoresby Sund and on the East Greenland continental shelf. *Marine Geology* **111**, 37–53.

Eyles, C. H. 1988. A model for striated boulder pavement formation on glaciated shallow marine shelves: an example from the Yakataga Formation Alaska. *Journal of Sedimentary Petrology* **58**, 62–71.

Eyles, C. H. 1994. Striated boulder pavements. *Sedimentology* **88**, 161–173.

Fyfe, G. J. 1990. The effect of water depth on ice-proximal glaciolacustrine sedimentation: Salpausselkä I, southern Finland. *Boreas* **19**, 147–164.

Gore, D. B. 1992. Ice-damming and fluvial erosion in the Vestfold Hills, East Antarctica. *Antarctic Science* **4**, 227–234.

Hansom, J. D. 1983. Ice-formed intertidal boulder pavements in the sub-Antarctic. *Journal of Sedimentary Petrology* **53**, 135–145.

Johnson, P. G. & Kasper, J. N. 1992. The development of an ice-dammed lake: the contemporary and older sedimentary record. *Arctic and Alpine Research* **24**, 304–313.

King, L. H. 1994. Till in the marine environment. *Journal of Quaternary Science* **8**, 347–358.

Kovacs, A. & Mellor, M. 1974. Sea ice morphology and ice as a geologic agent in the southern Beaufort Sea. In: Reed, J. C. & Slater, J. E. (Eds) *The Coast and Shelf of the Beaufort Sea*. Arctic Institute of North America, Arlington, VA, 113–164.

Larsen, E., Longva, O. & Follestad, B. A. 1991. Formation of De Geer moraines and implications for deglaciation dynamics. *Journal of Quaternary Science* **6**, 263–277.

Matthews, J. A., Dawson, A. G. & Shakesby, A. 1986. Lake shoreline development, frost weathering and rock platform erosion in an alpine periglacial environment, Jotunheimen, southern Norway. *Boreas* **15**, 33–50.

McCabe, A. M. & Eyles, N. 1988. Sedimentology of an ice-contact glaciomarine delta, Carey Valley, Northern Ireland. *Sedimentary Geology* **59**, 1–14.

Nichols, R. L. 1953. Marine and lacustrine ice-pushed ridges. *Journal of Glaciology* **2**, 172–175

Nielsen, N. 1991. A boulder beach formed by waves from a calving glacier; Eqip Sermia, West Greenland. *Boreas* **21**, 159–168.

Peacock, J. D. & Cornish, R. 1989. *Glen Roy Area: Field Guide*. Quaternary Research Association, Cambridge.

Powell, R. D. 1990. Glacimarine processes at grounding-line fans and their growth to ice-contact deltas. In: Dowdeswell, J. A. & Scourse, J. D. (Eds) *Glacimarine Environments: Processes and Sediments*. Geological Society Special Publication No. 53, 53–73.

Powell, R. D. & Molnia, B. F. 1989. Glacimarine sedimentary processes, facies and morphology of the south-southeast Alaska shelf and fjords. *Marine Geology* **85**, 359–390.

Reimnitz, E. & Barnes, P. W. 1974. Sea ice as a geologic agent on the Beaufort Sea Shelf of Alaska. In: Reed, J. C. & Slater, J. E. (Eds) *The Coast and Shelf of the Beaufort Sea*. Arctic Institute of North America, Arlington, VA, 301–353.

Sissons, J. B. 1977. Former ice-dammed lakes in Glen Moriston, Inverness-shire. *Transactions of the Institute of British Geographers* **2**, 224–242.

Sissons, J. B. 1978. The parallel roads of Glen Roy and adjacent glens, Scotland. *Boreas* **7**, 183–244.

Sissons, J. B. 1982. A former ice-dammed lake and associated glacier limits in the Achnasheen area, central Ross-shire. *Transactions of the Institute of British Geographers* **7**, 98–116.

Stoker, M. S. & Holmes, R. 1991. Submarine end-moraines as indicators of Pleistocene ice-limits off northwest Britain. *Journal of the Geological Society, London* **148**, 431–434.

Thomas, G. S. P. 1984. A late Devensian glaciolacustrine fan-delta at Rhosemor, Clwyd, North Wales. *Geological Journal* **19**, 125–141.

Zilliacus, H. 1989. Genesis of De Geer moraines in Finland. *Sedimentary Geology* **62**, 309–317.

12
Interpreting Glacial Landscapes

In the previous chapters we have examined the processes and products of both glacial erosion and deposition. The conditions under which each type of landform or sediment is formed have also been reviewed. In this chapter we assemble this information and examine the pattern of landform–sediment distribution produced by a glacier. It is this distribution which controls the nature of the glacial landscape. The glacial landscape is formed from an assemblage of different landforms and sediments, the distribution of which is controlled by the variation within the glacier, both in space and time, of such variables as ice velocity, ice thickness, basal thermal regime, mass balance, geology, and basal topography. Since the glacial landscape of much of the northern mid-latitudes was formed beneath the ice sheets of the Cenozoic Ice Age, the emphasis in this chapter is on ice sheets. There are two components to the landform–sediment distribution within an ice sheet: (1) the large-scale pattern of erosion and deposition, and (2) the distribution of individual landforms and sediments. We shall examine both of these components below.

12.1 THE PATTERN OF EROSION AND DEPOSITION WITHIN AN ICE SHEET

In Section 3.5 we showed how the velocity profile along a flow line within an ice sheet is a function of its mass balance, and rises from almost zero under the ice divide to a maximum beneath the equilibrium line (Figure 3.15). This profile can be used to demonstrate, at a conceptual level, the pattern of erosion and deposition found along the same flow line within the ice sheet.

If we think of erosion and deposition simply in terms of net sediment or rock transfer and ignore the individual processes involved, then it can be simplified and considered to be a function of changes in ice velocity. As velocity increases, extending flow will dominate which will tend to result in net erosion. In contrast, decreasing ice velocity will lead to flow compression and net deposition. The pattern of erosion and deposition within an ice sheet will, therefore, resemble that shown in Figure 12.1 for the Greenland Ice Sheet. This pattern will remain constant, although the magnitude of erosion and deposition will vary as the ice sheet grows and decays. This follows since a larger ice sheet has a larger accumulation zone, therefore more ice to discharge, and consequently a greater ice velocity. The

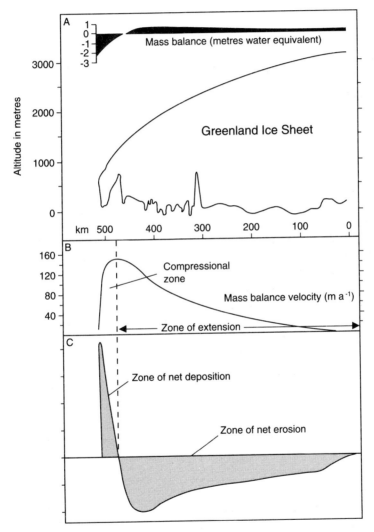

Figure 12.1 *Patterns of subglacial erosion and deposition within the Greenland Ice Sheet. This is based on the assumption that erosion occurs in areas of extending flow while deposition dominates in compressive flow regimes and that the volume of erosion must equal the volume of deposition. [Diagram reproduced with permission from: Boulton (1987) In: Menzies, J. & Rose, J. (Eds)* Drumlin Symposium, *Balkema, Fig. 5, p. 37]*

accuracy of the pattern of erosion and deposition in Figure 12.1 is limited by assumptions made about its construction: (1) the basal topography is assumed to be flat, of uniform geology and easily eroded in all directions; (2) the ice sheet is warm-based throughout; (3) erosion equals deposition within the ice sheet and meltwater does not transfer sediment beyond the ice margin; and (4) the ice divide remains in a fixed location. Despite these assumptions the pattern of erosion and deposition in Figure 12.1 provides a first approximation to that found within a real ice sheet and can be used to demonstrate how a flat landscape is modified by erosion and deposition during the growth and decay of an ice sheet.

Figure 12.2 shows a time–distance diagram for a hypothetical transect through an ice sheet. This diagram shows the span or width of the ice sheet at nine successive time steps and the pattern of erosion and deposition present beneath the ice sheet at each time step. The left-hand side of the ice sheet is located in a more maritime climate than the right-hand half and consequently experiences a larger mass

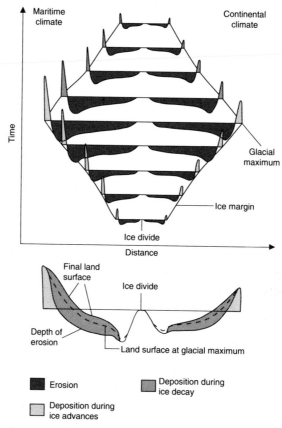

Figure 12.2   *Time–distance diagram for a transect through an isothermal mid-latitude ice sheet (i.e. warm-based throughout). The expansion and contraction of the ice margin and the pattern of erosion and deposition within it at nine time steps is shown. The shape of the landsurface that would result from the growth and decay of this ice sheet is shown below*

balance gradient and greater ice flow velocity. In contrast, the right-hand side experiences a continental climate and therefore has a lower mass balance gradient and rate of ice flow. As a consequence the pattern of erosion and deposition is slightly different either side of the ice divide (Figure 12.2). By integrating the values for net erosion and deposition for all nine time steps at each point covered by the ice sheet, it is possible to predict what the land surface looks like at the glacial maximum (Time 5), assuming that the original surface beneath the ice sheet was horizontal (Figure 12.2). In the same way it is possible to predict what the final surface looks like at Time 9 (Figure 12.2). From Figure 12.2, the following observations can be made about the changing form of the landscape beneath this hypothetical ice sheet.

A. The total volume of erosion and deposition is greater in the maritime half of the ice sheet due to the greater mass balance gradient and ice velocity. Through mass balance, climate controls the net amount of erosion and deposition.
B. The area beneath the ice divide experiences little or no erosion.
C. Net erosion increases rapidly away from the ice divide and then declines towards the ice margin.
D. More net erosion is achieved during deglaciation than during ice sheet growth. This follows since as the ice sheet advances it overruns the sediment it has already deposited and must first erode this before it can act on the surface beneath. An advancing ice sheet is therefore constantly recycling its own sediment. During deglaciation erosion acts directly on the undisturbed sediment or rock surface and is then followed by deposition as the ice margin retreats.
E. On deglaciation, a layer of sediment is deposited over the eroded landscape. In reality this layer of sediment will be concentrated into landforms and will not necessarily form a continuous layer.
F. At the ice margin there are two layers of sediment: one deposited during ice sheet growth and one deposited during its decay.

In practice the assumptions on which this model is based are simplifications of reality: ice sheets may grow over a strong basal topography; ice sheets are not warm-based throughout, but have complex basal thermal regimes; meltwater transports sediment beyond the ice margin; geology is not uniform and lithological or structural variations have a dramatic effect on the susceptibility of different areas to erosion. However, field observations support the principal elements of the pattern of erosion and deposition identified in Figure 12.2: preglacial landscapes often survive under the ice divide; eroded landscapes are frequently buried under a depositional landscape (Box 12.1); most deposition occurs close to the ice margin; and maritime glaciers are generally associated with more erosion and deposition than continental ones. Figure 12.2 can therefore be considered as a first approximation of the pattern of erosion and deposition within an ice sheet. However, it does not tell us about the distribution of different types of landforms and sediments which compose this pattern of erosion and deposition.

BOX 12.1: THE STUDY OF BURIED GLACIAL LANDSCAPES

The erosional landform assemblages created by former ice sheets are frequently buried by later glacial deposition. In such situations, the mapping and interpretation of landforms of glacial erosion is difficult. One solution to this problem is to use sub-surface investigations such as boreholes and cores to investigate this buried landscape. Boreholes and cores are taken for a variety of purposes, such a mineral exploration, or for site surveys prior to construction work. The records of many of these boreholes are publicly available.

Sissons (1969) used existing borehole records as well as strategically placed hand cores to examine the eroded landscape buried beneath the thick glacial sediments of the Central Valley of Scotland. He used 1900 boreholes to map out the contours of the bedrock surface (the rock-head) beneath the glacial deposits. The spacing between the boreholes was sufficiently close to allow him to interpolate bedrock contours and examine the morphology of the buried erosional landscape. This landscape was found to consist of a series of valleys cut by both glacial erosion and meltwater. A similar exercise was undertaken by Howell (1973) in north-west England.

This work highlights the presence of buried erosional landscapes and of the use of publicly held borehole records to investigate their morphology.

Sources: Howell, F. T. 1973. The sub-drift surface of the Mersey and Weaver catchment and adjacent areas. *Geographical Journal* **8**, 285–296. Sissons, J. B. 1969. Drift stratigraphy and buried morphological features in the Grangemouth–Falkirk–Airth area, Central Scotland. *Transactions of the Institute of British Geographers* **48**, 19–50.

12.2 THE DISTRIBUTION OF LANDFORMS AND SEDIMENT WITHIN AN ICE SHEET

The distribution of individual landforms and sediments within an ice sheet is a function of a large range of variables at both a local and an ice sheet scale. Within the glacial landscape it is possible, however, to recognise a series of recurrent **landform–sediment assemblages** and to examine where these may occur within an ice sheet. These are sometimes referred to as **glacial landsystems** and effectively represent **landform–sediment facies**. These landform–sediment assemblages draw together the information presented in Chapters 5 to 11. The following erosional landform–sediment assemblages can be recognised:

1. **Selective linear erosion [E1]:** deep glacial troughs separated by upland areas which show no evidence of glacial erosion. This occurs in areas of strong topography which causes local variation in the basal thermal regime. Low areas, valleys, are overlain by thick ice and are therefore warm-based; intervening high areas are overlain by thin ice which is therefore cold-based. Rapid erosion occurs in areas of warm ice and no erosion occurs in the cold-based regions.

The thermal contrast between the glacial valleys and the intervening upland areas grows with continued erosion. On deglaciation, this type of landform assemblage contains deep glacial troughs and linear basins separated by areas of little or no erosion in which preglacial landforms and sediments may survive. This assemblage is often associated with ice sheets in areas with cold continental climates but strong relief.

2. **Landscapes of little or no erosion [E2]:** areas of preglacial landforms and sediments. These occur beneath cold-based ice where the processes of glacial erosion are largely ineffective and may also occur beneath ice divides where ice velocities are low. Within an ice sheet they are, therefore, associated with cold continental climates, areas of high relief or thin ice, and ice divides.

3. **Areas of areal scour [E3]:** these are composed of an assemblage of landforms including roches moutonnées, whalebacks, glacial troughs, rock basins, and striated surfaces. All rock surfaces show signs of glacial abrasion and plucking. These are associated with either warm-based ice or ice in thermal equilibrium. The intensity of the scour varies, from areas in which the preglacial land surface may influence the morphology of individual landforms to areas in which no preglacial legacy survives.

4. **Landforms of local glaciation [E4]:** in areas of local glaciation, valley glaciers and cirque glaciers cut troughs and cirques. These may contain roches moutonnées, whalebacks, rock basins, striated surfaces and other landforms associated with warm-based ice. Preglacial sediments and landforms survive beyond the glacial limits, although they may be modified by periglacial processes.

In addition, the following depositional landform–sediment assemblages can be recognised:

1. **Supraglacial landform–sediment assemblage [D1].** This assemblage is associated with high englacial and supraglacial debris contents either due to: (1) compressive flow and englacial thrusting which transfers large amounts of debris from the base of a glacier to its surface, a situation which might occur where ice flows against a regional bedrock scarp or where a belt of cold ice occurs at the ice margin behind which the ice is warm and therefore fast flowing; or (2) high englacial debris contents associated with large-scale freezing-on of debris due to a transition from warm-based to cold-based ice. Typically, large areas of kames and hummocky moraine result in which a complex stratigraphy of units of flow till, sand and gravel and supraglacial meltout till occur (Figure 8.14A). The topography may be controlled (structured when viewed in plan) or uncontrolled (chaotic when viewed in plan). If the topography is controlled then the pattern of ridges and mounds will reflect the debris structure within the former ice margin, such as the location and orientation of debris-rich thrust planes.

2. **Subglacial landform–sediment assemblage [D2].** This assemblage is associated with warm-based ice and may be best developed in areas of soft deformable sediment. The landform assemblage consists of megaflutes, drumlins, rogens and mega-scale glacial lineations. The sediment assemblage consists of defor-

mation till and lodgement tills which may be interbedded with subglacially deposited units of sand and gravel. Subglacial eskers may also be present.

3. **Maritime ice-marginal landform–sediment assemblage [D3]**. This assemblage occurs at warm-based ice margins located in maritime climates. Seasonal fluctuation of the ice margin is common and meltwater production is usually high. This landform assemblage is best developed at ice margins experiencing net retreat. Seasonal push moraines occur along the ice-front, often best developed where the glacier fore-field slopes towards the ice margin. Local areas of hummocky moraine may develop where the supraglacial sediment cover on the ice margin is high; for example, along the axis of medial moraines. Where the debris concentration within medial moraines is insufficient to retard ablation of the ice, the medial moraine will be marked on the glacier fore-field by a plume of boulders or an irregular layer of supraglacial moraine till. The lateral glacier margins may be marked by well-developed dump moraines. In mountainous areas, where the supply of supraglacial debris is high, larger areas of hummocky moraine and ablation moraines may occur. Along stationary parts of the ice margin outwash fans develop, elsewhere meltwater streams deposit outwash plains. Eskers formed from the infill of englacial tunnels or supraglacial channels may be exposed as the glacier retreats. The sedimentary facies produced by this landform assemblage is complex; sediments include lodgement tills, supraglacial melt-out tills, glaciofluvial sediments and non-glacial sediment incorporated into push moraines. These different sediments occur in close proximity and often show signs of glaciotectonics.

4. **Continental ice-marginal landform–sediment assemblage [D4]**. This assemblage occurs at warm-based ice margins located in continental climates. Seasonal fluctuations of the ice margin are uncommon, meltwater production may be low and the lower mass balance gradients mean that ice velocities are usually small. Although the glacier may be predominately warm-based, the actual ice margin may be either seasonally or permanently cold-based. This leads to strong compression at the glacier terminus. Ice-marginal moraines tend only to occur where the ice margin is advancing or stationary. If the glacier is experiencing steady retreat, few ice-marginal landforms will develop and an irregular spread of supraglacial melt-out or moraine till results. Ice-marginal moraines tend to be either large composite push moraines formed during a readvance/surge or more commonly of an ablational form. If thrusting occurs at the ice margin the morphology of the ablation moraine may be dominated by the geometry of the thrust planes. Small outwash fans may be present but are not usually dominant. A retreating ice margin will tend to give a relatively sparse landform assemblage except at points where ice was stationary or readvanced.

5. **Surging landform–sediment assemblage [D5]**. Certain valley glaciers and outlet glaciers of ice sheets may be subject to periodic surging. Such surges are associated with a distinctive landform assemblage. This landform assemblage has only been described from active glacial environments in which buried ice still contributes to the morphology of the landforms present. At these margins the following assemblage has been recorded: in front of the margin of the glac-

ier surge a large composite push moraine is formed, which may be ice-cored. Behind this moraine, stagnant ice occurs. On ablation, crevasse-squeezed ridges are revealed, as well as a complex topography of hummocky moraine and kames formed by the melt-out of the stagnant ice and by glaciofluvial deposition on the melting ice. The presence of crevasse-squeezed ridges is considered by many to be indicative of a glacier surge.

6. **Cold-based or polar ice landform–sediment assemblage [D6].** Glaciofluvial processes are largely absent beneath cold-based ice. Ice-marginal sedimentation is dominated by the deposition of dump moraines or ablation moraines formed by the concentration of supraglacial and englacial debris at the ice margin by englacial thrusting and shearing at the ice margin. The melt-out or sublimation of englacial debris may produce a topography of hummocky moraine, the surface of which may reflect the pattern of folding and shearing and therefore the thickness of debris within the ice. Little or no reworking of debris occurs. Sediments consist of basal and supraglacial melt-out till and sublimation till.

7. **Glaciofluvial landform assemblage [D7].** These develop along major meltwater discharge routes within ice sheets, particularly in areas where the geometry of the ice margin or the underlying relief concentrates or confines the meltwater. A stationary or slowly retreating ice margin is also important. At these locations meltwater discharge is sufficient to rework all or most of the glacial sediment delivered to the ice margin, and the landform–sediment assemblage is dominated by glaciofluvial processes. Whole valleys or large areas of the ice margin may be dominated by glaciofluvial landforms. Large outwash fans develop over the ice front to give a complex glaciofluvial landscape as the ice retreats. Meltwater channels, outwash fans, kame terraces, kames, eskers and braided eskers are common. Sediments consist primarily of glaciofluvial sands and gravels.

8. **Glaciolacustrine assemblage [D8].** Where ice margins dam water against reverse slopes or terminate in large proglacial lakes a distinct assemblage of landforms and sediments develop. The principal control over the presence of this assemblage is the topography and the geometry of the ice margin. It may occur at all types of thermal ice margin. Landforms include meltwater channels, lake shorelines, De Geer moraines, ice-contact outwash fans, ice-contact deltas, delta moraines, fluvial deltas, eskers and moraine banks. Sediments consist of sands and gravels; laminated, often rhythmitic, silts and fine sands; and diamictons deposited by the 'rain-out' of ice-rafted debris or by sediment gravity flows.

9. **Glaciomarine fjord assemblage [D9].** At glaciers that terminate in glacial troughs (fjords) or other constricted depressions which are submerged by the sea, a glaciomarine landform–sediment assemblage develops. The geometry of the landforms and sediment bodies is restricted by the relief of the fjord. Subaqueous push moraines, outwash fans, moraine banks, ice-contact deltas and eskers may all be present. Above the water surface the lateral margins of the fjord glacier may be marked by lateral moraines and meltwater channels. Shorelines and deltas formed by tributary rivers may also be present.

Sediments consist of sands and gravels, laminated silts and fine sands, and diamictons deposited by the 'rain-out' of ice-rafted debris or by sediment gravity flows.

10. **Glaciomarine continental shelf landform–sediment assemblage [D10].** Where large parts of an ice sheet terminate on the continental shelf the geometry of the glaciomarine landforms and sediments is not usually constrained by relief. Large moraine banks and delta moraines may develop. If the ice margin is located close to the shelf edge then the geometry of the sediment units may be limited and fans may develop at the shelf edge. Sediments consist of sands and gravels, laminated silts and fine sands, and diamictons deposited by the 'rain-out' of ice rafted debris or by sediment gravity flows.

The landform–sediment assemblages described represent a broad summary of the recurrent associations of landforms and sediments found at the margin or beneath glaciers. As with any general summary, exceptions will exist, particularly since local controls may influence the combination of landforms that occur at any one location. More importantly, the distinction between one assemblage and the next may often be difficult to define. Despite these qualifications the landform–sediment assemblages described provide a means of examining the broad distribution of landforms within an ice sheet.

Within an ice sheet different landform assemblages or facies occur in different locations and may become superimposed on one another as the ice sheet grows and decays. For example, a maritime ice-marginal assemblage may be superimposed on a subglacial system or on a landscape of areal scour. Similarly, a glaciomarine continental shelf assemblage may be superimposed on an assemblage of subglacial landforms and sediments. Interpreting vertical sediment sequences or landform assemblages requires the careful application of Walther's principle (see Section 8.1.7). Superimposed landform–sediment assemblages or facies provide a picture of the lateral pattern of glacial environments beneath the glacier that deposited them. This picture is complicated where landform–sediment assemblages of different ages and therefore the product of different glaciers are juxtaposed. In these cases the careful application of the stratigraphical tool kit is required to determine the relative order of events—something which is not always possible to resolve unambiguously.

From the above discussion of each landform–sediment assemblage it should be apparent that their distribution within a glacier or ice sheet is a function of such variables as: (1) the basal thermal regime and its variation within the glacier; (2) the mass balance gradient; (3) sea-level; (4) relief; and (5) geology.

Basal thermal regime determines the type of erosional landscape that may exist within an area (Box 12.2). For example, warm-based ice is associated with areas of areal scour and dictates the presence or absence of the subglacial depositional landform–sediment assemblage. It may also have a role in the location of areas covered by the supraglacial landform–sediment assemblage. The mass balance gradient, which is a function of climate, controls ice velocity and the sensitivity of the ice margin to seasonal variation in ablation and ice flow. For example, an ice margin in a maritime climate, with a high mass balance gradient, will be domi-

BOX 12.2: LANDSCAPES OF GLACIAL EROSION AND BASAL THERMAL REGIME

Probably the most important attempt to examine the distribution of landscapes of glacial erosion and examine their glaciological controls was undertaken by Sugden (1977, 1978). This work involved three phases: (1) the reconstruction of the basal thermal regime of a former ice sheet, in this case the Laurentide or North American Ice Sheet; (2) mapping of the pattern of erosional landforms and the intensity of glacial erosion within the limits of this former ice sheet; and (3) comparison of the pattern of erosion with the pattern of basal thermal regime. Sugden (1977) reconstructed the Laurentide ice sheet and its basal thermal regime using a simple glaciological model and input parameters derived either from the published literature or by analogy with present-day ice sheets in Antarctica and Greenland. The pattern of basal thermal regime consisted of warm ice sheet core surrounded by a zone of cold ice, which in turn was followed in some areas by another warm-based zone. Sugden (1978) then mapped out the pattern of erosion using satellite images. He recognised areas of selective linear erosion; areas of little or no erosion; areas of areal scour; and areas of local glacial erosion. He obtained an index of the intensity of glacial erosion by calculating the density of lake basins; a high density of basins corresponding to a high rate of erosion. This is based on the assumption that most lake basins are excavated by glacial erosion. A zone of intense glacial erosion was recognised in a broad belt which curved across the continent to the south of Hudson Bay. Sugden (1978) then compared the pattern of basal thermal regime with the pattern of erosion. He found a broad coincidence between the thermal zones and the pattern of erosion. Areas of areal scour coincided with warm-based portions of the ice sheet, while zones of little or no erosion were located beneath areas of cold-based ice. The zone of intense erosion identified by the lake basin density index coincided with a zone of thermal transition between warm ice and cold ice, the ideal condition for intense glacial plucking due to widespread freezing-on of glacial debris as meltwater flowed from the warm-based area of the ice sheet to the cold zone. Variations from this broad pattern of agreement could be attributed to topography or the effects of geology (Sugden 1978). Although the pattern of basal thermal regime predicted by Sugden (1977) has been questioned by more recent modelling exercises, this work is important as it established the link between basal regime and landscapes of glacial erosion.

Sources: Sugden, D. E. 1977. Reconstruction of the morphology, dynamics and thermal characteristics of the Laurentide ice-sheet at its maximum. *Arctic and Alpine Research* **9**, 21–47. Sugden, D. E. 1978. Glacial erosion by the Laurentide ice-sheet. *Journal of Glaciology* **20**, 367–391.

nated by the maritime ice-marginal landform–sediment assemblage, which consists of seasonal push moraines, dump moraines, flutes, and outwash fans. Mass balance and thermal regime may combine to determine which glaciers are prone to surging behaviour and therefore likely to contain the surging landform–sediment assemblage.

Sea-level is also important because it controls the distribution of glaciomarine landform–sediment assemblages. Local sea-level is controlled by both the eustatic (or global) sea-level and the local land elevation. Eustatic sea-level is controlled on the timescale of a glacial cycle by global ice volume. As the volume of water contained in ice sheets increases across the globe, sea-level will fall as water is stored within the ice sheets. Local land elevation is determined by the isostatic depression caused by an individual ice sheet. An ice sheet is heavy and causes the crust to be depressed, reducing land elevation—a phenomenon known as isostatic depression or loading. As the ice sheet grows the crust is depressed, and as it decays the crust rises, although the response of the crust is slow and therefore lags behind the growth and decay of the ice sheet. For a lowland ice sheet the proportion of the ice margin in contact with the sea may depend on the phasing between local changes in crustal elevation (isostatic loading) and global variations in eustatic sea-level (global ice volume). For example, an ice sheet that is decaying in phase with a global decrease in ice volume, such that the crust is rising as the ice sheet decays and as eustatic sea-level rises, may have a consistent proportion of its ice margin in contact with the sea. In contrast, if an ice sheet decays out of phase with a global decrease in ice volume, such that the crust is still depressed as eustatic sea-level rises, then the proportion of its ice margin in contact with the sea will rise dramatically (see Section 10.2). Isostatic loading may also play a part in controlling the distribution of the glaciolacustrine landform–sediment assemblage. Isostatic loading by an ice sheet not only depresses the crust beneath the ice sheet but also the proglacial area. The crust may also rise beyond the zone of depression to form a fore-bulge in compensation of the crustal depression beneath the ice sheet. Large proglacial lakes may develop between this fore-bulge and the glacier margin, causing much of it to experience glaciolacustrine conditions. This was particularly true for the North American Ice Sheet during the last glacial cycle which terminated close to the Great Lakes, causing them to expand into very large proglacial lakes.

Relief is also important in determining the distribution of landform–sediment assemblages within an ice sheet. For example, the fjord landform–sediment assemblages are restricted to areas of high relief in contact with the sea. Similarly, the presence of the supraglacial landform–sediment assemblage may be associated with the presence of regional bedrock scarps beneath the ice sheet, along which compressive flow may cause basal debris to rise towards the glacier surface. Relief also exerts a control on the basal thermal regime within an ice sheet: areas of variable relief beneath ice sheets with a cold continental climate may determine the presence of landscapes of selective linear erosion. In the warm-based troughs glacial erosion and the deposition of subglacial landforms may occur, while on the surrounding mountain-tops ice is cold-based and little or no erosion may occur, so the preglacial landscape may survive.

Finally, geology may also have an important role in determining the distribution of landform–sediment assemblages. For example, the subglacial assemblage is best developed where soft deformable sediments are present and, as was discussed in Section 9.2, subglacial eskers may be restricted to areas of hard free-draining bedrock (Box 9.6).

The distribution of the 14 landform–sediment assemblages within an ice sheet is therefore a function of a wide range of variables, determined primarily by climate, relief and geology. In order to illustrate how these landform–sediment assemblages may be combined within an ice sheet to produce glacial landscapes, a series of hypothetical examples are used below.

Figure 12.3 shows the distribution of landform–sediment assemblages formed by an ice sheet with a horizontal bed and a simple pattern of basal thermal regime. The model shows the evolution of the landscape over five time steps during the decay of the ice sheet. The final disintegration of the ice sheet and the associated landforms and sediments are not included within the model. The pattern of basal thermal regime is simple but provides a first approximation, based on theoretical models, as to what we might expect to find beneath a former mid-latitude ice sheet (Figure 3.13). The broad pattern of erosion and deposition is based on that in Figure 12.1. Figure 12.3 consists of a transect through a former ice sheet which runs from a warm maritime climate on the left to a continental polar climate on the right. This is similar to the climatic range that would have been experienced by a north–south transect through the North American Ice Sheet or a west–east transect through the British or Scandinavian ice sheet during the last glacial cycle.

The final landscape produced after deglaciation, within this model, can be divided into three areas, each of which is described below:

A. A central zone of little or no erosion [E2] beneath the former ice divide in which the preglacial landscape may survive.
B. The landscape beneath the maritime or left-hand half of the ice sheet. To the left of the central zone of little or no erosion [E2] one first encounters a zone of areal scour [E3] with little deposition. Further south, the proportion of deposition on this eroded land surface [E3] increases and a maritime ice-marginal landform–sediment assemblage develops [D3], which may be locally superimposed upon the subglacial landform–sediment assemblage [D2]. Both these assemblages tend to occupy valley floors and low-lying areas, while the erosional landscape dominates on the higher areas [E3]. This reflects the concentration of fast-flowing ice, and therefore glacial debris, along valleys. Further to the left, most of the erosional landscape [E3] is buried beneath ice-marginal and subglacial landforms [D2 and D3]. Locally areas of the glaciofluvial landform–sediment assemblage [D7] may occur along major discharge routes of the ice sheet or where meltwater discharge is concentrated into confined areas by topography. Close to the ice margin both the maritime ice-marginal landform–sediment assemblage [D3] and the subglacial assemblage [D2] bury a preglacial landscape which may show little evidence of erosion [E2]. Beyond the ice margin an assemblage of proglacial glaciofluvial landforms and sediments occurs.

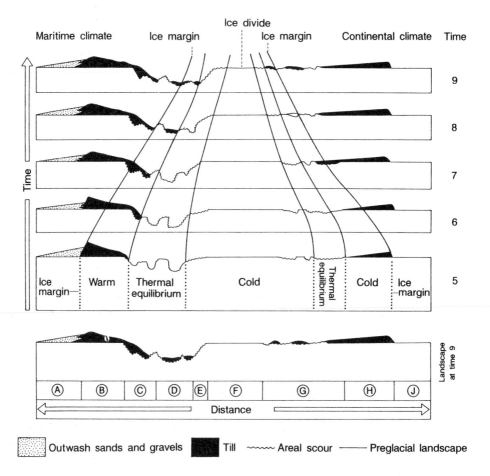

Ice divide

Maritime climate   Ice margin   Ice margin   Continental climate   Time

Time

| Ice margin— | Warm | Thermal equilibrium | Cold | Thermal equilibrium | Cold | Ice —margin |

Landscape at time 9

| (A) | (B) | (C) | (D) | (E) | (F) | (G) | (H) | (J) |

Distance

Outwash sands and gravels   Till   ~~~ Areal scour   —— Preglacial landscape

Figure 12.3 *Conceptual model of the landscape evolution beneath a mid-latitude ice sheet with a simple basal thermal regime, uniform geology and an initially flat land surface. The left-hand half of the ice sheet is located in a maritime area, while the right-hand half experiences a more continental climate. The shape of the landsurface beneath this ice sheet is shown at five time steps during deglaciation. The final landscape and the landform–sediment assemblages that would result are as follows. (A) Glaciofluvial landforms. (B) Maritime ice-marginal landforms [D3] and subglacial landforms [D2] above a buried preglacial landsurface [E2]. (C) Maritime ice-marginal landforms [D3] and subglacial landforms [D2] above a buried landscape of areal scour [E3]. (D) Maritime ice-marginal landforms [D3] and subglacial landforms [D2] in valley floors. Landscapes of areal scour in upland areas [E3]. (E) Landscape of areal scour [E3]. (F) Little or no erosion, survival of preglacial landsurface [E2]. (G) Landscape of areal scour in low-lying areas [E3], while preglacial landscape may survive in upland areas [E2]; local deposition of a cold landform assemblage [D6] may occur. (H) Cold landform assemblage [D6] occurs over a buried preglacial landsurface [E2]. (J) Minor glaciofluvial landforms*

C. The landscape beneath the continental or right-hand half of the ice sheet. Moving to the right of the central zone of little or no erosion [E2] beneath the ice divide, the landscape contains areas of both areal scour [E3] and little or no erosion [E2]. The amount of deposition is relatively small and consists of land-forms and sediments from the cold-based or polar ice-marginal assemblage [D6]. Towards the margin this eroded landscape may become more pro-nounced, but is buried to a much greater extent by glacial landforms and sedi-ment [D6]. Closer to the former ice margin the sediments and landforms of the cold-based or polar ice-marginal landform–sediment assemblage [D6] are found over an undisturbed preglacial landscape [E2]. Minor glaciofluvial land-forms and sediments occur beyond the ice limit.

The glacial landscape at any one point is, therefore, the product of several superimposed landform–sediment assemblages. The exact distribution of each depositional assemblage is controlled by local topography, the geometry of the ice margin and the rate of ice flow. The glacial landscape should, therefore, be viewed as a superimposed assemblage of individual landforms and sediments.

Figure 12.4 shows a second hypothetical glacial landscape. In this case several complicating variables have been added to the scenario. These are: (1) the intro-duction of geology in the form of soft deformable sediment to the south of the ice divide, and (2) the introduction of a marine ice margin to the north. Again, the final landscape produced after deglaciation can be divided into three areas, each of which is described below.

A. A central zone of little or no erosion [E2] beneath the former ice divide in which the preglacial landscape may survive.
B. The landscape beneath the maritime half of the ice sheet. Moving left from the central zone of little or no erosion [E2] one first encounters a zone of areal scour [E3]with little deposition. Further to the left the proportion of deposition on this eroded landsurface increases [E3] and a maritime ice-marginal landform–sediment assemblage develops [D3], which might be locally super-imposed upon a subglacial landform–sediment assemblage [D2]. As one moves onto the soft deformable bed, the basal contact becomes one formed by sub-glacial deformation and may be drumlinised [D2]. This surface is covered first by a subglacial landforms and sediments, on top of which the maritime ice-marginal landform–sediment assemblage [D3] will be locally present. The sub-glacial landform–sediment assemblage becomes increasingly important towards the maximum extent of the former ice margin and may include a com-ponent of deformation till [D2]. Close to the ice margin the subglacial deforma-tion facies changes from excavational (erosion) to accretion. The maritime ice-marginal landform–sediment assemblage is also present [D3]. Beyond the ice margin a proglacial glaciofluvial landform–sediment assemblage occurs.
C. The landscape beneath the continental half of the ice sheet. Moving to the right of the zone of little or no erosion [E2] beneath the ice divide, the landscape is one of areal scour [E3] and areas of little or no erosion [E2] in which the preglacial surface may survive. The amount of deposition is small and what

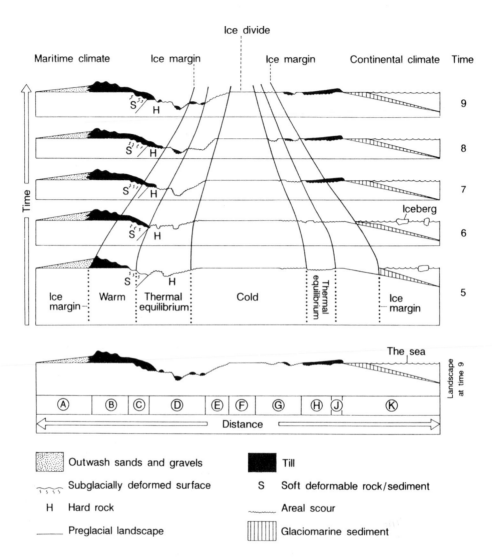

exists consists of landforms and sediments from the cold-based or polar ice-marginal assemblage [D6]. Towards the margin this eroded landscape becomes more pronounced, but is buried to a much greater extent by glacial landforms and sediment. The total volume of sediment present is relatively low however. Closer to the former ice margin the sediments of the cold-based or polar ice-marginal landform–sediment assemblage [D6] are found over an undisturbed preglacial landscape [E2]. Finally a zone of glaciomarine sedimentation is encountered which starts inside the maximum extent of the former ice sheet and extends for a considerable distance beyond it [D9/10]. Much of the sediment transported by the ice sheet is deposited in this zone, which explains the reduced sediment volume deposited on land.

A variety of further complications can be introduced into these scenarios to produce glacial landscapes which replicate more closely the pattern of landforms and sediments observed in a particular area. For example, one could introduce a strong preglacial topography into the scenario. Figure 12.5 shows the glacial landscape produced by an ice sheet with a continental climate and a preglacial topography. The presence of a strong topography modifies the pattern of basal thermal regime, creating areas of warm-based ice in the topographic lows and areas of cold-based ice on the raised areas. This results in not only an area of selective linear erosion [E1], but also in the deposition of the subglacial landform–sediment assemblage [D2] and the continental ice-marginal landform–sediment facies [D4] within the glacial troughs, adjacent to which there are areas of little or no glacial erosion [E2]. Locally, as the ice retreats away from topographic highs, ice-dammed lakes and the associated landform–sediment assemblage may occur [D8].

In reality, the glacial landscape produced by a large ice sheet may be much more complex and locally variable than the above examples suggest. This is particularly true since the examples are limited by three simplifications: (1) they involve a single ice sheet with no inheritance from earlier glacial episodes; (2) the pattern of basal thermal regime is not only simple but is stable as the ice sheet decays; and (3) the ice divides remain in the same place. In practice, the former mid-latitude ice sheets of the Cenozoic Ice Age were much more complex. Each ice sheet, apart from the first, grew and decayed over the remains of evidence from earlier glaciers (Box 12.3). They may have had locally very complex patterns of thermal regime, particularly in areas of pronounced topography, which were not necessarily stable during decay (Box 12.4). There is also evidence to suggest that in some cases ice divides may change position or vary in relative size and importance (Box 2.2). As a

---

Figure 12.4 *Conceptual model of the landscape evolution beneath a mid-latitude ice sheet with a simple basal thermal regime, basic geological variation including an area of soft deformable sediment, a marine margin and an initially flat land surface. The left-hand half of the ice sheet is located in a maritime area, while the right-hand half experiences a more continental climate. The shape of the landsurface beneath this ice sheet is shown at five time steps during deglaciation. The final landscape and the landform–sediment assemblages that would result are as follows. (A) Glaciofluvial landforms. (B) Subglacial landforms [D2] and maritime ice-marginal landforms [D3] above a buried preglacial landsurface [E2]; widespread subglacial deformation of a constructional form. (C) Subglacial landforms [D2] and maritime ice-marginal landforms [D3]; sedimentary base may show evidence of erosion by subglacial deformation (excavation deformation). (D) Maritime ice-marginal landforms [D3] over landscapes of areal scour [E3]; deposition is concentrated into low-lying areas and erosional landscapes may dominate in upland areas. (E) Landscape of areal scour [E3] with local deposition. (F) Little or no erosion; survival of preglacial landsurface [E2]. (G) Landscape of areal scour in low-lying areas [E3], while preglacial landscape may survive in upland areas [E2]. (H) Cold landform assemblage [D6] occurs over a buried landscape of areal scour [E3]. (J) Cold landform assemblage [D6] occurs over a buried preglacial landscape [E2]. (K) Glaciomarine landform–sediment assemblage [D10]*

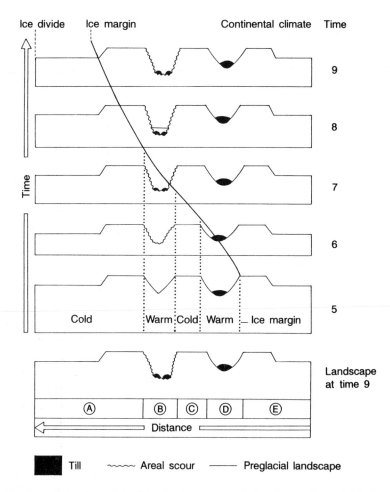

Figure 12.5   *Conceptual model of the landscape evolution beneath a mid-latitude ice sheet with a simple basal thermal regime, uniform geology, and a pronounced basal topography. Only the right-hand half of the ice sheet is shown, located in a continental climate. The shape of the landsurface beneath this ice sheet is shown at five time steps during deglaciation. The final landscape and the landform–sediment assemblages that would result are as follows. (A) Little or no erosion of the preglacial landscape [E2]. (B) Extensive valley modification by selective linear erosion [E1]; a subglacial landform–sediment assemblage [D2] with continental ice-marginal landforms [D4] may occur on the valley floor. (C) Little or no erosion of the preglacial landscape [E2]. (D) Little valley modification, but subglacial landforms [D2] plus continental ice-marginal landforms [D4] may occur on the valley floor. (E) Minor glaciofluvial landforms and sediments*

consequence, the observed landform pattern is in many locations much more complex than that suggested here. However, the above examples serve to illustrate: (1) the broad patterns of landform–sediment distribution that occur within an ice sheet; (2) the importance of the landscape as the product of a superimposed

BOX 12.3: COLD-BASED DEGLACIATION AND THE PRESERVATION OF GLACIAL LANDFORMS

In recent years attention has focused upon the apparent survival of early glacial landforms and deposits beneath the last Scandinavian Ice Sheet. Parts of Norway and Sweden appear to contain a remarkable assemblage of superimposed landforms belonging to different and successive ice sheets. The last Scandinavian Ice Sheet is known to have covered most of Scandinavia, yet beneath parts of the ice sheet large areas of the glacial landscape appear to pre-date it. Glacial landforms formed by earlier ice sheets, dated by organic remains, appear to have survived beneath this ice sheet. The hummocky moraine illustrated below, and the organic sediments in the kettle holes associated with it, are believed to pre-date the last ice sheet yet are located within its known limit. They are intimately associated with biogenic evidence which pre-dates the last glacial cycle. Extensive fields of drumlins formed when the ice sheet was at its maximum extent appear to have been unaffected by deglaciation and the associated changes in ice flow direction. Similarly periglacial landforms which pre-date the last glacial cycle have also been observed in areas known to have been covered by the last ice sheet. The survival of these preglacial landforms and of the landforms formed during the glacial maximum is only possible if the Scandinavian Ice Sheet experienced an entirely cold-based deglaciation. Numerical models suggest that as ice sheets grow and decay a warm-based marginal zone passes across the landscape, as illustrated in Figure 3.11. However, this cannot have been the case for the Scandinavian Ice Sheet if the evidence is correctly interpreted. The cause, mechanics and significance of such

a cold-based deglaciation is not yet understood, but it probably reflects the climatic and topographic conditions under which deglaciation occurred.

Source: Kleman, J. 1994. Preservation of landforms under ice sheets and ice caps. *Geomorphology* **9**, 19–32. [Photography: R. Lagerbäck]

BOX 12.4: DEGLACIATION AND CHANGING PATTERNS OF BASAL THERMAL REGIME

Mooers (1990) studied the landforms produced during the decay of two ice lobes, the Rainy Lobe and the Superior Lobe, on the southern margin of the ice sheet that existed in North America during the last glacial cycle. The two lobes advanced to their maximum along the St Croix moraine at approximately the same time. The St Croix moraine is characterised by frontal outwash plains and broad areas of kames and hummocky moraine. The moraine is typical of that produced by a glacier with a mixed thermal regime and a cold-based ice margin. The decay or recession of the Rainy ice lobe is marked by a similar series of ablation moraines and belts of hummocky moraine. No subglacial landforms are present. In contrast, the decay of the Superior ice lobe is documented by a series of well-defined and closely spaced recessional moraines. These moraines are superimposed on drumlins, eskers and tunnel-valleys.

Mooers (1990) suggests that the landform record of these two ice lobes reflects different basal thermal regimes. At their maximum extent both lobes appear to have been of mixed regime with a cold ice margin and a warmer interior. Thrusting towards the ice margin is reflected by the ablation moraines and the widespread presence of kames and hummocky moraine. The thermal regime of the Rainy Lobe changed little during decay, but the Superior Lobe appears to have undergone a rapid evolution to a warm-based glacier in which subglacial processes operated.

This example illustrates how local spatial and temporal variations in basal thermal regime are recorded in the glacial landscape and contribute to its complexity.

Source: Mooers, H. D. 1990. A glacial-process model: the role of spatial and temporal variation in glacier thermal regime. *Geological Society of America Bulletin* **102**, 243–251.

sequence of landform–sediment assemblages; and (3) a starting point from which the landform history of individual areas can be considered.

12.3 SUMMARY

Ice sheets generate a distinct pattern of glacial erosion and deposition. The principal elements of this are that: (1) little or no erosion occurs beneath the ice divide;

(2) erosion and deposition is greater beneath ice sheets in maritime locations; (3) net erosion increases rapidly away from the ice divide and then declines towards the ice margin; (4) more erosion is achieved during deglaciation than during ice sheet growth; (5) deposition rises steadily towards the ice margin; and (6) during deglaciation a layer of sediment is deposited over the eroded landscape.

Within the glacial landscape at modern glaciers we can identify a series of 14 landform–sediment assemblages or facies, the distribution of which within an ice sheet is controlled by such variables as: (1) the basal thermal regime and its variation within the glacier; (2) the mass balance gradient; (3) sea-level; (4) relief; and (5) geology. The glacial landscape is the product of the superposition of different landform–sediment assemblages and should be interpreted as such. It is possible to illustrate at a conceptual level the distribution of landforms and sediments within an ice sheet. This may not, however, do justice to the actual complexities present within the glacial landscape present within former glacial areas. This is due to the inheritance of glacial landforms and sediments from earlier episodes of glaciation.

Unravelling the complex landform and sediment record left by Cenozoic and pre-Cenozoic glaciers is one of the most fundamental problems of glacial geology. To begin to read this story from the record of landforms and sediment left in the landscape we must apply the tools of glacial stratigraphy: (1) establishing the relative order of landforms and sediments, and (2) interpreting those landforms and sediments in terms of the glacial environments in which they formed. This book has concentrated on the interpretation of the landforms and sediments. As we have seen in the preceding chapters, each landform and sediment holds a clue about the glacier that formed it: a clue to the size of the former glacier, to the direction of ice flow, to its basal thermal regime, or to its mass balance and therefore to the prevailing climate. This interpretation is based on the concept of uniformitarianism or actualism; that modern glacial processes can be observed and their products studied in order to interpret the glacial landscape produced by ancient glaciers. With this book we have provided the basic information with which to identify and interpret glacial landforms and sediments. We have illustrated in this chapter how these landforms and sediments may be combined within the glacial landscape. You should now be able to read the clues within the glacial landscape etched and deposited by the glaciers of the Cenozoic Ice Age.

12.4 SUGGESTED READING

The pattern of erosion and deposition within an ice sheet is considered by Boulton & Clark (1990). The linkage between landforms of glacial erosion and basal thermal regime are explored in papers by Sugden (1974, 1977, 1978), Gordon (1979) and Glasser (1995). In contrast, little research has been done on the distribution of glacial landform–sediment assemblages within ice sheets. Eyles & Dearman (1981) recognise a series of glacial landsystems and Eyles *et al.* (1983) examine their distribution within Britain and North America. Mooers (1990) provides an interesting account of how changes in basal thermal regime may occur during deglaciation and are recorded in the glacial landscape. In recent years there has been a lot of work on the survival of landforms from preglacial and early glacial landscapes beneath the mid-latitude ice sheets of the last glacial cycle: Sugden (1968), Sugden & Watts (1977), Lagerbäck (1988), Rodhe (1988), Kleman (1992, 1994), Kleman *et al.* (1992), Dyke (1993) and Sollid & Sørbel (1994). These papers illustrate some of the complexities that exist within the glacial landscape.

Boulton, G. S. & Clarke, C. D. 1990. The Laurentide ice sheet through the last glacial cycle: the topology of drift lineations as a key to the dynamic behaviour of former ice sheets. *Transactions of the Royal Society of Edinburgh* **81**, 327–347.

Dyke, A. S. 1993. Landscapes of cold-centred Late Wisconsinian ice caps, Arctic Canada. *Progress in Physical Geography* **17**, 223–247.

Eyles, N. & Dearman, W. R. 1981. A glacial terrain map of Britain for engineering purposes. *Bulletin of the International Association of Engineering Geology* **24**, 173–184.

Eyles, N., Dearman, W. R. & Douglas, T. D. 1983. The distribution of glacial landsystems in Britain and North America. In: Eyles, N. (Ed.) *Glacial Geology.* Pergamon, Oxford, 213–228.

Glasser, N. F. 1995. Modelling the effect of topography on ice sheet erosion, Scotland. *Geografiska Annaler* **77A**, 211–215.

Gordon, J. E. 1979. Reconstructed Pleistocene ice sheet temperatures and glacial erosion in Northern Scotland. *Journal of Glaciology* **22**, 331–344.

Kleman, J. 1992. The palimpsest glacial landscape in northern Sweden—Late Weichselian deglaciation landforms and traces of older west-centred ice sheets. *Geografiska Annaler* **74A**, 305–325.

Kleman, J. 1994. Preservation of landforms under ice sheets and ice caps. *Geomorphology* **9**, 19–32.

Kleman, J., Borgström, I., Robertsson, A. M. & Kukkuesjiöld, M. 1992. Morphology and stratigraphy from several deglaciations in the Transtrand mountains, western Sweden. *Journal of Quaternary Science* **7**, 1–16.

Lagerbäck, R. 1988. The Veiki moraines in northern Sweden—widespread evidence of an early Weichselian deglaciation. *Boreas* **17**, 469–486.

Mooers, H. D. 1990. A glacial-process model: the role of spatial and temporal variation in glacier thermal regime. *Geological Society of America Bulletin* **102**, 243–251.

Rodhe, L. 1988. Glaciofluvial channels formed prior to the last deglaciation: examples from Swedish Lapland. *Boreas* **17**, 511–516.

Sollid, J. L. & Sørbel, L. 1994. Distribution of glacial landforms in southern Norway in relation to the thermal regime of the last continental ice sheet. *Geografiska Annaler* **76A**, 25–35.

Sugden, D. E. 1968. The selectivity of glacial erosion in the Cairngorm Mountains, Scotland. *Transactions of the Institute of British Geographers* **45**, 75–92.

Sugden, D. E. 1974. Landscapes of glacial erosion in Greenland and their relationship to ice, topographic and bedrock conditions. In: Brown, E. H. & Waters, R. S. (Eds) *Progress in Geomorphology.* Institute of British Geographers Special Publication 7, 177–195.

Sugden, D. E. 1977. Reconstruction of the morphology, dynamics and thermal characteristics of the Laurentide ice-sheet at its maximum. *Arctic and Alpine Research* **9**, 21–47.

Sugden, D. E. 1978. Glacial erosion by the Laurentide ice-sheet. *Journal of Glaciology* **20**, 367–391.

Sugden, D. E. & Watts, S. H. 1977. Tors, felsenmeer and glaciation in northern Cumberland Peninsula, Baffin Island. *Canadian Journal of Earth Sciences* **14**, 2817–2823.

Index